This volume fulfills a long-standing need of vertebrate paleontologists – whether amateurs attending their first excavation or professional preparators and curators – for a book that describes and explains modern paleontological techniques and practice. The authors of this volume are exceptional technicians in their field, and the book covers everything from field specimen collecting, through conservation methods, chemical preparation, molding, casting and painting, and mounting of vertebrate skeletons, to the final chapter devoted to the use of CT scans and X-ray methods.

Until now, most preparatory techniques and skills have been passed down by example and demonstration; rarely have they been standardized and set down in print. This book is an attempt to do that, and aims to enlighten workers on the most modern and successful methods that can be used in preserving and studying our fossil heritage. The volume instructs throughout on modern procedures and conservation techniques and educates workers toward an awareness of good conservation values. The editors and authors hope that it will find its place beside the three other common principles of good preparation: patience, concentration, and a good sense of humor.

Vertebrate paleontological techniques
Volume 1

Vertebrate paleontological techniques
Volume 1

edited by

Patrick Leiggi
Museum of the Rockies

Peter May
Research Casting International Ltd.

PUBLISHED BY THE PRESS SYNDICATE OF THE UNIVERSITY OF CAMBRIDGE
The Pitt Building, Trumpington Street, Cambridge, United Kingdom

CAMBRIDGE UNIVERSITY PRESS
The Edinburgh Building, Cambridge CB2 2RU, UK
40 West 20th Street, New York NY 10011–4211, USA
477 Williamstown Road, Port Melbourne, VIC 3207, Australia
Ruiz de Alarcón 13, 28014 Madrid, Spain
Dock House, The Waterfront, Cape Town 8001, South Africa

http://www.cambridge.org

First published 1994
First paperback edition 2004

A catalogue record for this book is available from the British Library

Library of Congress Cataloguing-in-Publication Data
Vertebrate paleontological techniques / edited by Patrick Leiggi,
Peter May.
p. cm.
Includes index.
ISBN 0 521 44357 1 (v.1) hardback
1. Fossils, Vertebrate–Collection and preservation.
2. Paleontology—Technique. I. Leiggi, Patrick. II. May, Peter
(Peter J.)
QE841.V388 1994
566'.075–dc20 94-4152
CIP

ISBN 0 521 44357 1 hardback
ISBN 0 521 45900 1 paperback

Dedicated to
Arnold Lewis and Gord Gyrmov

Contents

Contributors

William W. Amaral
Museum of Comparative Zoology, Harvard University, Cambridge, MA 02138

Kathy Anderson
The Mammoth Site, Inc., Hot Springs, SD 57747

Ann R. Bleefeld
American Museum of Natural History, Department of Vertebrate Paleontology, Central Park West at 79th Street, New York, NY 10024-5192

Kenneth Carpenter
Department of Earth Sciences, Denver Museum of Natural History, City Park, Denver, CO 80205

Dan S. Chaney
Department of Paleobiology, National Museum of Natural History, Smithsonian Institution, Washington, DC 20560

Sandy Clark
Department of Radiological Sciences, Sunnybrook Health Science Centre, Toronto, Ontario, Canada

John Congleton
Geology Department, Southern Methodist University, Dallas, TX 75275

Judy Davids
The Mammoth Site, Inc., Hot Springs, SD 57747

Walter B. Elvers
American Museum of Natural History, Department of Vertebrate Paleontology, Central Park West at 79th Street, New York, NY 10024-5192

Mark B. Goodwin
Museum of Paleontology, University of California, Berkeley, CA 94720

Frederick Grady
Department of Paleobiology, National Museum of Natural History, Smithsonian Institution, Washington, DC 20560

Jorg Harbersetzer
Forschungsinstitut Senckenberg, Senckenberganlage 25, D-6000 Frankfurt, Germany

Terry Hodorff
The Mammoth Site, Inc., Hot Springs, SD 57747

John R. Horner
Museum of the Rockies, Montana State University, Bozeman, MT 59717

Brian Iwama
Royal Ontario Museum, 100 Queens Park, Toronto, Ontario M5S 2C6, Canada

Alexander W.A. Kellner
American Museum of Natural History, Central Park West at 79th Street, New York, NY 10024-5192

Patrick Leiggi
Museum of the Rockies, Montana State University, Bozeman, MT 59717

Arnold Lewis
5244 Boxturtle Circle, Sarasota, FL 34232-3411

James Madsen
Dinolab, Box 9415, Salt Lake City, UT 84109

John G. Maisey
American Museum of Natural History, Department of Vertebrate Paleontology, Central Park West at 79th Street, New York, NY 10024-5192

Peter May
Research Casting International Ltd., 2342 Wyecroft Road, Unit G1, Oakville, Ontario L6L 6M1, Canada

Russell McCarty
Florida Museum of Natural History, University of Florida, Gainsville, FL 32608

Malcolm C. McKenna
American Museum of Natural History, Frick Curator, Department of Vertebrate Paleontology, Central Park West at 79th Street, New York, NY 10024-5192

James S. Mellett
Department of Biology, New York University, New York, NY 10003

Ian Morrison
Department of Vertebrate Palaeontology, Royal Ontario Museum, 100 Queen's Park, Toronto, Ontario M5S 2C6, Canada

Betty Quinn
Museum of the Rockies, Montana State University, Bozeman, MT 59717

Peter Reser
New Mexico Museum of Natural History, Box 7010, Albuquerque, NM 87194-7010

Raymond R. Rogers
Department of Geophysical Sciences, University of Chicago, 5734 South Ellis Avenue, Chicago, IL 60637

Ivy S. Rutzky
American Museum of Natural History, Department of Vertebrate Paleontology, Central Park West at 79th Street, New York, NY 10024-5192

Charles R. Schaff
Museum of Comparative Zoology, Harvard University, Cambridge, MA 02138

Sally Y. Shelton
Materials Conservation Laboratory, Texas Memorial Museum, 2400 Trinity, Austin, TX 78705

Darren Tanke
Royal Tyrrell Museum of Palaeontology, Box 7500, Drumheller, Alberta T0J 0Y0, Canada

Michael Tiffany
Museum of Paleontology, University of California, Berkeley, CA 94720

James W. Wilson
Museum of the Rockies, Montana State University, Bozeman, MT 59717

Foreword

Vertebrate paleontology, as well as other disciplines of paleontology, is a field of study where the accuracy of collection and preparation of specimens and data is the foundation that determines the ultimate quality of the science. Improper collection, data acquisition, preparation, and curation lead to inaccurate interpretation, much of which is exacerbated as time passes. Because these aspects of paleontology are so important to the final results of study, it is essential that the latest, most reliable methods be performed on a consistent basis.

Over the past few years preparators have been responsible for coming up with a variety of new techniques to better collect, clean, and preserve specimens. With the diversity of new techniques now being introduced into vertebrate paleontology, preparators are having to perform tasks that only a few years ago were reserved for more technically qualified personnel. Vertebrate preparation now includes osteohistological, CT scan, X-ray, and imaging techniques.

Because of the diversity in directions that the field is now taking, it is clear that few preparators will be able to accomplish all tasks equally well. *Vertebrate Paleontological Techniques, Volume 1,* is a comprehensive collection of papers dealing with specimen collection, preparation, conservation, and reproduction. This book will clearly be of tremendous help not only to professional and amateur preparators but also to professional paleontologists interested in expanding the directions of future research.

Dr. John R. Horner

Preface

Not all preparators have the luxury of being a link in a long chain; the knowledge gained by preparators from their predecessors is usually never disseminated beyond the immediate surroundings. These thoughts were the beginning of this volume.

The evolution of *Vertebrate Paleontological Techniques, Volume 1*, has been a long one. While working as a preparator for Jack Horner and Don Baird at Princeton University, Pat wondered why a techniques volume had never been specifically written for the science of vertebrate paleontology. This was in 1983. Later in the year, Pat met Chuck Schaff, who agreed that something should be written for our discipline. Nothing ever materialized.

In 1987 Pat discussed the idea with Bob Makela in the badlands of Montana. Bob was excited about the idea, and it looked as though the project was finally under way. Unfortunately, Bob died tragically in a truck accident early in the field season of 1987. Bob's lifelong contributions to field paleontology would have been a welcome addition to this volume. The section on Field Organization and Specimen Collecting in Chapter 4 of this volume is a tribute to his successes and professionalism.

In 1988 we found ourselves sitting in a tipi along the Two Medicine River on a rainy night in the badlands of Montana discussing the lack of information available to preparators. We had heard someone else had been writing a volume on techniques but it never seemed to materialize. We decided that night to attempt the task ourselves.

It began in 1989 with a massive mailout campaign to every member of the Society of Vertebrate Paleontology. The response was very positive if not overwhelming.

Vertebrate Paleontological Techniques, Volume 1, has been pared down from an enthusiastic 137 responses, abstracts, and manuscripts to a realistic number of 12 chapters. The authors are exceptional technicians in their field. Many of the 137 original contributors are not included in this volume. This was not because of any critical evaluation on our part but rather our desire for a more cohesive volume. We found that no two preparators ever think alike or agree on a single procedure or method. We did find, however, that preparators base their dedication and skills to the art of preparation on three common principles; patience, concentration, and a good sense of humor.

Instead of attempting to prepare the "mother of all volumes" on technique, we decided that it would be more appropriate to provide a vehicle for continuing the evolution of shared methods among professionals in the field of vertebrate paleontology. *Vertebrate Paleontological Techniques, Volume 2,* is already in progress.

We hope that this volume will be useful to students, interns, beginning preparators, and professionals alike. More importantly, if you ever consider this volume to be great bedtime reading material, please . . . see a doctor.

Acknowledgments

The editors wish to thank the following colleagues and friends for their kind help, support, and suggestions in making this volume possible: John Horner, Phillip Currie, Peter Dodson, David Weishampel, Chris McGowan, Gord Edmund, Rufus Churcher, James and Beatrice Taylor, Barbara Love, Kenneth Carpenter, Robert Harmon, Allison Gentry, John McMulkin, Jason Campbell, Scott Sampson, Mark Goodwin, Dan Chaney, Charles Schaff, Don and Beverly McKamey (quiet time at the McKamey ranch), Sheldon McKamey, Teresa May, Judith Weaver, Arthur Wolf.

For their patience and understanding we gratefully thank our wives,

Sheldon L. McKamey and Teresa M. May

Introduction

Patrick Leiggi and Peter May

For many years the science of vertebrate paleontology has been without an up-to-date, accurate technical volume. Manuals have been published – most recently, Feldmann, Chapman, and Hannibal's, *Paleotechniques,* (1989), which includes some chapters that relate to vertebrate paleontology – but on the whole they are not specific to the science. Volumes that directly relate to vertebrate paleontology are few and far between. The most recent was more of a manual than a book written by Howard Converse in 1984: *Handbook of Paleo-Preparation Techniques.* Earlier were Rixon's 1976 volume, *Fossil Animal Remains: Their Preparation and Conservation,* and the edited volume *Handbook of Paleontological Techniques,* by Kummel and Raup (1965). *Vertebrate Paleontological Techniques, Volume 1,* is the first offering in what we hope to be a series. It is an attempt to standardize techniques for the science of vertebrate paleontology.

When we began our careers as preparators in vertebrate paleontology, the technical procedures we were taught had been handed down from preparator to preparator. At the Royal Ontario Museum the majority of the dinosaur material was collected in the badlands of Alberta by Levi Sternberg, and prepared by him or under his direction. In the collections and warehouses of the Royal Ontario Museum many blocks are piled up that were collected as far back as 1918. If numbers had not been painted on the field jackets, it would be difficult to tell the difference between the field casts collected long ago and those collected as recently as last year. Like many other museums around the world, many magnificent prepared specimens rest in collection rooms. Most were prepared with nothing more than hammers and chisels for the coarse work and small picks for fine work. The skill of these pioneer preparators is evident in the work they performed and is revealed through the dust that has collected over the years, the quality of their workmanship. The technical processes developed by these early technicians for collecting, preparing, molding, casting, and mounting were passed on to professionals like Arnold Lewis and Gord Gyrmov.

These time-tested procedures remain the same today, but instead of hammers and chisels and horses and wagons, we now have pneumatic scribes and helicopters. The final result of the collecting and preparation of vertebrate information is still a rare and exquisite example of extinct animals that were, when alive, an integral part of our world's history. The preservation of the world's heritage is the job of every preparator who sets out to remove the dirt that surrounds the mystery hidden within the earth's sediments.

The past 40 years have introduced many new tools and materials to the science of vertebrate paleontology. *Vertebrate Paleontological Techniques, Volume 1* is a beginning in the process of standardizing techniques that can be applied by amateurs, beginning professionals, students, and experienced preparators alike. Many institutions are still using materials that are known to deteriorate over the years. Adhesives that break down in a year are still being applied to the joints of type specimens. Consolidants that last only five years before ultraviolet light causes breakdown in the stability of the applied materials are still being used. Valuable specimens are still being housed in collections in which the environment cannot be controlled.

We have called on the assistance of professional preparators who are not only skilled in procedure but also are aware of the conservation aspect of their work. Conservation aspects in archaeological technical practice is one field where consolidants and adhesives are analyzed for specimen deterioration and in the comparative safety of collections. Technical procedures are analyzed microscopically to ensure that nothing detrimental to the artifact will occur during the course of preparation. As in archaeology and other related fields, the need to educate preparators with the skills required for specimen preparation, and the introduction of values required to ensure the durability of prepared vertebrate fossil specimens, will provide an awareness of beneficial conservation practice.

We are aware that some techniques and materials described in this volume have not been fully tested in a paleontological application. We can only ask that preparators become more aware of conservation procedures and their application to vertebrate paleontological science. If vertebrate paleontological preparators can begin standardizing methods of preparation, specimen stability will be maintained. To accomplish this, we ask that all preparators add to the knowledge of others by publishing their experiences. We hope the information provided by the authors in this volume will be considered a tribute to the quality of preparation skills we have been taught and will continue to be shared with preparators, today and in the future.

1

Conservation of vertebrate paleontology collections

Sally Y. Shelton

The science of vertebrate paleontology and the science of conservation are converging to create a new research discipline. Vertebrate fossils are more susceptible to damage and deterioration than has often been assumed. The specific biochemical and geochemical structure and nature of the specimens, poor preparation materials and methods, and poor storage environments can act singly or in concert to effect significant deterioration. In addition, many new physical and chemical analytical techniques in use require that specimens be free of contaminants, including the wide range of polymeric materials used in preparation. Materials used to prepare or store vertebrate fossils should not cause permanent change in the fossils unless the primary immediate effect of the material (e.g., consolidation) is necessary to prevent certain loss of the specimen. Chemical compounds should, if at all possible, be reversible and of a known composition and concentration; the use of proprietary compounds should be a last resort. Accurate documentation of materials and methods used is as important as accurate field data and catalog information. Research and analysis in the field of paleontology conservation are necessary; peer-review publication at the highest levels of research publication is essential. Further investigation and training in the field will help guarantee the long-term stability and wide-range usefulness of vertebrate fossil material.

Paleontology and conservation

The word *conservation* has several meanings, not necessarily exclusive of each other. In this context it is used to refer to the science of preserving specimens in the best and most scientific way possible, so that both long-term stability and long-term value are ensured. The integrity of the vertebrate fossil specimen as a unique biogeochemical entity is of primary importance; all methods and materials used in its handling and storage should be selected to maintain that integrity where possible.

Conservation science is best developed in the fields of art and archaeology, less well in the natural history disciplines. An awareness of the fragility of natural history materials, the damage that can be done to them by poorly researched preservation materials and strategies, and the wealth of information at the cellular or molecular level that can be obtained from these specimens have sparked a surge of interest in the development of natural history conservation as a research discipline.

All aspects of vertebrate fossil and trace fossil study, from field and laboratory preparation to exhibit, research, education, and storage, can be examined from the standpoint of conservation. Such examination reveals that many approaches, although well intentioned, are rarely analyzed by the standards of scientific rigor that are applied to the results of studies on these specimens. In those cases in which the methods and materials in use actually reduce the stability, integrity, and research value of the specimens, the field of vertebrate paleontology has been poorly served.

In vertebrate paleontology conservation, the field of archaeological conservation has been used as a model for some of the most basic precepts of care. The sheer range of time, preservation modes, numbers of specimens, and research strategies in vertebrate paleontology, however, mandate the development of a dedicated research field to address the myriad problems not shared with archaeological investigation.

Recognizable vertebrate material in North America dates from the Lower Ordovician (500–550 million years ago) to the present. The range of preservation modes in vertebrate paleontology militates against a simple formulaic approach to conservation in this field. Skeletal elements may be unaltered bone, bone with the pore spaces filled with recrystallized minerals (but the organic matrix intact or preserved in some degree), or natural casts containing no original material. Original soft tissue has been preserved by freezing, bogs, and encasement in amber and anaerobic environments. Trace fossils (e.g., trackways) generally contain no body elements, yet provide information on behavior, kinesiology, and social structure that would otherwise be unavailable. The preservation of keratinaceous and collagenous protein structures (eggs, gastroliths, scales, otoliths, coprolites, and other body materials) has contributed enormously to an understanding of past life. Unusual preservation environments, including asphalt and oil shales, produce fossil vertebrate material that must be prepared in unusual ways.

The size of vertebrate fossil collections also raises significant concerns. Individual collections in the United States can range up to nearly 2 million specimens (Langston 1977), a significant number of which may be stored in field jackets or matrix awaiting preparation. It is not practical or feasible to design conservation strategies that focus only on single-specimen benchwork in large collections. Preventive conservation programs, which focus on the design of storage and handling strategies to minimize damage before it occurs, are more cost-effective and productive than are intensive individualized treatments of problems that have already occurred and may not be reversible.

Research fields based on or applicable to vertebrate paleontology become more numerous every year. Some, such as stratigraphy, systematic and cladistic analyses, and ecology, depend more on the preservation of gross morphology than on cellular or molecular structure. To date, the long-term effects of adding substances to consolidate, adhere, mount, or clean the specimen have been of little importance in these studies. Recent research indicates that the fraction of unaltered organic material in permineralized bone may be larger than was previously believed and that amino acids and other key biological molecules may be recovered essentially as they were in life. Histological, cellular, and molecular research requires that the skeletal material marked for analysis be free of contamination, including adhesives, consolidants, solvents, or substances introduced by even careful handling and storage. Techniques that are valuable for mounting an exhibit may be disastrous for research quality material. In short, the range of time, diversity, preservation modes, numbers of specimens, and uses demands that vertebrate paleontology be a flexible field, responsive to both collections and users.

A final point is the necessity of caring for vertebrate fossil specimens. Most collections are held in public trust in some form, and this imposes a duty of care. Failure to provide due care can leave the collection administrators in a position of legal liability if damage through negligence can be shown (Ullberg and Lind 1989). Apart from a legal duty, curators have a professional and ethical duty as well (Child, personal communication).

Agents of deterioration

One problem for vertebrate conservation is in the widespread misapprehension that vertebrate fossils are inherently stable, seldom react with their environments, and will last as long in collection or exhibit environments as they did in burial environments (Yochelson 1985). None of these statements is true. As Rixon (1961) writes: "Anyone who has been behind the scenes in any museum housing a large collection of fossils will know how brittle, friable, subject to chemical decay and to damage due to changes in humidity and temperature these specimens are." The rapid and extreme change in environment that occurs at the time of exposure often sets into motion a wide range of deleterious reactions. All the factors in a storage environment can affect the chemical compounds of the specimen adversely and synergistically (Goffer 1980). The substances used for short-term preservation of integrity may themselves be extremely reactive and short-lived, and often cause irreparable damage to the specimen as they age. Some of the minerals that fill pore spaces are unstable in the presence of changing humidity (e.g., pyrite and anhydrite); their reactions with the environmental conditions can destroy vertebrate fossils in a very short time. Poor storage conditions can accelerate reactions, particularly some efflorescences that occur in hydroxyapatite and calcium carbonate structures in the presence of wood acid vapors (FitzHugh and Gettens 1971). Fossils with a high fraction of organic material (such as collagens and keratin in unaltered organic structures) are subject to biodeterioration.

Michalski (1990) identifies nine major agents of deterioration that affect or can affect all objects and specimens in museum collections.

- *Physical neglect* – including all forms of structural damage through accident or carelessness, including damage or contamination of a specimen occurring in preparation and handling, as well as damage from unrestrained motion in an earthquake or other perturbation of the building.

- *Thieves and vandals* – particularly a concern in vertebrate fossil collections, which have a high commercial value in nonprofessional circles; some materials, including ivory, rhinoceros "horn," and amber, are particularly at risk for theft.
- *Fire* – either outright destruction or smoke damage, as well as the risk of losing key documentation; much damage is water related.
- *Water* – can cause efflorescence, swelling, dissolution, corrosion, delamination, and cracking.
- *Biodeterioration* – including mold and fungus growth, as well as attack by insects and rodents; this may be direct specimen attack or damage to documentation.
- *Air pollution* – may include both extraneous factors, such as acidic outdoor pollutants brought into the building, and intrinsic factors, such as outgassing of poorly selected paints and gaskets in cases, wood acid vapors, cigarette smoke, and solvent fumes.
- *Ultraviolet, visible, and infrared radiation* – can cause fading of key minerals as well as embrittlement of paper, film, and adhesives; undesirable IR levels can accelerate drying and cracking in sensitive materials; darkening and cross-linking of surface coatings and consolidants.
- *Temperature fluctuations and inappropriately high or low temperatures* – these generally act in concert with the effects of relative humidity variations.
- *Relative humidity fluctuations and inappropriately high or low humidity* – high humidities foster some forms of biodeterioration, particularly mold growth; fluctuations lead to the hysteretic response that causes cracking in bone, amber, and ivory; low humidities are deleterious to materials that were preserved in damp or wet burial environments; pyrite decay, efflorescences, and the destructive anhydrite/gypsum transition are all associated with changing relative humidities in the storage environment.

Preventive conservation in vertebrate paleontology attempts to minimize the damage that could be caused by each of these factors by taking measures to prevent their occurrence. Many of the preventive measures obviously center around the selection or modification of building, case, or other structural designs and materials. Others are affected by the amount and type of access allowed to specimens in collection or on exhibit. Although some factors can be mitigated easily (the UV fraction of light is not visible, not necessary for vision, and easily filtered from both windows and fluorescent tubes), others require advance planning for dealing with specific disasters. Many collections faced with the universal lack of adequate resources opt for a risk management strategy in which precautions are taken only against the most likely sources of deterioration affecting the given site or situation, while assuming the risk of less probable disasters.

It is accurate and prudent to assume that all specimens are in interaction with various environments at all times and that all specimens are breaking down as a result of such interactions, even if the rate is infinitesimally slow; but that, whereas some measures can be taken to retard the effects of entropy, the value of the collection lies in its use. The need for and effects of use are therefore necessary factors in devising a conservation plan – although use may accelerate deleterious reactions or even mandate destructive testing of part or all of a specimen. A useful conservation strategy must take the apparently opposing concerns of stability and use equally into account.

Vertebrate tissues and mode of preservation

The range of vertebrate diversity, tissues, and structures, as well as the range of preservation modes, precludes a simplistic approach to conservation in this field. It is best to understand the physical and chemical structure, preservation mode, and anticipated use of any specimen before planning its storage and handling. Although many fossil vertebrates are fairly durable in a controlled environment, some are far more fragile and require a more specialized system.

Unreplaced or sub-fossil bone, teeth, antler, ivory, and cartilage

Original mineralized or connective tissue materials are frequently found in the comparative collections maintained for reference and comparison. Original skeletal material may also be found in some Pleistocene collections from temperate or polar arid regions (e.g., dry caves and the Antarctic Dry Valley), as well as from permafrost environments. (Specimens found with significant preservation of original soft tissue are discussed in detail in a subsequent section.)

Bone is a complex structure. Its essential framework is a collagenous matrix invested with mineral salts, the most significant of which is hydroxyapatite [a calcium phosphate compound, $Ca_5(PO_4)_3OH$]. The degree and nature of mineral investment vary from class to class and, to some extent, within classes. The type and extent of mineral salt content determine many of the characteristics of bone (Eanes and Posner 1970). The outer layer of bone is the hard, compact cortex, with an inner spongy medulla. The degree of vascularization varies greatly among the vertebrate classes.

Some bone minerals, such as fluorine and manganese salts, are taken up during life and deposited during the constant process of osteoblastic remodeling; their presence and relative abundance are important in reconstructing environmental influences, diet, and age.

Original bone has its collagenous matrix more or less intact; this is subject to all the agents of biodeterioration if housed or handled inappropriately. Dry bone may become brittle and is easily damaged by poor storage and handling. In order to facilitate the drying of mammalian long bones, some preparers have drilled longitudinal holes in order to extract the marrow. This practice is exceedingly damaging; the unsupported bone undergoes torque during drying, and the morphology is permanently altered. Such bones should never be used for comparative anatomical uses, as their morphology may be deceiving.

The methods used to prepare Recent skeletons must be examined carefully. Use of a commercial enzyme-based detergent compound has been shown to cause total bone loss within 20 years (Shelton and Buckley 1990). In general, the use of caustic compounds such as ammonia, chlorine bleach, and ethanol baths is discouraged. Other methods for cleaning and degreasing bones without robbing them of their structural stability exist and are documented (Cumbaa 1983; Simmons 1987).

Teeth present some special problems. In mammals, enamel, which is highly mineralized, is apatitic, with a structural formula of $Ca_{10}(PO_4)6X_2$, in which X may represent fluoride, hydroxyl, carbonate, or bicarbonate ions (Zipkin 1970). Mammalian enamel is functionally impermeable, with very large hydroxyapatite content, whereas the underlying dentin, which has pore spaces, behaves more like bone (Parker and Toots 1980). Cementum and pulp structures share more characteristics with soft tissues. This causes differential responses to environmental changes, making teeth very susceptible to crazing and cracking. The teeth of other vertebrates are less complex, often consisting primarily of dentine. Teeth should never be frozen or subjected to sudden or severe changes in relative humidity or moisture content if cracking is to be avoided (Williams 1991). Because of their importance in establishing phylogenetic affinities, teeth should receive special attention in storage to prevent physical loosening and loss due to exposure to fluctuations in temperature and humidity.

Antler is a frontal bone outgrowth characteristic of cervids. It is the only vertebrate structure that is periodically shed and regrown (Goss 1983). Antler has no keratinaceous sheath (characteristic of horned ungulates) but is covered by a vascular membrane called "velvet" during active growth. Because it is grown so rapidly (complete regrowth between sheds is normally accomplished in a year), it is less well mineralized than bone and shows less cellular structure (O'Connor 1987); it does share with bone, though, an outer compact cortex and an inner spongy medulla. In general, antler can be stored in similar conditions to those suitable for the protection of bone.

Ivory is a dentine structure in which the mineralizing cells are not incorporated in the tissue but, rather, line up on the growing surface. This gives ivory an acellular structure (O'Connor 1987). Ivory is associated with proboscideans (similar structures in walruses and narwhals may also be included). It is particularly susceptible to environmental fluctuations. In a hysteretic response to fluctuations in humidity, ivory cracks concentrically and longitudinally. DNA information from Recent elephant ivory has been used to "fingerprint" individuals and herds so that illegal collecting can be traced to its origin (Cherfas 1989). Ivory is an economically valuable and politically sensitive material. It is, with rhinoceros "horn," one of the vertebrate tissues most likely to be stolen from a collection. These concerns suggest that ivory should not be stored with the rest of the collection, but housed in an area with proper climate control and good security.

Cartilage is another term for reinforced collagen with less mineral investment than bone. It can be found in connective tissues, fetal tissue membranes, incompletely ossified bone, and the skeletons of nonteleost fish. Biodeterioration is a primary concern with cartilage. Some cartilaginous materials are cleared, stained, and stored in glycerin.

In general, none of the materials discussed here should be fumigated; an alkaline fumigant can destroy amino acids very rapidly. Well-gasketted metal storage systems are vastly preferable to wooden cases. Skulls that contain teeth and/or ivory must never be frozen as a means of pest control, as this will cause crazing and cracking of the structures.

Eggs and otoliths

Eggshells are complex structures with mineralization by calcite in both birds and dinosaurs; the mineral investment of modern chelonian eggshells is aragonite, which is less stable in burial environments than calcite (Wyckoff 1972). Like other carbonate structures, eggshells should not be stored in materials made from wood or acidic wood products or in proximity to other acid-producing materials such as PVC, some PVA emulsion paints, insecticides, and the like. A white crystalline growth misidentified as mold in some eggshell collections is almost certainly an efflorescence resulting from attack by organic acid vapors.

Otoliths are calcium carbonate (generally aragonitic) structures in fish, analogous to the otoconia of other vertebrates (Secor, Dean, and Laban 1991). In teleost fish, they occur as three pairs of well-developed structures that function in balance and hearing. Their use in determining growth rates, species identification, and environmental influences is critically important (Panella 1971). Because they are calcium carbonate instead of calcitic, otoliths are extremely soluble in acid solutions or conditions that might not affect bone or calcite structures. Otoliths are generally stored dry or in 95% ethanol solutions (Hendrickson, personal communication) until they are needed for sectioning. If they are found dry at a paleontological or archaeological site, they should be stored in dry conditions.

Formaldehyde (or formalin, in its commercially available solution) is a notorious decalcifier. Its use should be strictly limited to immediate specimen fixation; neither formalin nor any other acidic solution should ever be used as a preservative for specimens intended to provide otolith or bone preparations (McMahon and Tash 1979; Hendrickson, personal communication). Otoliths will dissolve very quickly; bone will decalcify and leave collagenous matrix. Vertebrates that need to be preserved in fluid before being skeletonized should be stored in 70% undenatured ethyl alcohol.

Calcium carbonate is susceptible to attack by wood acid vapors; thus fossil carbonate structures should be stored in well-sealed metal or inert plastic storage systems. Storage in closed wooden cases or in containers made from acidic wood products can lead to an efflorescence of mineral salts, which is known to cause deterioration in a wide range of invertebrate and vertebrate calcium carbonate structures (Tennent and Baird 1985).

Unaltered proteinaceous structures

Keratins are structural protein compounds; they make up hair, baleen, horn sheath, beak sheath, feather, scale, nail, claw sheath, turtle scute, and hoof tissues. Keratins are very insoluble (Mills and White 1987). Their preservation in non-Recent specimens occurs under conditions similar to those that preserve collagen.

Keratin tissues respond poorly to environmental fluctutations, often splitting and delaminating. This type of physical change does not preclude their biochemical use, but it does impair their morphology. Some dense keratin structures such as horn sheath and baleen can be carefully restored to an approximation of their original morphology (Peever 1989), an invasive procedure that, though necessary to save some specimens, affects the biochemical structure of the keratin. Collagenous and keratinaceous structures that show signs of infestation can be safely bagged and frozen to eliminate pests.

Keratins and collagens are highly subject to biodeterioration. They should be stored in dry (50% humidity or less) conditions. Rhinoceros "horn," an agglutinated hair structure, is, like ivory, at a higher risk of theft than most specimens and requires security from theft while in storage.

Soft tissue [freezing, bogs]

Though rare, original soft tissue has been found in animals (generally of Pleistocene age) entombed in unusual burial environments. Important finds have been made in subpolar sites in permafrost sediments; and frozen specimens from Siberia and Alaska have been found for many years (Guthrie 1990). These animals may be coated with a layer of blue vivianite deposited in the burial environment. Such specimens must be preserved in an environment that will keep them from thawing and from oxidative changes. Their tissues are highly valuable for a variety of analyses. Frozen materials should never be fumigated or subjected to other chemical treatments.

Bog preservation frequently yields vertebrate remains in which the soft tissues are preserved by the action of tannic and other acids. These specimens are frequently decalcified by the preservation environment. Care of bog-preserved material is complex, requiring the prevention of oxidation and the maintenance of a stable case climate. These specimens require sophisticated care in their storage and handling if their integrity is to be maintained.

Amber

Amber is a term for any of a number of translucent hardened tree resins of varying age and structure (Mills and White 1987); the most characteristic component of Baltic amber is succinic acid (Langenheim 1965), but the composition varies widely. Amber is a well-known mode of preservation, most commonly of a wide range of invertebrates.

Vertebrate preservation in amber, though rare, does happen. Lazell (1965) reports the discovery of two partial but diagnostic Oligocene or Miocene *Anolis* specimens in amber from Chiapas, Mexico. In the Dominican Republic, an Eocene frog (Poinar and Cannatella 1987) and an *Anolis* (Rieppel 1980) in amber have been documented. Fragmentary remains of hair or feathers in amber have also been reported occasionally.

Vertebrate specimens in amber are exceedingly important as sources of environmental, biomolecular, and ecological information. Samples from at least one specimen exposed to the surface by cutting and polishing have been removed for study (Cannatella, personal communication). The specimen should never be removed from the amber, and the amber should be stored in a dark environment free of temperature and humidity fluctuations. Cracking can be a problem in amber exposed to low relative humidities or to fluctuations; a relative humidity of 55% is recommended (Trusted 1985). The amber may require cutting and polishing to reveal the specimen, and the polished surfaces react rapidly to exposure to ammonia, hydrogen sulfide, formic acid, and acetic acid (Waddington and Fenn 1988). For these reasons, exposed storage or storage in acidic storage materials (including wood and acidic wood products) is unacceptable. Immersion in mineral oil or treatment with a wide range of now unacceptable surface coatings has been recommended in the past (Beck 1982), but these are neither necessary nor desirable if the specimen is kept in the right environmental conditions. Amber is highly marketable and thus should be particularly protected from theft.

Permineralized bone

Bone with pore spaces filled and certain elements of the apatite framework replaced with minerals from solution in a burial environment is one of the most universally familiar modes of vertebrate fossilization. Unfortunately, the formation and structure of these fossils have not always been well understood. Permineralized bone is not the same as a natural cast (see discussion in the next section). Evidence is mounting that a significant fraction of the original material may remain, as does the original hydroxyapatite crystal orientation (Hare 1980; Mervis 1991; Gillette personal communication). These fossils should never be treated as if their mineral components have completely replaced the original structure; chances are that mineralization proceeds as a void-filling process rather than as molecule-by-molecule replacement.

The minerals that fill in a permineralized bone vary widely, and the specific mineral composition determines much of the nature of the fossil specimen. Each specimen represents a unique combination of life and taphonomic events. Some are relatively stable and require little attention; others are extremely unstable and may break down rapidly in a storage environment (Waller 1980).

In the burial environment, silica and aluminosilicates are common void-filling minerals, as are manganese, barium, lead, iron, calcite, and gypsum compounds. Expansion of these minerals may cause crushing of the original cellular or crystal structure to some degree, but this is a physical change, without chemical interaction. Fluorine and yttrium may be found in the apatite structure; they tend to replace hydroxy and calcium but leave the orientation of the crystals unimpaired. Bone and dentin are permineralized nearly twice as rapidly as teeth (Parker and Toots 1980).

One of the most unstable minerals found in permineralized bone is microcrystalline pyrite, FeS_2. Under conditions of high humidity or saturation, the microcrystalline form of pyrite breaks down in an expansive oxidation reaction that liberates corrosive sulfuric acid compounds. Despite older references to bacteria as a cause of this problem, it now appears that the *Thiobacillus ferrooxidans* bacterium is merely opportunistic, though possibly catalytic, and arrives after the reaction has begun. Neutralization of the reaction products by ammonia gas treatment (Waller 1987) has been shown to have short-term effectiveness for affected material.

Pyrite decay cannot be reversed and can be controlled only by storage in stable dry conditions ($<$ 30% RH). Uncontrolled, it will destroy the specimen, as has happened to several notable fossils. The marcasite morph of pyrite, contrary to older accounts, is almost never implicated in this condition. Older recommendations for treatment with bactericides and disinfectants are ineffective and should not be used.

Bone that has been void-filled with gypsum or anhydrite [$CaSO_4.2(H_2O)$ or $CaSO_4$] may respond very poorly to extreme or prolonged fluctuations in relative humidity. As its name suggests, anhydrite is the anhydrous morph of gypsum, and the phase shift between the two involves the take-up or release of water. This causes an associated volumetric change that may be sufficient to crack and destroy the bone.

Occasionally, mineral salts will rupture the surface of a fossil specimen. This may represent either the migration of soluble salts from the interior of the specimen (another artifact of the burial environment) or the interaction of the surface with acid wood vapors. The latter is most commonly seen in calcium materials, in which acetic acid and formic acid react with the calcite to form calcium formate and calcium acetate. Calcium chloride (calclacite) and calcium nitrate salts have also been identified. This has been referred to as Byne's disease in malacology but is more accurately called an efflorescence.

Bone that has been void-filled with silicates is very stable; it does not appear to be easily susceptible to breakdown or corrosion. In some cases the expansive recrystallization of silicates may deform or crush cell structure from within, a process that can easily be confused with the effects of external pressures. Permineralization generally increases the weight of the bone greatly. The heaviness and brittleness of fossil bones in this condition may exacerbate storage and handling problems.

Natural casts and molds

When a structure buried in sediment decays, it may leave a hollow in the surrounding sediments (the mold), which in turn may be filled in completely (the cast) by materials extraneous to the original specimen. There is no original organic material and no evidence of cell structure or hydroxyapatite crystal orientation left in the specimen (though there may be trace chemicals in the immediate matrix). The nature of the matrix sediments and replacing minerals determines the nature of the specimen.

Natural molds and casts are less severely affected (in terms of research value) by chemical contamination than are permineralized bones. They can be treated as mineral entities, subject to the same standards of care as mineralogical holdings. It is always wise to establish which minerals are present, so that care of the less stable substances can be planned. Mineral identification may involve some degree of destructive testing.

Natural casts are rarer in vertebrates than in invertebrates and are found less often than permineralized bone. If there is doubt, it is best to treat all fossil bones as if they have an original organic fraction until their true nature can be determined.

Trackways and trace fossils

Trackways and trace fossils are valuable specimens. Their role in determining kinesiology, ethology, and social structure of extinct vertebrates, as well as environmental conditions at the time of formation, is unparalleled (Lockley 1986). In general, these fossils do not include body fossils of any kind but, rather, the impressions made by the physical structure, movement, or activities of the animals.

Trackways, as their name implies, are the footprints, tail marks, and other such signs of the movement of an animal or group of animals across a soft substrate. The prints are preserved in the substrate as it hardens to rock. In some cases, such as the Paluxy River dinosaur trackway, evidence of invertebrate bioturbation is preserved as well, furnishing valuable clues to the paleoecology of the area. Trace fossils have never been transported from their site of formation, maintaining their close relationship with the original depositional conditions (Frey 1975).

Trackways are sometimes removed from their site – an extensive, exhausting, and extremely damaging operation that is strongly discouraged unless the site is facing certain obliteration or destruction. Thus tracks are often left in situ (discussed in a subsequent section).

The nature of the stone determines the care of trackways. Sandstone and shale trackways may be more vulnerable to short-term erosive effects than are limestone and dolomite trackways. All are susceptible to weathering, in particular the severe effects of freeze–thaw cycles. One dinosaur trackway on exhibit has apparently been damaged by rising damp and the reactions of the plaster used as a joint filler (Shelton 1993). Limestone and dolomite trackways can be rapidly dissolved by acid precipitation. Trackways at the National Museum of Wales are suffering from delamination and crumbling caused by desiccation (Buttler, personal communication).

Other trace fossils may include skin impressions (Ratkevich 1976; Sternberg 1990), burrows, and gnawings or other toothmarks. Their storage and handling are similar to those for track fossils.

Asphalt and oil shales

Preservation in hydrocarbon-based deposits has yielded some of the most significant collections in vertebrate paleontology. Asphalt deposits are formed when native hydrocarbon deposits seep to the surface or are exposed; they generally have lost most or all of their volatile fraction (Mills and White 1987). Oil shales contain large amounts of keratogen oil.

Hydrocarbon-impregnated matrix has yielded some of the finest and most detailed fossils known. It appears that the hydrocarbons infiltrate the bone without effecting true replacement of organic or hydroxyapatite structures. The extensive vertebrate fauna of the La Brea tar pits is notable for its retention of collagen protein microstructure, which has been used for radiocarbon dating of the specimens and site (Shaw, personal communication).

Preparation of hydrocarbon-impregnated fossils generally involves the removal of as much of the saturating hydrocarbon material as possible. The use of strong solvents (such as 1,1,1-trichloroethane for the La Brea material) has not impaired the retrievability of microstructure material.

Coprolites, gastroliths, and calculi

Coprolites, or preserved fecal material, may be represented as either dried organic substances or natural casts. Dried coprolites, often found in arid Pleistocene cave deposits, are valuable sources of information on diet, ecology, parasitology, and climate. They must be kept dry, preferably in a sealed polyethylene bag or box with a preconditioned desiccant, to prevent damage by moisture and oxidation. Coprolites that are natural casts seldom show any remnant of internal structure, although their external morphology is a useful indicator. They are treated in the same way as are other natural casts.

Gastroliths, or "stomach stones," are often found in association with sauropod fossils (Ratkevich 1976). These are usually very smooth and geochemically stable; they were swallowed by the animal as an aid to digestion. Gastroliths should be stored with the specimen from which they came, in the same storage conditions.

Calculi of the biliary and urinary systems are very infrequently found in Recent and Pleistocene material. These are acellular deposits with a characteristic ringed deposition pattern in cross section. They must be kept dry and sealed for future study.

Materials and methods

A wide range of materials has been used in the treatment of fossil vertebrates (Howie 1984). In general, the published research on polymeric materials has been exceptionally slow in reaching the vertebrate paleontology preparation community. Knowledge of the short- and long-term effects of many adhesives, consolidants and coatings has not been well disseminated. A comparison of amateur and professional publications (Matthews 1962; La Plante 1977; Casanova and Ratkevich 1981; MacFall and Wollin 1983; Converse 1984) reveals almost no difference in the materials and methods recommended, although the aims of collecting differ widely between the two groups. Most writers recommend proprietary products without any reference to composition, testing, stability, or undesirable reactions. This cavalier attitude is not appropriate for research collections, and tends to establish the use of inferior materials that introduce unnecessary contamination in addition to their necessary function. Scientific rigor in materials selection is no less important than scientific rigor in other aspects of vertebrate paleontology research.

A specimen that has been saturated with consolidants, adhesives, and coatings (or even one that has been stored in a collection or exhibit for some time) is not the same biogeochemical entity that it was before such additions or exposure. It cannot be validly compared to a specimen from the same burial environment (or even from the same individual) that is newly exposed. This difference may not be significant in studies based on external morphology, but it may be enormous in studies based on cellular or molecular structure. It is prudent to collect uncontaminated samples in all field excursions, particularly in those cases where consolidation of the majority of the material is necessary for its retrieval.

Certain problems inherent in the retrieval of vertebrate fossils have been noted as long as fossils have been objects of interest. The effects of weathering frequently bring the fossil to the attention of a collector as well as destroy its integrity, so that collection is made more difficult. Also, specimens in clay soils may begin to crack visibly within hours after exposure to a drier climate. Element-by-element retrieval of a vertebrate specimen is often forgone in favor of block quarrying, which preserves the orientation of the elements relative to each other and also enables some retention of the original matrix. Measurements of in situ moisture content to indicate the best storage environment can help (Child, personal communication).

The primary concerns in retrieving vertebrate fossils have always been maximum recovery of all elements and any associated material of interest, adhesion of broken parts, consolidation of fragile or weathered material, and accurate documentation and meaningful organization of the find (Rixon 1976). Although the methods and materials used in pursuit of these objectives have varied widely, the goals remain essentially unaltered.

A more detailed discussion of the characteristics of adhesives and consolidants is provided in Chapter 2.

Documentation

Documentation of the treatment history of a specimen, from its original exposure through its life in storage, is one of the most critical of the many complex spheres of paleontology conservation – and one most often overlooked. Collectors who would never fail to document the geographic and stratigraphic locality, date of collection, and identification of a specimen will often consistently fail to record its physical and chemical treatment history. These gaps in the specimen record may have grave consequences if addition of a compound provokes an unforeseeable chemical reaction.

Inks for numbering should be selected for stability and colorfastness (Williams and Hawks 1986). Traditionally, vertebrate fossils are directly numbered on a clean bone surface without diagnostic features. Numbering on printed patches requires the use of a stable paint; correction fluid is not recommended. For very small fossils, the capsules or vials in which they are stored may be labeled instead.

The paper stock for labels should be durable, acid-free, and relatively light-stable. Synthetic paper stock made from polyethylene is becoming popular for both fluid and dry vertebrate collections. This material is nonreactive and stable; it cannot be used with a laser printer but has been successfully used with dot-matrix and impact printers (Gisbert, Palacios, and Garcia-Perea 1990). Older labels may require special care or removal to archival storage (Hawks and Williams 1986).

Forms for routine recording of vertebrate paleontology conservation data are in use at several institutions (Fitzgerald 1988; Seymour 1988; Howie, personal communication). These forms should be part of the specimen's permanent record and must be available to anyone conducting work on treated material.

Field materials and methods

Vertebrate fossil specimens must be collected or documented in order to be scientifically valuable. Archaeological approaches are not necessarily valid here, as vertebrate paleontology fieldwork tends to be formation specific rather than site specific (Committee on Guidelines for Paleontological Collecting 1987). An exposed fossil is subject to all the forces of weathering and will almost certainly be destroyed if left in situ (except in certain rare instances, as noted in the discussion on in situ sites). The techniques used in removing the fossil are the first form of treatment it receives.

The exposure and removal of a fossil specimen from its matrix are two of the most significant factors in determining its future stability and usefulness. Careful work, thorough documentation, and selection of stable materials all contribute to the long-term integrity and usefulness of the specimen.

Removing the matrix material places the specimen at an unavoidable risk of damage. Some methods used to facilitate this process, including dynamite, picks, and backhoes, greatly increase the amount of cracking and breakage in the material. Manual methods, such as probes and brushes, are vastly preferable, but they may not be feasible in a very hard matrix or in instances where time is at a premium. Preparation in the field should be minimal (Croucher and Woolley 1982).

Rigid jacketing was developed in response to the intensive collecting of Cope and Marsh, and may have been devised by Williston or Sternberg (Shor 1971). The earlier use of element-by-element removal of fossil skeletons placed the bones at severe risk of damage (Scott 1939; Sternberg 1990). Jacketing the bones in the field enabled collectors to preserve the original configuration of the bones while providing protection against the rigors of removal and transportation. Strips of cloth (commonly burlap) are soaked in a solution that hardens as it dries, such as flour paste (Simpson 1982) or plaster. A separating material must be used to ensure that the plaster does not adhere directly to the specimen. Separators between the bone and plaster have included Japanese rice paper or other papers with good wet strength. Films and foils are not appropriate separators, as they do not sufficiently conform to the specimen.

The preservation of orientation, degree of articulation, and other taphonomic factors make block removal the best approach to the removal of large vertebrates or large assemblages from the field. This may result in a block of significant mass and weight. All factors of removal and transportation should be worked out in advance, so that the risk of dropping and breaking is minimized.

Documentation of the methods and materials used is extremely important (Sease 1988). Such records should be kept in a log separate from personal field notes, and the information should be transferred to the permanent specimen record at the collection or institution.

Screening for small fossils is a useful strategy that may be performed with dry matrix or with the assistance of water. Stacked screens of progressively smaller mesh apertures catch small specimens as they are shaken or washed through. Care must be taken to prevent loss of material through slipping or tilting of the screens.

Laboratory preparation materials and methods

Preparation is necessary for exposing the features of the fossil specimen. The preparator should know the fine structure, preservation mode, and potential use of the specimen before determining the best approach (Feldmann 1989). In preparation, it is important to make a clear distinction between preservation and restoration. Some older approaches are based on treating all specimens as if they were destined for exhibition. Modeling and infilling of missing structures, painting to mask demarcation lines between original and void-filled areas, extensive adhesion and consolidation, permanent articulation in a life posture, and similar approaches have sometimes been used to restore specimens before their best use is determined.

Restoration is a complex field that relies on a wide range of techniques that in themselves are often irreversible. It is probably best to restore only those specimens that have been clearly targeted as immediate display material, and then only if their original condition is so fragmentary or fragile that they cannot be meaningfully or safely exhibited otherwise. (The use of casts for exhibition is far safer insofar as protecting the integrity of specimens; the changes to the specimen that may be necessary for casting are often far less damaging than those of restoration.)

Restoration as such is not necessary for, and may easily reduce the usefulness of, primary research material. In developing an awareness of the damage caused by injudicious restoration, vertebrate paleontology follows the example set in archaeological conservation:

> Lack of technological sophistication and devilishly good intentions often destroyed . . . significant information which an object (or specimen) contained. Restorers of the old school often made a specimen into what they thought it should be, and neglected what it actually was. The newer, sounder approach settles instead for preservation. (Kopper 1982:104)

Even the best preparation will not necessarily save a badly damaged or deteriorated specimen (Yochelson 1989). Specimens that are to be prepared by mechanical or chemical means should be evaluated for stability and with their prospective use (research, exhibition, or education) in mind.

Mechanical and chemical preparation

In vertebrate paleontology more than in invertebrate, the matrix is often completely removed from the specimen during preparation. Removal of the matrix surrounding a fossil may be accomplished by mechanical or chemical means. Mechanical preparation involves the use of some type of percussive or concussive force, from probes and needles or hammer and chisel to dental tools, air abrasive units, and saws. All specimens must be supported and cushioned so that the shock waves from the tool used do not destroy the specimen (Wagstaffe and Fiedler 1968). Many must be consolidated to prevent loss in this process. For this reason, some workers prefer to delay preparation of materials that may be needed for chemical analysis. Some imaging techniques will reveal the exact position and orientation of the fossils embedded below the matrix surface, reducing the risk of poorly placed blows.

Nonabrasive methods are generally preferred for materials in unconsolidated or soft matrices. Use of air abrasive units has proved effective in reducing particularly hard and uniform matrices that defy other means of removal (Stucker, Galusha, and McKenna 1965).

Chemical preparation is often used when the fossil and its matrix differ sufficiently in chemical composition that a caustic substance (generally an acid) will preferentially attack the matrix without causing excessive damage to the fossil. This is often used for particularly hard matrix encasements or delicate specimens. Chemical disruption by water or hydrogen peroxide may be used to break soft matrices down, and sequestrants such as water softeners or chelating agents will form stable mineral complexes that can be safely removed (Whybrow and Lindsay 1990). Various acids, including acetic or formic acid, for preparing vertebrate specimens from carbonate rocks (Wagstaffe and Fiedler 1968; Green 1970) have been used. In combination with washing and screening, acid preparation yields fossils that are free of matrix and available for study. The nature of the acid and its safety in use must be considered carefully, and the reaction must be stopped at the moment that the matrix removal is sufficient in order to prevent specimen damage. Some specimens may require drying and consolidation as they are exposed by the acid to protect the new surface. Acid-prepared fossils may not be usable for protein studies. Hydrochloric, hydrofluoric, and oxalic acids, recommended in some earlier works, are too dangerous to both the specimen and the worker (Rixon 1961; Green 1970) and produce too many undesirable salts to be useful (Wagstaffe and Fiedler 1968).

Collection and storage

Vertebrate fossil collections are, in theory, the best means of preserving these remnants of vanished biotas for study (Rudwick 1985). Most, if not all, of a specimen's history in a collection will be spent in storage, either in the collection itself or in the specialized storage environment of an exhibit. Storage conditions, however, frequently are unstudied or are ranked as low priorities for upgrades. Selection of storage materials, systems design, and integration must be made with the specimen's stability and long-term usefulness in mind (Howie 1978).

Storage systems for vertebrate fossils merit careful planning. In general, closed wooden cases are the poorest choice for storage furniture. They cannot easily be sealed adequately, and their outgassing accumulation of acid vapors is implicated in the development of efflorescences. In addition, any unsealed or poorly sealed case can admit fumes and particulates from outdoor or indoor sources. Gasketted steel cases with a powder paint are preferred. These can be well-sealed and release few or no deleterious fumes. Solvent-carried paint on a well sealed steel case may create a damaging internal microenvironment as the paint cures and outgasses. Ideally, cases painted with a solvent-carried paint should be left open, empty, to cure for at least a year before specimens are installed, but few collections can afford to let case space stand empty. It is better to invest in an alternative paint surface. Similarly, powder-painted steel drawers are far preferable to wooden ones.

Where wood cannot be avoided, it should receive some kind of surface treatment if it is to be used in uncontrolled environments. Wood in controlled environmental conditions (relative humidity 50% and 23°C) emits fewer acidic vapors than it does in environments at higher levels. Appropriate surface coatings, such as air-drying enamels, provide an effective vapor barrier that latex paints and shellacs do not (Miles 1986). Cured wood emits fewer fumes than does new wood, and hardwoods are more acidic than softwoods. Oak should never be used as a storage fabric. Specimens should never be in direct contact with any case fabric, and should be arranged so that they do not move when the drawer does. Rolling and bumping can cause severe damage.

The use of compactor systems has received attention over the past few years. Compactors can expand the effective use of collection space by reducing the number of open aisles. However, the weight of many vertebrate fossil collections and the sheer size of some specimens mandate the custom design of compactor systems in some cases. These should always be installed on the lowest floor possible. The number of fixed aisles can be specified in the design. Of the available systems, the manual is recommended over the electric; there are fewer jolts in starting and stopping and no risk of a power failure closing off part of the collection. After an earthquake, the compactor cases in one natural history collection were found to have moved a few inches on their rails, sparing the specimens on their shelves any of the rattling and breakage associated with freestanding cases under the same circumstances (C. Patterson, personal communication).

The materials in or on which the fossils are stored should be selected for their stability and lack of acidity. Inert polyethylene materials such as ethafoam are often used as drawer liners, and acid-free trays and boxes are highly desirable. Very fragile materials in a drawer can be further cushioned with bubble pack covered by ethafoam or acid-free tissue. Specialized systems for storing and labeling vials have been described by Fitzgerald (1986). No specimen should be stored on an unlined surface, and no specimen should bear its own weight without some form of padding or support.

Very large specimens cannot be enclosed in a drawer. A rack-and-pallet system, described by Fitzgerald (1989) and pioneered at the Canadian Museum of Nature, has been suggested as a means of safely storing and retrieving large fossils. The pallets themselves are padded. Environmentally sensitive specimens may be further protected by a hoop-and-bag system that provides a controlled microenvironment under an inert polyethylene bag, which is supported so that the plastic is not in direct contact with the specimen.

Ethafoam can be built up in layers to create custom "nests" for heavy specimens. This can also be done with paper-mâché and plaster, but these materials are inferior (the former for its acidity, the latter from its weight and brittleness). Ethafoam has also been used successfully as a vial stopper, replacing short-lived cork and plastic plugs.

Some aspects of storage environment can cause the degradation of sensitive materials (Padfield, Erhardt, and Hopwood 1982). Fossils known to contain a humidity-sensitive component should be stored at relative humidities that mirror those of the collection site. Efflorescences of calcium or other soluble salts are extremely damaging phenomena which, again, can be controlled by lowered relative humidities, as well as the removal of acidic atmospheric gases (FitzHugh and Gettens 1971). Pyritic decay, as has been discussed, also slows or is suspended when the specimen is stored at relative humidity levels of 30% or lower.

Plans for creating a bank of environmentally controlled steel cases have been drawn up by the Canadian Conservation Institute. Use of this or a similar system can create a stable microclimate in a building in which the internal climate cannot be easily controlled (Stolow 1966, 1977). Silica gel or other desiccants can also be used, but these should be spread out in a thin layer and periodically reconditioned. All of the desiccant particles must be in direct surface contact with the atmosphere to be effective; a beaker full of desiccant, frequently seen in museum collections, is only effective at the topmost layer. A long-term environmental monitoring program, including the use of calibrated recording hygrothermographs, is necessary to establish the environmental levels and fluctuations, as well as the effectiveness of mitigation strategies (Macleod 1978; Thomson 1988).

Some fossil vertebrates (notably specimens from the Morrison Formation, Hagerman Lake Beds, Cypress Hills Formation, and Karoo Sequence) are radioactive in varying degrees. Because these are sources of radon, their storage and handling should receive special attention. Radon tends to concentrate in indoor areas and can pose a pulmonary health hazard at high levels. Storage in separate rooms with adequate ventilation may be necessary, as only large volumes of air are effective in reducing the concentration to a safe level. The use of separate storage areas, radon detectors, and reasonable precautions in storage and display of these specimens is recommended (Carman and Carman 1989). Simple film exposure tests for the presence of radioactivity can be made without a Geiger meter or scintillation counter (Blount 1990). Dose rate meters, background radiation meters, and alpha detectors (for radon) are also recommended.

Collections of nonbone material (matrix, invertebrates, pollens, etc.) may be stored in a variety of ways. Unprepared matrix material is not generally stored with cataloged specimens; its weight and bulk alone may dictate off-site storage. It should never be stored outdoors, however, where exposure and weathering can cause loss and undesirable changes. Unconsolidated matrix can be separated into polyethylene bags for storage, so that the weight is distributed for ease in handling. Hard matrix may be broken down before bagging or boxing.

Cataloged invertebrates and plant materials may be deposited in another collection or may be retained in the vertebrate collection, depending on the preferences of the institution and its collections. Specimens from a single site must always be cross-listed for easy retrieval of all ecological components of an area. Storage conditions are the same as for other fossil material.

Documents – site maps, photographs and negatives, photographic slides, film, microscope slides, and paper – should never be stored with fossil specimens. Their storage requirements are very different, and such concentration of materials increases the risk of total loss of information in a fire or other disaster. Photographs, maps, valuable original documents (including holographic materials and historic labels), and other paper stocks should be stored in the archival boxes made for this purpose. Exposed and unexposed film stock (including negatives and slides) should always be stored in a refrigerator away from chemicals. The colorfastness of photographic slides is notoriously short-lived; cool temperatures prolong the color quality. The storage of microscope slides may be dictated by the mounting medium used, if known; many materials have been used for mounting and sealing microscope slides, and it is important to protect these against failure (Garner and Horie 1984). In general, they are best kept in metal storage units in dark and cool conditions. A collection that has not set up an archival arrangement for these types of materials, so that they receive safe and separate but easily retrievable storage, should consider establishing such a system.

Specimens prepared for use in scanning electron microscopy (SEM) may be coated to increase their effective conductivity for analysis by high-KV and high-magnification microscopes. This coating is often gold or a gold–palladium alloy, but may be another metal or carbon. The mounting medium and coating material should always be documented. Most SEM specimens are safely stored with other specimens in the collections without removal of the coating or separation from the stub. Both coatings can be very harmful to specimens. It is suggested that type specimens that require SEM study not receive the coating if they can be studied with a lower- magnification, lower-KV microscope (Golden 1989). Some workers have found that a cast in cellulose acetate may work just as well as an original for SEM study (Claugher and Taylor 1990).

Accurate labeling, accessible catalog and inventory systems, and reliable guides to the arrangement of a collection are important features of a storage system. They can minimize the amount of unnecessary specimen handling, which is always a source of deterioration.

In situ sites

Some notable vertebrate fossil specimens (both body and track fossils) are never removed from their original site but are exposed and displayed in situ. Some famous examples in the United States include Dinosaur National Monument (Utah), Badlands National Park (South Dakota), Big Bend National Park (Texas), Agate Fossil Beds National Monument (Nebraska), Ashfall State Park (Nebraska), Florissant Fossil Beds National Monument (Colorado), Berlin Ichthyosaur State Park (Nevada), Dinosaur Valley State Park (Texas), Hot Springs Mammoth site (South Dakota), and Dinosaur State Park (Connecticut). Other major sites include Dinosaur Provincial Park (Alberta), Lark Quarry Environmental Park (Queensland, Australia) and the Dashantu site (Sichuan, China). Their in situ preservation enhances their public accessibility, but it can also engender some problems.

Vertebrate fossils left in situ are subject to weathering year-round. Freeze–thaw cycles are particularly damaging, as is exposure to acid precipitation (Amoroso and Fassina 1983). Protection of some kind is sometimes afforded by small domes or open or closed buildings. Few if any of the sites protected by buildings are equipped with any kind of environmental control system, so that climatic fluctuations, although lessened, are still influential. Also, the specimens are more difficult to study in matrix, and can only be exposed to a certain extent without losing the structural support of the matrix.

At Dinosaur National Monument (DNM), one of the most severe problems is the system of cracks in the exposed quarry surface. These have admitted water to the surface from the outside on several occasions, and it is feared that they are enlarging over time. Monitoring of these may suggest the extent of the problem and may necessitate strategies to anchor the irreplaceable surface. Specimens that do detach are removed to a collections area after their original position is carefully documented; no attempt is made to readhere them, although epoxy resins were used for this in the past. Occasional consolidation with PVA solutions is carried out. PVA is not appropriate for all sites or rock types, however. Other than the cracking of the quarry face, the most important problem at DNM appears to be security and the always-difficult balance between public access and protection (Chure, personal communication).

Several in situ sites, in fact, have been the targets of thieves and vandals. Casts have replaced original body fossils in both the Badlands and Big Bend sites as a direct response to these problems. (In the former case, casts were stolen as well.) The Dinosaur Valley trackway on the Paluxy River has been the target of unauthorized attempts to make plaster casts (which caused damage to the track surface) and of a notorious series of incidents in which human footprints were chiseled into the trackway. Tracks at the Lark Quarry dinosaur trackway in

Queensland, a fragile sandstone trackway with a thin limonite cover, were finally, ingeniously, anchored after a series of setbacks in the preservation of the trackway, including human souveniring and the deleterious effects of kangaroos attracted to the shade provided by the roof over the site (Agnew 1984). These tracks were coated with a solvent-based silicone resin, and the cracks were infilled with polyethylene foam rod and masking silicone elastomer–sand mixture. This system goes beyond cosmetic repair; it slows the expansion of cracks and the theft of broken pieces and is essentially inert and easily repairable (Agnew et al. 1989).

Consolidation of in situ material is definitely not recommended as a conservation strategy. Most fossil-bearing matrix is porous to some extent, and many consolidants form water-repellent but still permeable films that penetrate only a few centimeters of the surface (Rossi-Manaresi 1982; Waller, personal communication). Thin surficial consolidation will not only fail to prevent breakage and spalling, but may actually accelerate than as larger pieces are forced off by the forces of rising damp and weathering. The depth of in situ deposits makes effective consolidation unlikely, and many coatings produce deleterious breakdown products. On-site management, structural protection, implementation of environmental control within a sheltering structure, or removal from extremely threatening conditions are the only effective ways to maintain the integrity of in situ sites (Coles 1986). Making casts for laboratory study or for exhibit is also recommended (Thulborn 1990), provided that the method chosen does not do short- or-long-term damage to the track surface (Conrad 1989; Ishigaki and Fujisaki 1989; Maceo and Riskind 1989; Obata et al. 1989).

Some references suggest the use of ink, felt-tip tracing, or staining of in situ trace fossils to enhance their visibility for photography or study (Farrow 1975). These approaches are no longer recommended, as they introduce an irreversible source of matrix contamination.

Analytical research approaches

Destructive testing

Destructive testing involves the partial or complete destruction of a specimen in order to study an aspect of its nature or structure not apparent from its external morphology. This is a necessary and vital use of vertebrate paleontology collections. Type and figured specimens are, in general, not subjected to destructive testing, but the limited amount of fossil vertebrate material may force the use of such key specimens. Testing and its results should always be documented in the permanent specimen record, and residual material (e.g., thin sections or slides) should be appropriately stored under the original specimen number.

Thin sections are used for microscopic studies of the histological and mineralogical structure of a section (transverse, sagittal, or longitudinal) of a specimen. Preparing a thin section usually involves cutting the section; cleaning, vacuum drying, and impregnation with a polymer consolidant; attachment to a glass slide with epoxy or polyester resin; and finally, grinding the section to near-transparency. This section affords a view of the internal structure of the specimen and, as such, is of immense scientific value. Because the specimen is permanently changed, and because some parts are destroyed, sectioning is not recommended for type specimens unless the scarcity of the material and the significance of the research dictate otherwise. The sections are equivalent to any other specimens in their need for identification, care, and storage.

Microprobe studies are based on the analysis of thin sections coated with carbon; these studies yield information on the exact chemical nature of the specimen as well as some structural information. These serve as clues to diet and other environmental influences (Toomey, personal communication). Use of the microprobe involves essentially the same amount of destructive preparation as thin sections. The remaining slides should be cared for similarly.

Many techniques for dating specimens or matrix, including radiocarbon, potassium–argon, and argon-40/argon-39, necessitate the destruction of a sample to a powder for analysis (Goffer 1980). The information obtained should be a permanent part of the specimen record under a permanent specimen number. There is no residual material to be retained after such a process. Fission-track dating requires the preparation of a thin section rather than crushing.

X-ray diffraction and atomic absorption studies also involve crushing a specimen to a powder, which is then subjected to analysis. From this, the biochemical or geochemical nature of the specimen can be assessed. The powder is not retained at the end of the analysis.

Wyckoff (1972 p. 32) comments: "Most proteins are so unstable that it has been generally assumed that they could not persist for geologically significant periods of time, and until recently little serious effort was made to recover them from fossils." Analysis of fossil collagen, proteins, and amino acids is one of the most exciting developments in the field of vertebrate paleontology, but requires that the greatest care in handling be taken. Any source, even water or unprotected handling, can result in contamination of the results (Gillette, personal communication). Isolating proteins is an increasingly technologically complex analysis that does destroy the sample used.

Immunologically active glucoproteins have been found in Pleistocene material (Wyckoff 1972; Rothschild and Turnbull 1987). Collagen identifiable as such at a microstructure level has been extracted from La Brea Pleistocene vertebrates (Wyckoff et al. 1963; Wyckoff and Doberenz 1965), at varying degrees of degradation (Doberenz and Wyckoff 1967). It appears that the variable rate of survival of amino acid, protein, or collagen structures depends more on the diagenetic conditions than on the inherent stability or instability of the materials. Exposure to moisture and high heat will destroy collagenous protein (Wyckoff 1972).

Some dinosaur bones have proved to contain identifiable proteins (Miller and Wyckoff 1968). The isolation of these structures may be complicated by a complex diagenesis, in which matrix sediments may show a high degree of protein presence (Armstrong and Tarlo 1966), indicating leaching and diffusion from the specimen. Microbial contamination can also introduce foreign amino acids, which are difficult to separate. Recently, a multidisciplinary study on a vertebra of the sauropod *Seismosaurus* has revealed the presence of proteins in the bone that were not found in the matrix, suggesting that they may in fact prove to be proteins from the 150 million-year-old individual (Gurley et al. 1991; Gillette et al., in press).

Material targeted for protein studies should be newly exposed bone without contamination of any kind. This includes contamination from common field materials (adhesives, consolidants, coatings, plaster, and water) as well as contamination that can result from handling or from storage in a museum environment (Gillette, personal communication). If it is not clear whether or not a specimen in the field will be studied in this way, fieldworkers should remove an uncontaminated sample for future studies before proceeding with consolidation, adhesion, or plastering.

In all destructive tests, copies of printout information should be part of the permanent record. The information obtained should be recorded in a permanent field or file under the permanent specimen number. The specimen should receive this number before processing, so that the information can be easily tracked.

Nondestructive testing

Imaging technologies have made it possible in some cases to study a specimen sectionally without invasive procedures (e.g., cutting). This is particularly valuable for the study of type, rare, or fragile material (Conroy and Vannier 1989). Studies in computer imaging, volumetric visualization, and nuclear magnetic resonance spectrometry are based on different applications of imaging technology.

In these examples of nondestructive testing, the specimen remains unaltered in appearance and gross morphology. The results of testing (printouts or films) should become part of the permanent record. Film stock requires particularly dark, cool storage for maximum image retention; such documents should receive separate storage. Printouts from a direct computer printer linkup should be copied and filed with the specimen record.

Duplication, exhibition and restoration

Casts

Casting vertebrate fossils for research, education, and exhibition purposes is a vital part of the activities of most preparation laboratories. Casts extend the availability of scarce, fragile, or valuable fossils (such as type or figured specimens), which cannot ethically be lent by scientific collections. Molding carries certain inherent risks to the specimen, and the molds and casts themselves may have relatively short shelf lives.

Both molds and casts in plaster have been made at least since 1904, when replicas of the Carnegie *Diplodocus* specimen were distributed to a number of European rulers (Krishtalka 1989). The setting of plaster is an exothermic reaction; thus specimens vulnerable to damage from heat should not be molded in plaster. Plaster is porous and absorbent, and so must be kept in environmentally controlled conditions to prevent damage by moisture uptake, freezing, or mold growth. Plaster is also very brittle, and reinforcing elements of metal or cloth may corrode or decay and cause severe damage (Beale, Craine, and Forsythe 1977).

Specimens that are to be molded often require both cleaning and consolidation in order to survive the process (Rigby and Clark 1965). Documentation of this should be made so that these specimens are not used for chemical analysis in the future. If the specimen is desirable for both analytical work and casting, a sample should be removed first for future work and stored so that it will not be contaminated by handling or by poor storage environments. Older impregnation methods, such as Bakelite dissolved in alcohol, shellac, polyvinyl acetate emulsions, or cellulose nitrates, are not recommended. Compounds that are in widespread use today should be selected for their relative nonreactivity and inertness over time as much as for their effectiveness.

Materials for making molds, which will necessarily be in direct contact with the specimen, should be selected carefully. Oil-free clays should always be used in preference to oil-based clays. Natural waxes, such as beeswax and paraffin, may leave a surface residue that can be difficult to remove without inflicting damage. Paper-mâché, silicone rubber, latex, fiberglass, and plaster are commonly used molding materials (Baird 1955; Rigby and Clark 1965; Watson and Grant 1988). Polyvinyl chloride compounds should be avoided altogether in vertebrate paleontology; they tend to liberate hydrochloric acid as they deteriorate (Mills and White 1987). Many compounds require the use of a separator substance, which should not leave a damaging residue.

Casts have similarly been made from a wide range of materials, from organic substances (gelatin and agar) and plaster to synthetic substances (polyurethane and epoxy resins, fiberglass, and other plastics). The best casting materials are those that clearly retain the morphology of the fossil and have long-term stability.

Galvanoplastic casts are used for replicating specimens fossilized on a single bedding plane with low relief (Zangerl 1965). A latex peel is taken from the specimen and used as the basis for the complex casting process.

Molds should not be stored with the specimens they duplicate. Some materials may degrade and release harmful reaction products. In addition, molds and casts serve as a valuable form of documentation of specimen morphology. In the event that the specimen is stolen or lost, molds and casts may be the only vestiges of it left. They should be stored in environmentally controlled conditions in steel cases, and must always be clearly labeled with the name and number of the original specimen. Stable casts, particularly of specimens that are in other collections and thus not available for study, should be considered equally important as the original material and are often stored with the original specimens (Langston 1977). At the first sign of deterioration, however, they should be removed to more controlled storage.

Exhibition

No collection is valuable if it is never used, and exhibition is one common use of fossil material. Most museums, because of their charter, accreditation, or mission statement, are obliged to exhibit some part of their holdings to the public on a regular basis. Exhibits can be one of the most educational and interactive aspects of vertebrate fossil specimen use.

An exhibit is a specialized form of visible storage, generally under far less protective conditions than are provided by good collection storage (Hodges 1982). The specimen is generally at more risk from all forms of damage and deterioration, particularly vandalism and theft, air pollution, light and UV radiation, and temperature and relative humidity fluctuations. Specimens for exhibition should never be those that are rare, fragile, or of typological or voucher significance.

Casts are frequently substituted for original specimens on exhibit. This has caused some controversy, as some museum professionals believe that the experience of educative encounters with original materials is part of the uniqueness of a museum visit. Rising market values, awareness of deterioration, and a reluctance to compromise the safety of the original specimens, however, may mandate the use of casts. It is far more appropriate to paint, infill, and effect other essentially restorative or cosmetic procedures for effective display on casts than on original specimens.

Some exhibition procedures of the past have caused tremendous damage to vertebrate fossils. Since the early part of the twentieth century, many vertebrates have been exhibited in postures that mimic the poses and articulation during life. The use of external armature to support the bones in position eventually gave way to the use of internal armature. This approach requires drilling the bones with longitudinal or transverse holes to accept structural rods, as well as screws, plates, wires, clamps, and nonreversible adhesives. Removing the specimen or any of its individual elements for study may be prohibitively difficult. Coatings applied in the past, notably shellac and beeswax, are difficult to remove; they act as particulate traps, which may increase the deterioration of the specimen if the particles are acidic.

Exhibiting original fossil specimens that are sensitive to fluctuations in relative humidity may require the use of humidity-controlled cases. Although desiccants such as silica gel may be used for this, a more efficient approach is the construction of humidity control modules (Michalski 1982; Sease 1990). Larger specimens that cannot be easily enclosed in a more controlled environment may have to be removed from exhibit if they show significant environmental damage. In such a case, the use of casts is strongly recommended for exhibit; the original specimen requires more controlled storage conditions.

New exhibits should be designed with the input of curators, exhibit designers, and conservators in order to provide the most effective displays at the least risk to the specimen. Support materials made from inert polymers (e.g., polymethyl methacrylate) are superior to metal supports for smaller specimens, and, because they are clear, can be used as external supports without intruding visually. Larger mounts may require several support methods to ensure their safety and stability. Lightweight casts have an advantage in this regard. Fossil vertebrates should never be drilled for internal armature support; this is an outdated approach that entails unnecessary damage.

Recommendations

Natural history collections in general are threatened by a variety of economic and academic factors (West 1988). To survive into the next century, vertebrate paleontology collections must be of demonstrable value to new research concepts as well as the anchor for previous research results. The care taken in storage and handling vertebrate fossil specimens may make a considerable difference in their longevity, stability, and usefulness.

Preparators and collections managers traditionally have been viewed as "subprofessional personnel" (Patton 1972), but this attitude is changing in many systematics disciplines as the roles of all collections workers become more complex. Formal training and continuing education are becoming as necessary in collections management as the discipline of paleontology itself.

Preparators and curators must work together to determine the best approach to the preservation of fossil vertebrates on a case-by-case basis; no formulaic approach can answer the needs of the expanding research fields based on the collections. It is vital that specialized training at a formal level be provided so that preparation and conservation approaches are standardized, improved, and brought into synchronization with the rapidly emerging research approaches that demand more sensitive and sophisticated (and less intrusive) preparation of materials. The call to establish collections management as a discipline in itself (Simmons and Manning 1991) has never been more necessary, as academic pressures increasingly restrict the ability of curators to provide specimen care or training in collection work.

Most preparators learn techniques on the job, an unsatisfactory approach that creates a field based on oral history or anecdotal evidence. All too quickly this leads to the entrenchment of inferior approaches. The paleontology curriculum sponsored by the Denver Museum of Natural History (Stucky, personal communication) is an excellent example of effective, college-credit training in methods and materials that yields better prepared and documented material, an increased rate of donations, and a rise in effective community involvement. Such an approach could be extended to provide formal coursework for students planning to enter collections-related employment of any kind, perhaps following the curriculum of the Collections Care Pilot Training Program sponsored by the Bay Foundation. The amount of new information available in this field cannot be adequately communicated by work experience alone.

Peer-review publication of structural studies, materials research, and methods applications must be encouraged for the dissemination of vital information. Informal networks are valuable sources of information, but they lack academic credibility and must be augmented by formal venues.

Materials for the treatment and storage of vertebrate fossil specimens should be selected with preventive conservation in mind. In situations in which environmental control is difficult or impossible to achieve, the microenvironment created by a well-sealed case containing no acidic products creates an environmental buffer zone for the specimens.

Adhesives, consolidants, solvents, and other materials that necessarily come into direct contact with the fossil material should be carefully studied. Fossil specimens may be of more value if they are not contaminated with such substances. Substances that are used should not simply be bought off the shelf without knowledge of their structure and long-term characteristics. A material that can be reversed easily as needed is generally preferable to one that cannot. Information on their composition is available from the manufacturer in the form of a materials safety data sheet (MSDS). More detailed information on stability and longevity is easy to obtain from the Canadian Conservation Institute's product information service (available on-line through the Conservation Information Network). Because many manufacturers will not divulge proprietary information, the CCI's service to the field is incalculable.

No intrusive materials, such as adhesives and consolidants, should be used unless they are necessary to stabilize a damaged specimen that will be otherwise lost. A stable specimen does not need treatment, and a specimen that is broken but stable should have its parts safely grouped together in storage, not readhered, until such time as its study, illustration, or exhibition makes reassembly necessary. The principle of leasts (the least amount needed to do the job, applied over the least amount of time needed to be effective) should govern the application of intrusive materials. Overuse does not extend or ensure the specimen's stability and may even work against it. Most such materials were not designed for the uses to which they are put in vertebrate paleontology, and the features that make them valuable in industrial applications may not be the same features needed by paleontologists.

The need for rigor in materials and methods research parallels that in all other research fields. The approaches used by hobbyists and amateur collectors are simply inappropriate for research collections in many cases. Documentation of the specimen's storage and handling history is vital to establishing effective research.

A conservation approach to vertebrate paleontology specimen care is crucially important to the field as it develops new lines of research and information. Vertebrate paleontology anchors many disciplines; it, in turn, is anchored by the nature and quality of its specimen collections. The long-term survival and stability of those specimens as an irreplaceable and useful scientific resource must be the ultimate goal of collections management and conservation.

Acknowledgments

The credit for developing and leading the field of geological conservation belongs to F. M. P. Howie of the Natural History Museum (London), G. R. Fitzgerald and R. R. Waller of the Canadian Museum of Nature, and C. J. Collins of the Geological Conservation Units, Cambridge University. I thank them for their generous and patient mentorships. For information, assistance, and advice on this chapter, my gratitude to the following: D. C. Cannatella, C. J. Durden, D. L. Hendrickson, J. S. Johnson, E. L. Lundelius, Jr., R. H. Rainey, W. G. Reeder, R. S. Toomey, and E. Yarmer of the University of Texas; J. Simmons, University of Kansas; D. Chure, Dinosaur National Monument, Utah; D. Gillette, Division of State History, state of Utah; C. Shaw, Page Museum; G. Brown, University of Nebraska; G. McDonald, Cincinnati Museum of Natural History; C. Patterson (presently at the Denver Art Museum) and R. Stucky, Denver Museum of Natural History; and R. E. Child and C. Buttler, National Museum of Wales. Finally, thanks once more to the sponsors and coordinators of the Natural History Collections Care Pilot Training Project (Bay Foundation and the Natural History Museum of Los Angeles County) for their groundbreaking contributions to effective conservation education. This is publication No. N.S.-39 of the Texas Memorial Museum.

References

Agnew, N. 1984. The use of silicones in the preservation of a field site – the Lark Quarry dinosaur trackways. In N. S. Brommelle, E. M. Pye, P. Smith and G. Thomson (Eds.), *Adhesives and consolidants* (pp. 87–91). Preprints of the contributions to the Paris Congress, 2–8 September 1984. London: International Institute for Conservation of Artistic and Historic Works.

———, H. Griffin, M. Wade, T. Tebble, and W. Oxnam. 1989. Strategies and techniques for the preservation of fossil tracksites: an Australian example. In D. D. Gillette and M. L. Lockley (eds.) *Dinosaur tracks and traces* (pp. 397–407). Cambridge: University Press.

Amoroso, G. G., and V. Fassina. 1983. *Stone decay and conservation: Atmospheric pollution, cleaning, consolidation and protection.* New York: Elsevier.

Armstrong, W. G., and L. B. Halstead Tarlo. 1966. Amino-acid components in fossil calcification. *Nature* (London) *210*:481–482.

Baird, D. 1955. Latex micro-molding and latex-plaster molding mixtures. *Science 122*:202.

———. 1978. The burnt dope technique and other intertidal ploys. *The Chiseler* (Carnegie Museum) *1*(2):16–17.

Beale, A., C. Craine, and C. Forsythe. 1977. The conservation of plaster casts (pp. 18–26). Preprints of papers presented at the fifth annual meeting of the American Institute for Conservation of Historic and Artistic Works.

Beck, C. 1982. Authentication and conservation of amber: Conflict of interests. In N. S. Brommelle and G. Thompson (eds.), *Science and technology in the service of conservation* (pp. 104–107). Preprints of the contributions to the Washington Congress, 3–9 September 1982. London: International Institute for Conservation of Artistic and Historic Works.

Bird, R. T. 1985. *Bones for Barnum Brown.* V. Theodore Schreiber, Ed. Fort Worth, TX: Texas Christian University Press.

Blount, A. 1990. A low-cost radioactivity test for geologic specimens. *Collection Forum 6*(1): 8–11.

Carman, M. R., and J. D. Carman. 1989. Health considerations of radon source fossil vertebrate specimens. *Collection Forum 5*(1):5–10

Casanova, R., and R. P. Ratkevich. 1981. *An illustrated guide to fossil collecting* (3rd ed.). Happy Camp, CA: Naturegraph Publishers.

Cherfas, J. 1989. Science gives ivory a sense of identity. *Science 246*:1120–1121.

Claugher, D., and P. D. Taylor. 1990. Electron microscopy. In D. E. G. Briggs and P. R. Crowther, (eds.) *Palaeobiology: A synthesis* (pp. 508–511). Oxford: Blackwell.

Coles, J. M. 1986. The preservation of archaeological sites by environmental intervention. *In-situ archaeological conservation: Proceedings of meetings* (pp. 32–55). Instituto Nacional de Antropologia e Historia de Mexico and the Getty Conservation Institute.

Committee on Guidelines for Paleontological Collecting. 1987. *Paleontological collecting.* Board on Earth Sciences, Commission on Physical Sciences, Mathematics, and Resources, National Research Council. Washington, DC: National Academy Press.

Conrad, K. 1989. The use of alginate in track replication. In D. D. Gillette and M. L. Lockley, (Eds.) *Dinosaur tracks and traces* (pp. 397–407). Cambridge: Cambridge University Press.

Conroy, G. C., M. W. Vannier, 1989. The Taung Skull revisited: New evidence from high resolution Computed Tomography: *South African Journal of Science 85*(1), pp. 30–32.

Converse, H. H. 1984. *Handbook of paleo-preparation techniques.* Gainesville: Florida State Museum.

Croucher, R., and A. R. Woolley. 1982. Fossils, minerals and rocks: Collection and preservation. London: British Museum (Natural History) and Cambridge University Press.

Cumbaa, S. L. 1983. Osteological preparation techniques used by the Zooarchaeological Identification Centre. In D. J. Faber (Ed.), *Proceedings of the 1981 workshop on care and maintenance of natural history collections. Syllogeus 44*:29–35.

De Witte, E. 1975. Soluble nylon as a consolidation agent for stone. *Studies in Conservation 20*:30–34.

Doberenz, A. R., and R. W. G. Wyckoff. 1967. Fine structure in fossil collagen. *Proceedings of the National Academy of Sciences 57*:539–541.

Eanes, E. D., and A. S. Posner. 1970. Structure and chemistry of bone mineral. In H. Schraer (Ed.), *Biological calcification: Cellular and molecular aspects* (pp. 1–26). New York: Appleton-Century-Crofts.

Farrow, G. E. 1975. Techniques for the study of fossil and Recent traces. In R. W. Frey (Ed.), *The study of trace fossils: A synthesis of principles, problems, and procedures in ichnology* (pp. 537–554). New York: Springer-Verlag.

Feldmann, R. M. 1989. Selection of appropriate preparation techniques. In R. M. Feldmann R. E. Chapman, and J. T. Hannibal (Eds.), *Paleotechniques* (pp. 24–29). Paleontological Society Special Publication 4.

Feller, R. L., and D. B. Encke. 1982. Stages in deterioration: the examples of rubber cement and transparent tape. In N. S. Brommelle and G. Thomson (Eds.), *Science and technology in the service of conservation* (pp. 19–23). Preprints of the contributions to the Washington Congress, 3–9 September 1982. London: The International Institute for Conservation of Artistic and Historic Works.

Fisher, D. W. 1965. Collecting in sedimentary rocks. In B. Kummel and D. Raup (Eds.), *Handbook of paleontological techniques* (pp. 150–155). San Francisco: Freeman.

Fitzgerald, G. R. 1986. The card-vial system for the storage of small vertebrate fossils. In J. Waddington and D. M. Rudkin (Eds.), *Proceedings of the 1985 workshop on care and maintenance of natural history collections* (pp. 105–198). Life Sciences Miscellaneous Publications, Royal Ontario Museum.

———. 1988. Documentation guidelines for the preparation and conservation of paleontological and geological specimens. *Collection Forum* 4(2):38–45.

———. 1989. A form-fitted pallet for the storage of large fossils. *Geological Curator* 5(2): 72–76.

FitzHugh, E. W., and Gettens, R. J. 1971. Calclacite and other efflorescent salts on objects stored in wooden museum cases. In R. H. Brill (Ed.), *Science and archeology* (pp. 91–102). Cambridge, MA: MIT Press.

Frey, R. W. 1975. The realm of ichnology: its strengths and limitations. In R. W. Frey (Ed.), *The study of trace fossils: A synthesis of principles, problems, and procedures in ichnology* (pp. 13–38). New York: Springer-Verlag.

Garner, R., and C. V. Horie. 1984. The conservation and restoration of microscope slides of mosses. *Studies in Conservation* 29:93–99.

Gillette, D. D. (Ed.). 1989. Chemistry of dinosaur bone. New York, in press. Proceedings of a workshop at Los Alamos National Laboratory.

Gisbert, J., F. Palacios, and R. Garcia-Perea. 1990. Labeling vertebrate collection with Tyvek(R) synthetic paper. *Collection Forum* 6(1):35–37.

Goffer, Z. 1980. *Archaeological chemistry*. New York: Wiley.

Golden, J. 1989. Golden oldies: Curating SEM specimens. *Collection Forum* 5(1):17–26.

Goss, R. J. 1983. *Deer antlers: regeneration, function, evolution*. New York: Academic Press.

Green, M. 1970. Recovering microvertebrates with acetic acid. *South Dakota Geological Survey Circular 40*.

Gurley, L. R., J. G. Valdez, W. D. Spall, B. F. Smith, and D. D. Gillette. 1991. Proteins in the fossil bones of the dinosaur, "*Seismosaurus*." *Journal of Protein Chemistry* 10(1):75–90.

Guthrie, R. D. 1990. *Frozen fauna of the Mammoth Steppe: The story of Blue Babe*. Chicago: University of Chicago Press.

Hare, P. E. 1980. Organic geochemistry of bone and its relation to the survival of bone in the natural environment. In A. K. Behrensmeyer and A. P. Hill (Eds.), *Fossils in the making: Vertebrate taphonomy and paleoecology* (pp. 208–219). Chicago: University of Chicago Press.

Hodges, H. 1982. Showcases made of chemically unstable materials. *Museum* 34(1):56–58.

Hopwood, W. R. 1979. Choosing materials for prolonged proximity to museum objects (pp. 44–49). Preprints of papers presented at the 1979 annual meeting of the American Institute for Conservation, of Historic and Artistic Works.

Howie, F. M. P. 1978. Storage environment and the conservation of palaeontological material. *The Conservator* 2:13–19.

———. 1979. Museum climatology and the conservation of palaeontological material. *Special Papers in Palaeontology* 22:103–105.

———. 1984a. Materials used for conserving fossil specimens since 1930: A review. In N. S. Brommelle, E. M. Pye, P. Smith, and G. Thomson (Eds.), *Adhesives and consolidants* (pp. 92–97). Preprints of the contributions to the Paris Congress, 2–8 September 1984. London: International Institute for Conservation of Artistic and Historic Works.

———. 1984b. Conservation and storage: geological material. In M. A. Thompson (Ed.), *Manual of curatorship* (pp. 308–322). London and Boston: Butterworths.

Ishigaki, S., and T. Fujisake. 1989. Three dimensional representation of *Eubrontes* by the method of moire topography. In D. D. Gillette and M. L. Lockley, (Eds.), *Dinosaur tracks and traces* (pp. 421–425). Cambridge: Cambridge University Press.

Koob, S. 1990. The continued use of shellac as an adhesive: Why? In N. S. Brommelle, E. M. Pye, P. Smith,and G. Thomson (Eds.), *Adhesives and consolidants* (p. 103). Preprints of the contributions to the Paris Congress, 2–8 September 1984. London: International Institute for the Conservation of Artistic and Historic Works.

Kopper, P. 1982. *The National Museum of Natural History.* New York: Abrams.

Krishtalka, L. 1989. *Dinosaur plots and other intrigues in natural history.* New York: Morrow.

Langenheim, R. L. 1965. Collecting amber. In B. Kummel and D. Raup (Eds.), *Handbook of paleontological techniques* (pp. 184–188). San Francisco: Freeman.

Langston, W. L., Jr. (chairman). 1977. *Fossil vertebrates in the United States: The next ten years.* A report of the Society of Vertebrate Paleontology, Advisory Committee for Systematics Resources in Vertebrate Paleontology.

La Plante, T. C. 1977. *The weekend fossil hunter* (pp. 65–78). New York: Drake.

Lazell, J. D. 1965. An *Anolis* (Sauria, Iguanidae) in amber. *Journal of Paleontology 39*(3):379–382.

Lockley, M. 1986. Dinosaur tracksites. *Geology Department Magazine* (University of Colorado at Denver), special issue 1.

Maceo, P. J., and D. H. Riskind, 1989. Field and laboratory moldmaking and casting of dinosaur tracks. In D. D. Gillette and M. L. Lockley (Eds.) *Dinosaur tracks and traces* (pp. 419–420). Cambridge: Cambridge University Press.

MacFall, R. P., and J. Wollin. 1983. *Fossils for amateurs: A guide to collecting and preserving invertebrate fossils.* New York: Van Nostrand Reinhold Company.

Macleod, K. 1978. Relative humidity: Its importance, measurement and control in museums. *Canadian Conservation Institute Technical Bulletin.*

Matthews, W. H. 1962. *Fossils: An introduction to prehistoric life.* New York: Barnes and Noble.

McMahon, T. E., and J. C. Tash. 1979. Effects of formalin (buffered and unbuffered) and hydrochloric acid on fish otoliths. *Copeia 1979*(1):155–156.

Mervis, J. 1991. For fun, Los Alamos team goes digging for dinosaurs. *The Scientist 5*(17): 1, 8–9.

Michalski, S. 1982. A control module for relative humidity in display cases. In N. S. Brommelle and G. Thomson (Eds.), *Science and technology in the service of conservation* (pp. 28–31). Preprints of the contributions to the Washington Congress, 3–9 September 1982. London: International Institute for Conservation of Artistic and Historic Works.

———. 1990. An overall framework for preventive conservation and remedial conservation (pp. 589–591). Preprints of papers presented at the annual meeting of the American Institute for Conservation of Historic and Artistic Works.

Miles, C. 1986. Wood coatings for display and storage cases. *Studies in Conservation 31*: 114–124.

Miller, M. F., and R. W. G. Wyckoff. 1968. Proteins in dinosaur bones. *Proceedings of the National Academy of Sciences 60*:176–178.

Mills, J. S., and R. White. 1987. *The organic chemistry of museum objects.* London: Butterworths.

Obata, I., H. Maruo, H. Terakado, T. Murakami, T. Tanaka, and M. Matsukawa. 1989. Replicas of dinosaur tracks, using silicone rubber and fiberglass-reinforced plastics. In D. D. Gillette and M. L. Lockley, (eds.) *Dinosaur tracks and traces* (pp. 397–407). Cambridge: Cambridge University Press.

O'Connor, T. P. 1987. On the structure, chemistry and decay of bone, antler, and ivory.

Proceedings of a conference held by the United Kingdom Institute for Conservation, Archaeology Section (p. 6–8).

Padfield, T., D. Erhardt, and W. Hopwood. 1982. Trouble in store. In N.S. Brommelle and G. Thomson (Eds.), *Science and technology in the service of conservation* (pp. 19–23). Preprints of the contributions to the Washington Congress, 3–9 September 1982. London: International Institute for Conservation of Artistic and Historic Works.

Panella, G. 1971. Fish otoliths: daily growth layers and periodical patterns. *Science 173*: 1124–1127.

Parker, R. B., and H. Toots. 1980. Trace elements in bone as paleobiological indicators. In A.K. Behrensmeyer and A. P. Hill (Eds.), *Fossils in the making: Vertebrate taphonomy and paleoecology* (pp. 197–207). Chicago: University of Chicago Press.

Patton, T. 1972. Current challenges to curators and administrators of collections. In W. Langston, Jr., et al. (Eds.), *Fossil vertebrates in the United States* (pp. 5–13). A report of an ad hoc committee of the Society of Vertebrate Paleontology on the status of fossil vertebrate conservation in the United States.

Paul, C. 1989. *The natural history of fossils* (pp. 10–45). London: Weidenfeld and Nicholson.

Peever, M. 1989. A treatment for bison hornsheaths. *Collection Forum* 5(1):32–37.

Poinar, G. O., and D. C. Cannatella. 1987. An Upper Eocene frog from the Dominican Republic and its implications for Caribbean biogeography. *Science 237*:1215–1216.

Ratkevich, R. P. 1976. *Dinosaurs of the Southwest*. Albuquerque: University of New Mexico Press.

Rieppel, O. 1980. Green anole in Dominican amber. *Nature* (London) *286*:486–487.

Rigby, J. K., and D. L. Clark. 1965. Casting and molding (pp. 389–413). In B. Kummel and D. Raup (Eds.) *Handbook of paleontological techniques*. San Francisco: Freeman.

Rixon, A. E. 1961. The conservation of fossilized material. *Museums Journal 61*:205–207.

———. 1976. *Fossil animal remains: their preparation and conservation*. London: Athlone.

Rossi-Manaresi, R. 1982. Scientific investigation in regard to the conservation of stone. In N. S. Brommelle and G. Thomson (Eds.), *Science and technology in the service of conservation*. Preprints of the contributions to the Washington Congress, 3–9 September 1982. London: International Institute for Conservation of Artistic and Historic Works.

Rothschild, B. M., and W. Turnbull. 1987. Treponemal infection in a Pleistocene bear. *Nature 329*(6134):61–62.

Rudwick, M. J. S. 1985. *The meaning of fossils: Episodes in the history of palaeontology*. Chicago: University of Chicago Press.

Scott, W. B. 1939. *Some memories of a palaeontologist* (pp. 171–173). Princeton, NJ: Princeton University Press.

Sease, C. 1981. The case against using soluble nylon in conservation work. *Studies in Conservation 26*:102–110.

———. 1988 A conservation manual for the field archaeologist. *Archaeological research tools* (Volume 4). Los Angeles: University of California, Los Angeles, Institute of Archaeology.

———. 1990. A new means of controlling relative humidity in exhibit cases. *Collection Forum* 6(1):12–20.

Secor, D. H., J. M. Dean, and E. L. Laban. 1991. Manual for otolith removal and preparation for microstructural examination. Electric Power Research Institute and Belle W. Baruch Institute for Marine Biology and Coastal Research.

Seymour, K. 1988. Computerized specimen and preparation/conservation worksheets for fossil vertebrates. *Collection Forum* 6(2):46–50.

Shelton, S. Y., and J. S. Buckley. 1990. Observations on enzyme preparation effects on skeletal material. *Collection Forum* 6(2):17.

Shor, E. N. 1971. *Fossils and flies: The life of a compleat scientist, Samuel Wendell Williston (1851–1918)*. Norman: University of Oklahoma Press.

Simmons, J. E. 1987. Herpetological collecting and collections management. *Herpetological Circular 16*. Society for the Study of Amphibians and Reptiles.

———. and A. Manning. 1991. Collections management. *Association of Systematics Collections Newsletter* 19(4):46.

Simpson, G. G. 1982, *Attending marvels: A Patagonian journal*. Chicago: University of Chicago Press.

Sternberg, C. H. 1990. *The life of a fossil hunter*. Bloomington: Indiana University Press.

Stolow, N. 1966. Fundamental case design for humidity sensitive museum collections. *Museum News Technical Supplement 11*.

———. 1977. The microclimate: a localized solution. *Museum News 56*(2):52–62.

Stucker, G. F., M. J. Galusha, and M. C. McKenna. 1965. Removing matrix from fossils by miniature sandblasting. In B. Kummel and D. Raup (Eds.), *Handbook of paleontological techniques* (pp. 273–275). San Francisco: Freeman.

Tennent, N., and T. Baird. 1985. The deterioration of Mollusca collections: Identification of shell efflorescences. *Studies in Conservation 30*:73–85.

Thomson, G. 1988. *The museum environment*. London: Butterworths.

Thulborn, T. 1990. *Dinosaur tracks*. New York: Chapman and Hall.

Trusted, M. 1985. *Catalogue of European ambers in the Victoria and Albert Museum*. London: Victoria and Albert Museum.

Ullberg, A. D., and R. C. Lind, Jr. 1989. Consider the potential liability of failing to conserve collections. *Museum News 68*(1):322–333.

Waddington, J., and J. Fenn. 1988. Preventive conservation of amber: Some preliminary investigations. *Collection Forum* 4(2):25–31.

Wagstaffe, R., and J. H. Fiedler. 1968. The preservation of natural history specimens (Vol. II, pp. 231–242). New York: Philosophical Library.

Waller, R. 1980. The preservation of mineral specimens (pp. 116–128). Preprints of papers presented at the eighth annual meeting. American Institute for Conservation of Historic and Artistic Works.

———. 1987. An experimental ammonia gas treatment for oxidized pyritic mineral specimens (pp. 623–630). Preprints of the annual meeting of the International Council of Museums, Committee for Conservation.

Watson, W. Y., and S. Grant. 1988. Flat fossil molds with silicone. *Collection Forum* 4(1): 8–9.

West, R. 1988. Endangered and orphaned natural history and anthropology collections in the United States and Canada. *Collection Forum* 4(2):65–74.

Whybrow, P. J., and W. Lindsay. 1990. Preparation of macrofossils. In D. E. G. Briggs and P. R. Crowther, (Eds.), *Palaeobiology: A synthesis* (pp. 499–502). Oxford: Blackwell.

Williams, S. L., and C. A. Hawks. 1986. Inks for documentation in vertebrate research collections. *Curator* 29(2):96–108.

———. 1986. Care of specimen labels in vertebrate research collections. In J. Waddington and D. M. Rudkin (Eds.) *Proceedings of the 1985 workshop on care and maintenance of natural history collections*. Life Sciences Miscellaneous Publications, Royal Ontario Museum, pp. 93–108.

Williams, W. L. 1991. Investigation of the causes of structural damage to teeth in natural history collections. *Collection Forum* 7(1):13–25.

Wyckoff, R. W. G. 1972. *The biochemistry of animal fossils*. Bristol: Scientechnica Publishers.

———, and A. R. Doberenz. 1965. The electron microscopy of Rancho La Brea bone. *Proceedings of the National Academy of Sciences* 53: 230–233.

Wyckoff, R. W. G., E. Wagner, P. Matter III, and A. R. Doberenz. 1963. Collagen in fossil bone. *Proceedings of the National Academy of Sciences 50*:215–218.

Yochelson, E. L. 1985. *The National Museum of Natural History: 75 years in the natural history building*. Washington, DC: Smithsonian Institution Press.

———. 1989. Some perspectives on preparation. In R. M. Feldmann, R. E. Chapman, and J. T. Hannibal (eds.), *Handbook of paleontological techniques* (pp. 413–420). San Francisco: Freeman.

Zangerl, R. 1965. Galvanoplastic reproduction of fossils. In B. Kummel and D. Raup (Eds.), *Handbook of paleontological techniques* (pp. 413–420). San Francisco: Freeman.

Zipkin, I. 1970. The inorganic composition of bones and teeth. In H. Schraer (Ed.), *Biological calcification: Cellular and molecular aspects* (pp. 69–103). New York: Appleton-Century-Crofts.

2

An evaluation of adhesives and consolidants recommended for fossil vertebrates

Sally Y. Shelton and Dan S. Chaney

One of the major activities in vertebrate fossil preparation is the treatment of broken, crushed, or fragile specimens with a polymer to repair breaks and consolidate small fragments to conserve the specimen for future scientific study. Many organic and semisynthetic and synthetic polymers pose unacceptable hazards for long-term conservation of specimens. Among these are shellac, beeswax, starch pastes, hide or bone glues, and cellulose nitrate, which have been recommended and used on fossil vertebrate specimens in the past for their short-term advantages. The hazards to fossil vertebrate specimens that may result from these and other inadequately tested adhesives and consolidants have not been widely recognized or acknowledged in vertebrate paleontology. Permeable coatings providing some vapor-barrier capability are recommended instead, in combination with effective environmental control of the storage or display system. Two thermoplastics dissolved in acetone are here recommended: Polyvinyl acetate (VINAC B-15) is effective for general conservation; and two acrylics, polyethyl methacrylate/polymethyl methacrylate copolymer (Acryloid B-72) or polybutyl methacrylate (Acryloid B-67), are best for protection of specimens requiring acid preparation. These are not the only useful polymers, but they are some of the most versatile.

Vertebrate fossil preparation is the treatment of broken, crushed, and/or fragile specimens. This joining of pieces does not restore the bone to its original condition. Instead, it merely maintains the broken pieces in a configuration that approaches that of the original unbroken element as closely as possible. Long-term conservation of fossil vertebrates has not been studied in depth until relatively recently. The fragility and environmental vulnerability of many specimens are all too often overlooked (unless the damage to the specimen is so severe that it poses an overt and immediate threat, as in pyritic decay or salt efflorescence). Formal venues for research and publication in the preparation of fossil vertebrates seldom exist, and materials research is usually a low priority at most laboratories. The unfortunate result is that the use of inferior materials tends to be taught by example and entrenched in the oral history of preparation, with no rigorous scientific analysis of the materials used. Many adhesives, recommended and used in the past for their short-term advantages, have been found to lack long-term stability and to deteriorate over time. The results are either adhesive joint failure and often irreparable damage to the specimen or cross-linking, which causes shrinkage and greatly reduces the reversibility of the material. Other ways in which polymers deteriorate include yellowing and loss of plasticizers.

From some of the earliest references in vertebrate paleontology, it is evident that collecting and gluing have long been simultaneous activities (Buckland 1866). The materials used for reattaching broken parts, consolidating crushed material to prevent further dimensional warping, or coating fragile material for transport have varied widely over the years (Howie 1986). Organic substances derived from biological sources, such as shellac, beeswax, starch pastes, and hide or bone glues, were early favorites. The development of semisynthetic polymers led to the early use of unstable products, most notably cellulose nitrate, an inexpensive explosive (Selwitz 1988). In this century the rapid growth of synthetic polymer and plastics technology has resulted in a bewildering array of adhesive products. Many have been used on vertebrate fossils – and some with profoundly deleterious effects. Moreover, these products are designed and marketed for other purposes and are only secondarily used in paleontology; most are proprietary compounds whose components cannot be fully identified. For this and other reasons, in spite of the advice of other writers (e.g., Wolberg 1989), it is recommended that adhesives containing compounds not readily identifiable be avoided if at all possible until adequate information is available demonstrating that the material meets all requirements for an excellent adhesive.

Rixon (1976) and Howie (1984) have published reviews of most of the materials used for the mending and consolidation of fossil vertebrates, both those in contemporary use and those that have been applied in the past. However, the best recent work is by Horie (1987) – an invaluable reference for locating chemical and brand names of commercial polymeric materials, especially those cited in British or European references and for questions on older publications or compounds.

This chapter summarizes basic information concerning commonly used adhesives, including some products that were not available for testing at the time of Rixon's (1976) work. Classes of products, rather than particular brands or manufacturers' names, are discussed with regard to their short- and long-term properties. The products recommended are those that have proven to have long-term stability and some degree of reversibility, in addition to their ease of use. More detailed and specific information can be found in the publications of Horie (1987) and Howie (1984).

Properties of adhesives and consolidants

The state of preservation of the specimen, its relative stability, and its potential for future research are all factors in deciding how to conserve the specimen. Extracting, preparing, and exhibiting vertebrate fossil material depend on maintaining the integrity and orientation of the specimen. Breakage, crushing, or softness may make these procedures prohibitively difficult or even impossible.

Unfortunately, the ideal compound for adhesion and consolidation, one that could be used for all applications and removed completely without causing any physical or chemical change to the specimen, does not exist. The most desirable, both for conservation and safety, properties of a polymeric material in fossil vertebrate preparation include the following (in no order of priority):

- *Long-term stability,* that is, no change in structure or composition – unstable polymers may cross-link, a process that causes shrinkage of the polymer and damage to the adhering bone surface; unstable polymers may become brittle, causing the joint to fail over time and posing a risk to the specimen.
- *Water-clear* – a yellow or black resin is obviously undesirable, even if it is effective in other ways, because of the discoloration of the joints and the risk of long-term staining of the specimen; this includes both the original color and the colors to which it may change over time.
- *Utility* – consider the particular requirements of a project: Does the polymer meet the criteria for a consolidant and/or an adhesive? Is it acid resistant if that is required for safe preparation of the specimen?
- *Solubility* – in common and reasonably safe solvents such as water, acetone, ethanol, or methanol – the use of any solvent or other chemical compound should be carefully considered. A material safety data sheet, or MSDS, is invaluable in evaluating the selection and safe use of these potentially hazardous substances.
- *Vapor-barrier capability* – a useful adhesive coating, though not rigid and impermeable like glass, will nevertheless provide some degree of protection from environmental fluctuations for sensitive materials.
- *Reversibility* – in conservation this term refers to the ideal condition of being able to remove all traces of a compound if needed. Preparators more commonly use the term to mean that an adhesive bond can be weakened and loosened with a solvent, without attempting to remove all of the material. True reversibility is seldom possible; undoing an adhesive bond is quite possible.
- *Resistance to acid attack,* for those specimens requiring acid preparation – one of the major reasons for the use of polymeric materials on vertebrate fossils.
- *Availability in 100% solid form without additives,* including plasticizers, stabilizers, or emulsifiers – this enables the preparator to make a controlled and precise solution of the desired adhesive material without untraceable proprietary contaminants.
- *Cost* – may not be a consideration but the cost of any adhesive or consolidant used on a specimen will amount to pennies in comparison to what has been and will be spent to collect and house the specimen; there is no correlation between price and suitability of a consolidant.
- *Glass transition point,* or T_g – temperature at which a plastic starts to flow, however slowly; providing information about the brittleness of the plastic at working conditions; low T_g's ($< 20°C$) will result in plastics that, under elevated temperatures, will not hold specimens together adequately, and due to their softness, incorporate dust into the consolidant; too high a T_g will result in a plastic that is much too brittle and in solution will not penetrate well.

Adhesive substances are commonly used to reattach broken pieces to each other, either surficially (adhesion) or through solventborne penetration (consolidation) or to coat the surface. Many of these substances are not easily removed or are, in fact, irreversible; they can introduce a foreign substance to the specimen that can cause more deterioration, and if the treatment is not documented the adhesive may be unidentifiable in later years. If the specimen is later used for sophisticated technological analysis (Gurley et al. 1991; Gillette 1992, personal communication), these contaminants can interfere with the results and perhaps solutions to other scientific questions. Moreover, in spite of recommendations in older references (e.g., Converse 1984), it is neither necessary nor desirable to glue or consolidate all specimens. Pieces may be bagged or boxed together to maintain their affiliation until a decision on their best use has been reached. Also, the use of polymeric materials to provide a coating or sealer for exposed surfaces is often ill-advised. Some of the most damaging environmental reactions, including salt efflorescence and pyritic decay, will continue under the coating (Winkler 1973), causing larger segments of the affected surface to fall off.

Specific materials

The advantages and disadvantages of some commonly used adhesives are summarized in this section, in groups defined by origin.

Natural organic materials

The oldest adhesives are those derived from botanical or zoological sources, including natural waxes, shellac, gum arabic (Wagstaff and Fiedler 1968), and protein glues (Howie, 1986). These natural organic adhesives share several problems that make them undesirable in vertebrate fossil preparation: They are generally not stable over time, and they are particularly susceptible to biodeterioration (e.g., mold, insects, rodents). Further, a specimen so treated in the past may be more damaged by attempts to remove the organic adhesive than by the adhesive itself. Stabilizing the old adhesive can be difficult and time-consuming.

Shellac, in particular, was used in an ethanol solution as the field consolidant of choice for many years. A natural resin produced by a scale insect (Mills and White 1987), it has been widely used for coating or consolidating archaeological and paleontological materials. Shellac is extremely unstable over time, and its shrinking can cause severe surface and joint damage (Koob 1984a). Though soluble to some extent, it is difficult to remove fully, and it has a tendency to turn yellow. Shellac hold dust and other particulates on the surface. These particulates are often acidic and cause pitting and erosion, especially in environments of elevated humidity, and can obscure many key features of specimens. Shellac, though once valued for its availability and ease of application, should not be used in vertebrate paleontology owing to its poor aging qualities, cross-linking, and discoloration. Other compounds can provide the same bonding characteristics without the inherent disadvantages.

Beeswax has been used to provide a glossy coating for bones, and protein glues have been used for adhesion of Recent and lightweight fossil specimens.

Rixon (1961) found that natural waxes are difficult or impossible to remove and, further, impede the use of other materials.

Semisynthetic adhesives

One of the most popular semisynthetic adhesives is cellulose nitrate, the polynitrate ester of the natural plant polysaccharide cellulose. It is a semisynthetic polymer first manufactured for the production of "gun cotton" as a military explosive (Selwitz 1988). Stabilization of cellulose nitrate with camphor led to the development of celluloid and engineering plastics, lacquers, and coatings.

Cellulose nitrate polymers have been popular in fossil vertebrate preparation for some time (Grenn 1924). They are easy to use, quick-setting and -drying, durable, and removable to some degree; in addition, they are very soluble in acetone (almost the universal solvent of vertebrate paleontology) and better at consolidation than shellac in alcohol solutions.

Cellulose nitrates, however, lack long-term stability. Cellulose nitrate resins have been proven to break down when exposed to light, moderate heat, relative humidity or other moisture sources, and atmospheric gases (Koob 1982, 1986). This instability results in discoloration, brittleness, weakness, and the formation of nitric, formic, and oxalic acids. The acids so formed have the potential to attack the surface of the specimen or surrounding specimens and materials. Covering old cellulose nitrate joints or coatings with a new polymeric film is not effective. Cellulose nitrate adhesive is acidic and will continue to degrade. Failure and embrittlement of the adhesive often cause it to pull away from the specimen, resulting not only in failure of the joint but also removal of the bone surface to which the adhesive was in contact.

Cellulose nitrates do not cross-link, but they degrade over time and lose their adhesive ability. They will shrink with loss of plasticizer and, in the process, they tend to turn yellow (which often causes staining). The instability problems with cellulose nitrates have been reported, and the compounds have not been recommended for conservation applications for some time. These reports seem not to have not been seen or heeded by the majority of vertebrate paleontologists and their preparation staffs. Vertebrate paleontology (and perhaps paleontology as a whole) has lagged behind other fields (e.g., archaeology) in awareness of this problem.

There are several well known and widely used adhesives that are cellulose nitrate based. Glyptal, developed for use in coating electrical wires, has found widespread use in paleontology as a coating, at least in some countries (Brink 1957). Wilson (1965) recommended using Glyptal in combination with polyvinyl acetate emulsions, but this is not now recommended. Model airplane glue and Ambroid are others to be avoided for this reason. Because of the multiple disadvantages of cellulose nitrate compounds, their use should be avoided in vertebrate paleontology preparation. The short-term advantages are far outweighed by the long-term disadvantages.

Synthetic adhesives

The first wholly synthetic plastic produced for commercial use was Bakelite, a thermoset, phenolic, urea–formaldehyde resin that was applied in solution (Case

1925) for use in vertebrate paleontology. However, its immediate advantages were outweighed by its tendency to turn brittle and its irreversibility (Howie 1984). Several synthetic resins, though, provide excellent adhesion and consolidation of fossil bones. Two of the most widely used are polyvinyl acetate and polybutyl methacrylate solutions. Though Rixon (1961) valued polyvinyl acetate emulsion ("white glues") as a consolidant because of the inertia of the suspended solids, an acrylic (Acryloid B-72 Rohm and Haas) solution is preferred for dry material (Koob 1984b). Polyvinyl acetate emulsions, in addition to being difficult to reverse, are a source of acetic acid, which will attack the specimen. Polyvinyl acetate in an acetone solution has been found effective for general conservation (cf. Storch 1983).

Polyacrylates are often selected because of their stability and resistance to yellowing (Koob 1986). They do not cross-link and do not appear to degrade readily. Their solubility varies; most are soluble in acetone or other aromatic solvents. Acrylic copolymers are useful in applications where reversibility is a primary concern. The acrylic copolymer Acryloid B-72 (Rohm and Hass) has found widespread use as an adhesive and a consolidant and is considered a stable and dependable synthetic polymer. Acrylic emulsions and colloidal dispersions appear to be widely useful for a wide range of conditions and materials (Koob 1984b).

Epoxy resins are compounded from an intermediate resin and a hardener, which initiates a cross-linking reaction over a period of days to minutes, depending on composition and temperatures (Mills and White 1987). Epoxies are widely used in industry but have limited use in conservation. They tend to turn yellow on exposure to heat or light, which may cause unsightly darkening of the adhesive bonds and staining of the adjacent material. Epoxies cannot be easily removed and are not recommended for direct application to research specimens, particularly geologic specimens (Macbeth and Strohlein 1965). Epoxy resins are useful in bonding heavy materials and specimens, and many preparators accept their inherent disadvantages. Where the joint is not apparent and no need for reversing the bond exists for the foreseeable future, epoxy resins may be useful. However, they do impair some research usefulness of the specimen.

Silicone resins have been used with success in the particularly problematic conservation of a dinosaur trackway (Agnew 1984; Agnew et al. 1989). The silicone was selected for its ability to repel water when used as a coating, the inertness of its breakdown products, and its ease of replacement. Samples of the trackway stone were tested with several compounds before the silicone resin was selected.

The cyanoacrylates (superglues) have not been fully tested as regards applicability in conservation. Repairs on glass made with these plastics are known to fail under conditions of elevated moisture. Their action is very rapid, but their long term stability is unknown, and they are extremely difficult to remove from a specimen (Whybrow and Lindsay 1990). More testing is needed. Their use in vertebrate paleontology is invaluable to preparation of micromounts to pins and to quick bonding of specimens. Their long-term characteristics are not known; thus their use must be documented in case of future problems.

One of the authors (Chaney) tested the effectiveness of several synthetic polymers in an acid-resistance test. Polymethyl methacrylate (B-67) and polybutyl methacrylate (B-72), both in acetone solutions, were best for protection in acid preparation. The test consisted of coating five sides of a limestone cube with a plastic that had been dissolved in acetone and air dried. The cube was then placed in a bath of 10% formic acid. Visual observations were made. Other than B-67 and B-72, the other plastics tested have not been fully examined for use in conservation work. Most are co- or tripolymers of proprietary composition, and the rest are of unknown composition, provided in a testing stage by several manufacturers.

Polyvinyl acetate solutions (not emulsions) and polybutyl methacrylate solution were best in controlled testing for specimen protection. These appear to have several advantages and few disadvantages. The solutions should be prepared from solid beads rather than purchased in pre-mixed compound form – a precaution that eliminates many industrial contaminants.

Recommendations

1. Adhesion and consolidation will continue to be necessary in vertebrate fossil preparation, but they need not dominate it. Use of adhesives and consolidants should follow careful planning and evaluation of the nature of the specimen, its immediate and future use, and its susceptibility to damage by treatment. Recommended materials should be synthetic and should match as closely as possible the characteristics discussed in this chapter.

2. Specimens intended for chemical analysis such as protein analysis or C^{14}, should never be treated with any substance (or even directly handled). It is always prudent to save at least some material from a site without treatment, so that future tests can be made on an uncontaminated sample.

3. Although it may be possible to identify the compounds used on specimens collected in earlier days, specific and precise tests are required (Haslam, Willis, and Squieral 1972; Rixon 1976). Even knowing the compound may not be sufficient to avert damage, as many older compounds are effectively irremovable. The specimen can be badly damaged by attempts to remove them. It may be best to make a high quality cast of the specimen and use the cast for further research if adhesive damage has made the specimen unusable.

4. Preparators in Canada and Great Britain have been encouraged to document their use of adhesives and consolidants. A form proposed by Fitzgerald (1988) lists several variables that should always be recorded when adhesives are used on research material. It is strongly suggested that each institution initiate a documentation program because work on the long-term effects of adhesives and consolidants is critically needed. Accelerated aging tests (Blackshaw and Daniels 1979) provide much valuable information, but such tests can only be carried out by careful, long-term observation and documentation. It would be valuable to know whether consolidant use affects optical imaging of fossil bones at the cellular level, for example. Only with research and publication can the standards for materials rise to meet the standards for research.

Laboratory use and modifications

Thermoplastics used in the laboratory need to have a variety of viscosities depending on their intended application. These materials should be stored so that they are ready to be dispensed when needed. A stock in the desired solvent of each type of thermoplastic should be maintained in the laboratory. A portion of the stock is transferred to an "application" container and thinned to the desired viscosity by the addition of solvent. Because the plastic beads have already been dissolved into the syrup, this further dilution takes little time. As the stock reaches about 25% of the original amount, more plastic beads and solvent are added to maintain a ready supply.

The application container will vary, depending on need and preference. One useful container for brush applications of very thin to thick solutions is a number of plastic bottles with rubber stoppers through which the handle of a paintbrush has been passed, providing ready access to a brush. For glue, the stock is poured into empty collapsible metal tubes; a variety of sizes are available for different applications.

Fumed silica can be added to PVA to increase bonding strength (Bryne 1984). In a solution of PVA 15% and acetone 85% add between 3% and 10% by weight fumed silica. This will result in a thixotropic gel that can be applied where needed from a collapsible metal tube. This has been found to create a brittle bond (Johnson, personal communication). More investigation is needed.

Supplies and suppliers

Collapsible metal tubes
Any number of sizes, shapes, colors, and materials are available We prefer aluminum nasal tip with no color (the paint comes off in acetone).

Unipac Supply Co., Box 98026, Pittsburgh, PA 15227, 412-885-2266

Acrylic resins
Acryloid B-72 and other useful acrylic resins are manufactured by Rohm and Haas. They are not available from the manufacturer. The manufacturer will provide a list of sources for these materials in your area.

Rohm and Haas Co., Independence Mall West Philadelphia, PA 19105

Polyvinyl acetate
Available in 100% bead form from at least two manufacturers.

Air Products and Chemicals VINAC B-15, 5 Executive Mall, Swedesford Rd., Wayne, PA 19087
Union Carbide AYAF, Old Ridgeway Rd., Danbury, CT 06817

Definition of Terms

These definitions are based on Hawley (1971), Mills and White (1987), and Williams (1988).

adhesion: the join of two surfaces by valence action, interlocking action or both.

adhesive: any substance capable of bonding other substances together by surface attachment.

coating: a surface skin application of an adhesive.

consolidant: an adhesive, often in a volatile carrier solvent, applied to soak through a specimen and force subsurface adhesion.

consolidation: holding together of an entire specimen or object in its original dimensional contours via the action of a consolidant.

cross-linking: the formation of bonds between otherwise unattached polymer chains; often a source of shrinking or degradation of the polymer; renders material insoluble.

emulsion: a dispersion of solid polymer particles in a fluid medium (often water).

glass transition temperature (T_g): temperature at which a high polymer changes from a brittle, vitreous state to a plastic state. The higher the T_g the more brittle the material at room temperature.

plastic: a base polymer plus necessary additives to adjust the properties of the compound to the desired value.

polymer: a large molecule formed by the linking together of many repeated units of the same small molecules (monomers) by normal covalent bonds; may be linear or nonlinear (branched).

solvent: a substance capable of dissolving another substance (solute) to form a solution.

thermoplastic: plastics that can be heated to a plastic state, molded, and then cooled to harden in that shape; theoretically this can be repeated and a new shape made.

thermoset: plastics that cannot be heated to a plastic state, then deformed. The desired shape is obtained by forming the initial polymer(s), then catalyzing it in position through heat, radiation, or a catalyst; often an exothermic reaction.

Acknowledgments

The study of adhesives and consolidants has been an ongoing project in the Department of Paleobiology at the National Museum of Natural History (Smithsonian Institution) for a number of years. Robert W. Purdy and Arnold D. Lewis spurred the initial examination and Frederick Grady has supported continued efforts in this area.

Thanks to J. Johnson, W. G. Reeder, R. H. Rainey, and E. Yarmer of Texas Memorial Museum; G. R. Fitzgerald and R. R. Waller of the Canadian Museum of Nature; C. V. Horie of the Manchester Museum; A. Doyle and F. M. P. Howie of the Natural History Museum (London): C. Collins of the Geological Conservation Unit, Cambridge University; and R. E. Child and C. Buttler, of the National Museum of Wales. This is publication No. N. S.-40 of the Texas Memorial Museum.

We would like to thank Dr. Steven Koob of the Sackler Gallery, Smithsonian Institution, Dr. Carolyn Rose of the National Museum of Natural History, and Gerald R. Fitzgerald of the Canadian Museum of Nature, all of whom reviewed early versions of this chapter, for their helpful comments.

References

Agnew, N. 1984. The use of silicones in the preservation of a field site – the Lark Quarry dinosaur trackways. In N. S. Brommelle, E. M. Pye, P. Smith and G. Thomson (Eds.), *Adhesives and consolidants* (pp. 87–91). Preprints of the contributions to the Paris Congress, 2–8 September 1984. London: International Institute for Conservation of Artistic and Historic Works.

———, H. Griffin, M. Wade, T. Tebble, and W. Oxnam. 1989. Strategies and techniques for the preservation of fossil tracksites: An Australian example. In D. D. Gillette and M. L. Lockley (Eds.), *Dinosaur tracks and traces* (pp. 397–407). Cambridge: Cambridge University Press.

Blackshaw, S. M., and V. D. Daniels. 1979. The testing of materials for use in storage and display in museums. *The Conservator* 3:16–19.

Brink, A. S. 1957. On the uses of Glyptal in palaeontology. *Palaeontologica Africana* 4:124–130.

Buckland, F. 1866. Curiosities of natural history (2nd ser.). London: Bentley [as cited in Howie 1986].

Byrne, G. 1984. In N. S. Brommelle E. M. Pye, P. Smith, and G. Thomson (Eds.) *Adhesives and Consolidants* (pp. 78–80). Preprints of the contributions to the Paris Congress, 2–8 September 1984. London: International Institute for Conservation of Artistic and Historic Works.

Case, W. 1925. The use of Bakelite in the preservation of fossil material. *Science 61*:453–454.

Converse, H. H. 1984. *Handbook of paleo-preparation techniques.* Gainesville: Florida State Museum.

Fitzgerald, G. R. 1988. Documentation guidelines for the preparation and conservation of paleontological and geological specimens. *Collection Forum* 4(2):38–45.

Grenn, W. A. 1924. Celluloid as a preservative. *Museum Journal* (London) 24:154.

Gurley, L. R., J. G. Valdez, W. D. Spall, B. F. Smith, and D. D. Gillette. 1991. Proteins in the fossil bones of the dinosaur, "*Seismosaurus.*" *Journal of Protein Chemistry 10*(1):75–90.

Haslam, J., H. A. Willis, and D. C. M. Squirrel. 1972. *Identification and analysis of plastics* (2nd ed.). London: Iliffe Books.

Hawley, G. G. (Ed.) 1971. *The condensed chemical dictionary* (8th ed.) New York: Van Nostrand Reinhold.

Horie, C. V. 1987. *Materials for conservation.* London: Butterworths.

Howie, F. M. P. 1984. Materials used for conserving fossil specimens since 1930: A review. In N. S. Brommelle, E. M. Pye, P. Smith and G. Thomson (Eds.), *Adhesives and consolidants* (pp. 92–97). Preprints of the contributions to the Paris Congress, 2–8 September 1984. London: International Institute for Conservation of Artistic and Historic Works.

———. 1986. Conserving and mounting fossils: A historical review. *Curator 29*(1):5–24.

Koob, S. P. 1982. The instability of cellulose nitrate adhesives. *The Conservator* 6:31–34.

———. 1984a. The continued use of shellac as an adhesive: Why? In N. S. Brommelle, E. M. Pye, P. Smith, and G. Thomson (Eds.), *Adhesives and consolidants.* Preprints of the contributions to the Paris Congress, 2–8 September 1984. London: International Institute for Conservation of Artistic and Historic Works.

———. 1984b. The consolidation of archaeological bone. In N. S. Brommelle, E. M. Pye, P. Smith, and G. Thomson (Eds.), *Adhesives and consolidants.* Preprints of the contributions to the Paris Congress, 2–8 September 1984. London: International Institute for Conservation of Artistic and Historic Works.

———. 1986. The use of Paraloid B-72 as an adhesive: Its application for archaeological ceramics and other materials. *Studies in Conservation 31*:7–14.

Macbeth, J. A., and A. C. Strohlein. 1965. The use of adhesives in museums. *Museum News Technical Supplement 7.*

Mills, J. S., and R. White. 1987. *The organic chemistry of museum objects.* London: Butterworths.

Rixon, A. E. 1961. The conservation of fossilized material. *Museum Journal* (London) *61*: 205–207.

Rixon, A. J. 1976. *Fossil animal remains: Their preparation and conservation.* London: Athlone.
Selwitz, C. 1988. Cellulose nitrate in conservation. *Research in Conservation 2*. Getty Conservation Institute.
Storch, P. S. 1983. Field and laboratory methods for handling osseous materials. *Conservation Notes 6*. Texas Memorial Museum, Austin.
Wagstaffe, R., and J. H. Fiedler. 1968. The preservation of natural history specimens, (Vol. II, pp. 231–242). New York: Philosophical Library.
Whybrow, P. J., and W. Lindsay. 1990. Preparation of macrofossils. In D. E. G. Briggs and P. R. Crowther, (Eds.) *Palaeobiology: A synthesis* (pp. 499–502). Oxford: Blackwell.
Williams, R. S. 1988. Polymers, plastics, rubbers, and elastomers in conservation. Lecture, Collections Care Pilot Training Program, Natural History Museum of Los Angeles County.
Wilson, R. L. 1965. Techniques and materials used in the preparation of vertebrate fossils. *Curator 8*(2):135–143.
Winkler, E. M. 1973. *Stone: Properties, durability in man's environment.* New York: Springer-Verlag.
Wolberg, D. 1989. Glues and other stick'ums and patch'ums: stabilizing compounds for strengthening fossils. In R. M. Feldmann, R. E. Chapman, and J. T. Hannibal (Eds.), *Paleotechniques.* Paleontological Society Special Publication 4:249–259.

3

Collecting taphonomic data from vertebrate localities

Raymond R. Rogers

Taphonomy, a field of study initially defined by Efremov (1940), explores the fate of organismal remains and traces. Its scope extends from events immediately prior to death through burial and diagenesis to fossil collection and curation. The analysis of taphonomic data can shed light on a wide array of paleobiological questions. For instance, the microscopic and chemical characteristics of bones and teeth (e.g., dissolution pits, scratch marks, trace element composition) can help identify specific agents of bioclast modification (Fisher 1981; Behrensmeyer, Gordon, and Yamagi 1986; Olsen and Shipman 1988; Tuross et al. 1989). On a somewhat larger scale, a comparison of taphonomic features from different depositional environments, tectonic settings, or time periods can reveal patterns and trends in the preservation and distribution of fossiliferous sediments (Dodson et al. 1980; Badgley 1986; Kidwell and Behrensmeyer 1988; Badgley et al. 1992).

This chapter discusses field methods for collecting taphonomic information from vertebrate sites. This treatment can by no means be exhaustive given the diverse ways in which vertebrates are preserved in the rock record. For this reason an annotated bibliography comprising numerous taphonomic case studies concludes the chapter.

Collecting taphonomic data

The type and quantity of taphonomic data and the relative ease with which these data can be extracted from the rock record vary, depending on the depositional environment and the "taphonomic mode" of occurrence (Behrensmeyer 1988). For example, an isolated bone fragment, a solitary intact skeleton, a hydraulically sorted vertebrate microsite, and a multi-individual bone bed will all yield very different data sets, even if all occur in a fluvial record. The goal of the taphonomist in the field is to sort through the available geologic and paleontological source material and collect the data necessary to reconstruct taphonomic history.

Taphonomic data can be broken down into two major categories – geologic data and (paleo) biologic data (see Table 3.1). Geologic data include the stratigraphy and sedimentology of a site. Biologic data comprise those taphonomic features directly related to the fossil bone assemblage (e.g., taxa represented, age

Table 3.1. *Types of geologic and paleobiologic data that should be collected for the taphonomic characterization of vertebrate localities.*

Geologic data	Paleobiologic data
Stratigraphic Unit (e.g., formation, member, bed. etc.)	Assemblage Data
Stratigraphic Position within Unit*	sample size
Lithology of fossiliferous stratum	taxa represented
color (fresh and weathered)*	number of individuals
thickness (lateral variation?)*	relative abundance
mineralogic composition	body size
grain size, sorting and shape	age and sex profiles
sedimentary structures*	articulation**
induration	skeletal sorting
cement	Quarry Data
pedogenic features*	size of accumulation (m^2, m^3)*
halos, invertebrate traces, mineral crusts, or micro-structures around bones**	density (bones/m^2, bones/m^3)*
Nature of Bounding Contacts*	spatial arrangement**
Geometry of Fossiliferous Stratum*	Bone Modification Data (side specific?)**
Nature of Surrounding Strata*	weathering**
Inferred Depositional Setting	polish**
	abrasion**
	scratch/tooth marks**
	root traces**
	borings (orientation?)**
	fragmentation/breakage**
	distortion (orientation?)**
	mode of fossilization

Data that should (or must) be recorded before excavation are indicated by two asterisks. Based upon taphonomic data sheets of Munthe and McLeod (1975) and Behrensmeyer (1991).

spectra, sorting, weathering, orientation, surface modification). Taphonomic narratives and data sheets often follow, at least in part, this bipartite division of data (Efremov 1940; Zangerl and Richardson 1965; Voorhies 1969; Munthe and McLeod 1975; Saunders 1977; Turnbull and Martill 1988; Rogers 1990). Although taphonomic data are often collected and presented under geologic and biologic subheadings, accurate taphonomic reconstructions can result only through the analysis of both data sets.

Geologic data

Reconstructing the depositional environment of a vertebrate locality should be a primary goal in any taphonomic study, because the physical, chemical, and biological processes that shape the rock record largely dictate the nature of included fossil material. Moreover, ecological parameters that influenced mortality and/or preservation may be revealed by paleoenvironmental analysis (Behrensmeyer 1991). The recognition of unusual lithologies in the rock record, such as ash beds, tektites, storm beds, and debris flow deposits, can further promote the reconstruction of taphonomic history (Fritz 1980; Voorhies 1985; Miller, Brett, and Parsons 1988). Several recent texts are devoted to the recognition and interpretation of facies and the reconstruction of depositional environments (Walker 1984; Reading 1986; Davis 1992).

Among the most fundamental of the data required to interpret geologic history and reconstruct depositional environments are measured stratigraphic sections. A measured section is simply a schematic description of the local rock record, but it provides a framework for assessing relative age relations between paleontological localities and stratigraphic horizons (e.g., ash beds, storm beds) and also provides control for sedimentologic and paleontological sampling. At sites with significant lateral variation in lithologies or bed thickness, several closely spaced sections may be measured and used to construct a two-dimensional cross-sectional view of lateral and vertical facies relationships. This may also be done on a regional scale, and long-range correlation may prove possible.

Detailed accounts of how to measure a stratigraphic section are available in Compton (1962) and Kottlowski (1965). Measuring a stratigraphic section, in its most general sense, entails delineating and describing lithologically coherent packages of strata, measuring their thicknesses, and characterizing their bounding surfaces. Stratigraphic sections are usually measured from bottom to top in order to follow natural stratal succession.

Measured sections should be routed through well-exposed, structurally undeformed rock exposures. An obvious objective when measuring a stratigraphic section for taphonomic purposes is that the section also pass through, or very near, the paleontological locality of interest. When an appropriate route is chosen, the strike and dip of strata are determined (see Compton 1962:28–33), and an appropriate measuring technique (e.g., eye heights, Jacob's staff, steel tape, plane table) is selected. In most situations, a Jacob's staff (a metrically calibrated wooden, metal, or plastic pole), combined with an Abney hand level or the clinometer of a Brunton compass, provides a fast and accurate means of measurement. Instructions for assembling a lightweight, collapsible Jacob's staff are outlined in Winker (1986).

The first step in measuring a stratigraphic section is delineation of the basal unit. Informal units comprising measured sections should be defined by changes in obvious lithologic criteria (composition, grain size, bedding, color, etc.), and thus often represent discrete depositional events. When the upper and lower contacts of a unit are determined, unit thickness can be measured, and a detailed description of unit lithology, geometry, and paleontology can be recorded in a field notebook.

Description of a rock unit should include color (fresh and weathered); lithologic composition; grain size, sorting, and shape; sedimentary structures; induration; type of cement; fossil content; and pedogenic features. Any type of structural deformation, such as folding, faulting, brecciation, or jointing, should also be noted. The nature of the contacts bounding a unit is extremely important to the local geologic history, and therefore deserves thorough scrutiny. Contacts are most often characterized as erosional (underlying beds or structures are truncated), sharp (no truncation but knife-edge change in lithology), or gradational. The geometry of a rock unit (e.g., lenticular, tabular, wedge-shaped) should be determined whenever possible. Field descriptions can be augmented by detailed petrographic and geochemical analyses if hand samples are collected.

The thickness of a unit can be obtained by orienting a Jacob's staff perpendic-

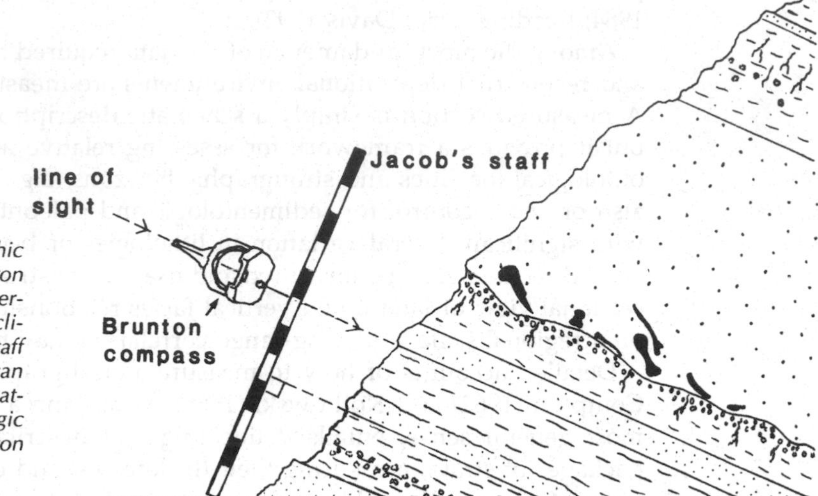

Figure 3.1. Measuring a stratigraphic section with a Jacob's staff and Brunton compass. Orient the Jacob's staff perpendicular to bedding using the clinometer of the compass. With the staff oriented properly, unit thickness can be read directly from the staff by locating bounding contacts of lithologic units through the sight of the Brunton compass.

ular to bedding at the base of a unit and sighting parallel to dip through the sight of the Abney hand level or Brunton compass. The clinometer (set for local dip) on the Abney hand level or Brunton compass is used to orient the Jacob's staff normal to bedding. Once the Jacob's staff is oriented properly, the upper contact of the unit is located through the sight of the hand level or compass, and unit thickness is read directly from the calibrated staff. When the basal unit of a measured section is thoroughly described and measured, the next unit can be delineated, and the process can proceed up section (see Figure 3.1).

Bone assemblage data

Taphonomic features directly related to, or measured from, the bone sample are here considered "bone assemblage data." These features include quarry area or volume, bone density (measured in bones per square meter or bones per cubic meter), sample size, number of individuals, taxonomic diversity, relative abundance of taxa, age and sex profiles, state of articulation and association, orientation, sorting, fragmentation, distortion, and various surface modifications (weathering, abrasion, scratch marks, toothmarks, gnaw marks, borings, root traces, etc.) (Table 3.1; see Behrensmeyer 1991). Many of these taphonomic features are best assessed after collection and preparation of the bone sample. Other variables, however, must be measured in the field while the bones remain in situ. Indeed, data pertaining to bone orientation, articulation, and association are invariably lost if bones are hastily extracted from encasing matrix.

A carefully drafted field map provides a permanent record of skeletal orientation, articulation, and original associations and thus should be primary data in any taphonomic field study. A field map can also demonstrate horizontal and vertical sorting and may reveal subtle features (e.g., taxonomic or size-class associations, density trends, localized bioturbation and/or fragmentation) that might otherwise go unnoticed. Accurate mapping of a site requires a system of reference, and this is normally provided by erecting a square grid system over the exhumed fossiliferous stratum. A meter-square grid system is suitable for most large-scale applications (e.g., bone beds); a smaller scale decimeter or centimeter-square grid system should be employed when mapping small, isolated skeletons or microvertebrate localities. A small-scale grid inserted within the framework of a larger grid system can augment accuracy.

A modicum of equipment is needed to construct a grid for field mapping. Stakes or large nails can be used as tie points, with string or wire stretched between points to define individual cells. Alternatively, a prefabricated grid of wood or plastic can be employed if the quarry is small. In either case, the plane of the grid system should be placed parallel to the bedding of the fossiliferous stratum.

Once a grid system is erected, mapping and excavation can commence. Under ideal circumstances, excavation will proceed across the quarry in preset increments, thus fostering analysis of the deposit in three dimensions (see Abler 1984 for a treatment of three-dimensional paleontological mapping). A separate map of the quarry surface should be plotted on graph paper for each increment exposed. A composite map illustrating in plan view all bones within a quarry can be generated by superimposing incremental maps on a light table.

Positions of bones can be conveniently measured from nodes (corner stakes) within the grid matrix using a measuring tape and Brunton compass. Alternatively, a plane table and transit can be utilized to locate bones accurately in three-dimensional space. (An introduction to plane table and transit techniques is presented in Compton 1962.) When a bone is located within the grid, its dimensions can be directly measured with ruler or tape, and the bone can be plotted on the map using a ruler and protractor. Map design may incorporate actual plan-view sketches of bones or schematic representations using lines (bones with a long axis) and circles (bones lacking a long axis) (see Figure 3.2). Collection numbers should be recorded on the field map for future reference.

Orientations of elongate skeletal elements should always be portrayed with trend, plunge, and polarity data. The trend of a bone is the compass direction of the vertical plane that includes the long axis of the bone. Bone plunge is the vertical angle between the long axis of the bone and a horizontal plane (see Figure 3.3). Bone polarity merely refers to whether one end of a bone is larger than the other (often the case in fragmentary limb bones). To measure the trend of a bone, orient the axial line of a horizontally leveled Brunton compass (see Compton 1962:21 for a description of a Brunton compass) parallel to the vertical plane that includes the long axis of the bone. Plunge is measured by orienting the edge of a Brunton compass parallel to the long axis of a bone and rotating the clinometer until the bubble in the clinometer level is centered (remember to adjust plunge measurements if bones occur in dipping strata). If the elongate axis of a bone is horizontal, trend is a bidirectional property (e.g., the bone trends N20E and S20W) and there is no plunge. If a bone is not perfectly horizontal, the bone trends in the direction of plunge (e.g., N20E, 10^0, where 10^0 indicates the angle of plunge). Trend, plunge, and polarity data should be recorded on the field map (space permitting) and in the field notebook.

Many other clues pertaining to taphonomic history can be lost or obscured if the sample is excavated before thorough taphonomic characterization. Subtle features of the sedimentary matrix encasing bones, such as alteration halos, mineral crusts, fine root traces, and localized invertebrate bioturbation (small burrows or

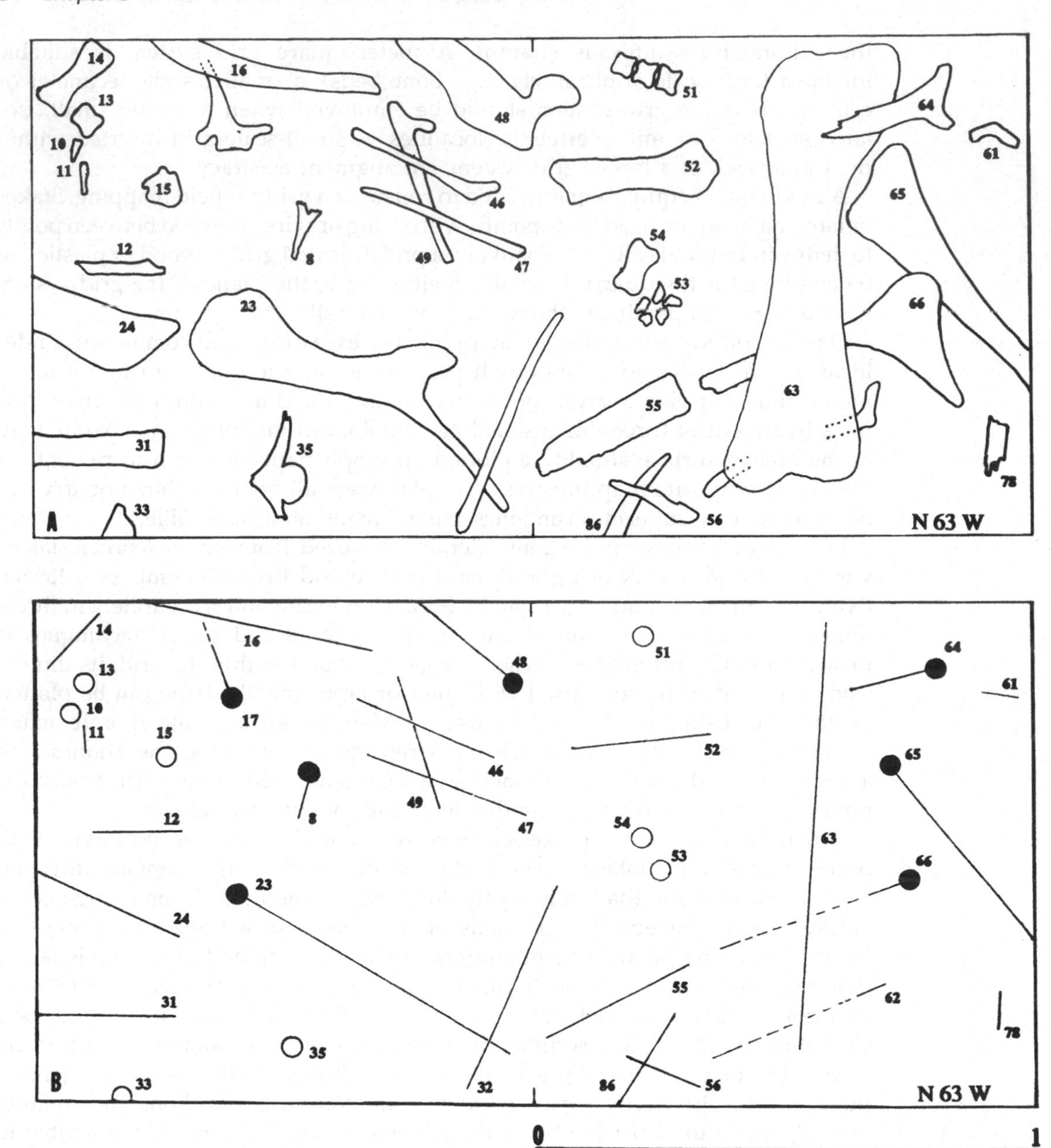

Figure 3.2. Bone bed maps. (A) Actual plan-view sketch of bone bed. (B) Schematic representation of same bone bed using lines and circles. Lines symbolize bones with a long axis and are scaled to bone length. Solid circles attached to lines indicate large ends of elongate (polar) bones. Open circles correspond to bones lacking a long axis or clear-cut polarity (adapted from Hunt 1978).

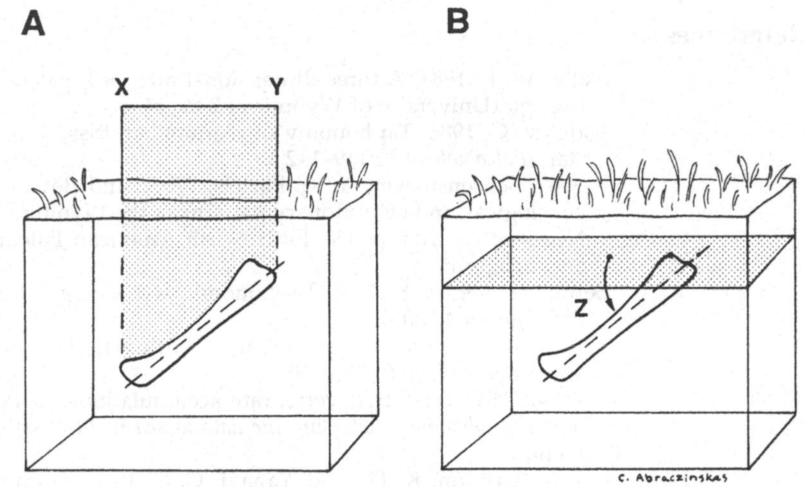

Figure 3.3. Block diagrams illustrating the trend and plunge of an elongate bone. (A) The trend of this bone is the compass direction (X–Y) of the vertical plane (stippled) that passes through the long axis of the bone. (B) The plunge of the same bone can be determined by measuring the angle (Z) between bedding (stippled plane) and the long axis of the bone. Because this bone is plunging, the trend is unidirectional toward X.

borings), may be destroyed during collection and preparation. Original bone-sediment relationships can be preserved if an intact fossiliferous block of oriented quarry matrix is collected (via a plaster jacket; see Chapter 4). The interpretation of one-sided surficial modifications, such as differential weathering (Behrensmeyer 1978; Rogers 1990), can also be confused if bones are collected without sufficient taphonomic documentation. The up-side of every skeletal element should be clearly indicated on the bone surface before extraction in the event that side-specific taphonomic features are recognized after preparation. Thorough photographic documentation of every locality is strongly encouraged.

Conclusion

Significant loss of taphonomic data can result if excavation proceeds before thorough taphonomic documentation. The basic methods outlined in this chapter should facilitate and promote the retrieval of field-based taphonomic data and thereby minimize data loss. Fundamental field-based taphonomic data include a locality map, with attendant bone orientation data, and in situ observations of bone–sediment or bone–ground surface associations. A measured stratigraphic section, with detailed sedimentologic and paleontological descriptions, is also essential.

The highly variable nature of vertebrate mortality and preservation renders an all-inclusive treatment of data collection techniques impossible, because every locality provides a unique array of taphonomic opportunities and logistical problems, and every investigator has his or her own aspirations. Thus methods of collecting taphonomic data must be designed with site-specific intentions. The methods presented in this chapter should be viewed as a means of collecting only the basic core of field-based taphonomic data. An annotated bibliography of taphonomic case studies is appended to this chapter to illustrate the variety of taphonomic studies possible, and the reader is referred to the methods sections of these studies for specific field strategies.

References

Abler, W. L. 1984. A three-dimensional map of a paleontological quarry. *Contributions to Geology* (University of Wyoming) *23*:9–14.

Badgley, C. 1986. Taphonomy of mammalian fossil remains from Siwalik rocks of Pakistan. *Paleobiology 12*:119–142.

———, Behrensmeyer, A. K., Bartels, W. S., and Bown, T. M. 1992. Preservational, paleoecological, and evolutionary patterns in the Wyoming-Montana Paleogene and Siwalik Neogene records (p. 13). Fifth North American Paleontological Convention, Abstracts and Program.

Behrensmeyer, A. K. 1978. Taphonomic and ecologic information from bone weathering. *Paleobiology 4*:150–162.

———, 1988. Vertebrate preservation in fluvial channels. *Palaeogeography, Palaeoclimatology, Palaeoecology 63*:183–199.

———, 1991. Terrestrial vertebrate accumulations. In P. A. Allison and D. E. G. Briggs (eds.), *Taphonomy: Releasing the data locked in the fossil record* (pp. 291–335). New York: Plenum.

———, Gordon, K. D., and Yanagi, G. T. 1986. Trampling as a cause of bone surface damage and pseudo-cutmarks. *Nature 319*:768–771.

Compton, R. R. 1962. *Manual of field geology*. New York: Wiley.

Davis, R. A., Jr. 1992. *Depositional systems: an Introduction to Sedimentology and Stratigraphy* (2nd ed.). Englewood Cliffs, NJ: Prentice Hall.

Dodson, P., Behrensmeyer, A. K., Bakker, R. T., and McIntosh, J. S. 1980. Taphonomy and paleoecology of the dinosaur beds of the Jurassic Morrison Formation. *Paleobiology 6*: 208–232.

Efremov, I. A. 1940. Taphonomy: A new branch of paleontology. *Pan-American Geologist* 74:81–93.

Fisher, D. C. 1981. Crocodilian scatology, microvertebrate concentrations, and enamel-less teeth. *Paleobiology 7*:262–275.

Fritz, W. J. 1980. Reinterpretation of the depositional environment of the Yellowstone "fossil forests." *Geology 8*:309–313.

Kidwell, S. M., and Behrensmeyer, A. K. 1988. Overview: Ecological and evolutionary implications of taphonomic processes. *Palaeogeography, Palaeoclimatology, Palaeoecology 63*:1–14.

Kottlowski, F. E. 1965. *Measuring stratigraphic sections*. New York: Holt, Rinehart and Winston.

Miller, K. B., Brett, C. E., and Parsons, K. M. 1988. The paleoecologic significance of storm-generated disturbance within a Middle Devonian muddy epeiric sea. *Palaios 3*:35–52.

Munthe, K., and McLeod, S. A. 1975. Collection of taphonomic information from fossil and recent vertebrate specimens with a selected bibliography. *Paleobios 19*.

Olsen, S. L., and Shipman, P. 1988. Surface modification on bone: Trampling versus butchery. *Journal of Archaeological Science 15*:535–553.

Reading, H. G. (Ed.). 1986. *Sedimentary environments and facies* (2nd ed.). Oxford: Blackwell.

Rogers, R. R. 1990. Taphonomy of three dinosaur bone beds in the Upper Cretaceous Two Medicine Formation of northwestern Montana: Evidence for drought-related mortality. *Palaios 5*:394–413.

Saunders, J. J. 1977. Late Pleistocene vertebrates of the Western Ozark Highland, Missouri. *Reports of Investigations No. 33*. Illinois State Museum.

Turnbull, W. D., and Martill, D. M. 1988. Taphonomy and preservation of a monospecific titanothere assemblage from the Washakie Formation (Late Eocene), southern Wyoming. An ecological disaster in the fossil record. *Palaeogeography, Palaeoclimatology, Palaeoecology 63*:91–108.

Tuross, N., Behrensmeyer, A. K., Eanes, E. D., Fisher, L. W., and Hare, P. E. 1989. Molecular preservation and crystallographic alterations in a weathering sequence of wildebeest bones. *Applied Geochemistry 4*:261–270.

Voorhies, M. R. 1969. Taphonomy and population dynamics of an early Pliocene vertebrate fauna, Knox County, Nebraska. *Contributions to Geology Special Paper No. 1*. University of Wyoming.

———. 1985. A Miocene rhinoceros herd buried in volcanic ash. *National Geographic Society Research Report 19*:671–688.
Walker, R. G. 1984. *Facies models* (2nd ed.). Geoscience Canada Reprint Series 1. Geological Association of Canada.
Winker, C. D. 1986. Plane-of-sight Jacob's staff for measuring sections oblique to dip. *Journal of Sedimentary Petrology 56*:564–565.
Zangerl, R., and Richardson, E. S. Jr. 1965. The paleoecological history of two Pennsylvanian black shales. *Fieldiana, Geology Memoirs 4*.

Bibliography

Several of the studies listed in this annotated bibliography of taphonomic case studies include detailed methods sections. Case studies lacking detailed treatments of methodology are also included to provide an indication of the type of data needed for taphonomic analysis.

Allison, P. A., Smith, C. R., Kukert, H., Deming, J. W., and Bennett, B. 1991. Deep-water taphonomy of vertebrate carcasses: A whale skeleton in the bathayal Santa Catalina Basin. *Paleobiology 17*:78–89.
A unique submarine-based study of deep-ocean, high-pressure whale taphonomy.
Andrews, P., and Ersoy, A. 1990. Taphonomy of the Miocene bone accumulations at Pasalar, Turkey. *Journal of Human Evolution 19*:379–396.
Taphonomy of a fragmentary Miocene mammal locality.
Andrews, P., Meyer, G. E., Pilbeam, D. R., Van Couvering, J. A., and Van Couvering, J. A. H. 1981. The Miocene fossil beds of Maboko Island, Kenya: Geology, age, taphonomy and palaeontology. *Journal of Human Evolution 10*:35–48.
Geologic and taphonomic investigation of two bone beds developed in nonmarine Miocene deposits with a general treatment of quarry mapping.
Badgley, C. 1986. Taphonomy of mammalian fossil remains from Siwalik rocks of Pakistan. *Paleobiology 12*:119–142.
Taphonomy of mammalian fossils in Late Miocene fluvial deposits. Good coverage of geologic data collection techniques.
Behrensmeyer, A. K. 1975. The taphonomy and paleoecology of Plio-Pleistocene vertebrate assemblages east of Lake Rudolf, Kenya. *Bulletin of the Museum of Comparative Zoology 146*:473–578.
Taphonomy of fossil vertebrates integrated with actualistic observations and experiments.
———. 1991. Terrestrial vertebrate accumulations. In P. A. Allison and D. E. G. Briggs (eds.), *Taphonomy: Releasing the data locked in the fossil record* (pp. 291–335). New York: Plenum.
Excellent review of vertebrate taphonomy, with a full treatment of the various types and implications of taphonomic data.
Clark, J., Beerbower, J. R., and Kietzke, K. K. 1967. Oligocene sedimentation, stratigraphy, paleoecology, and paleoclimatology in the Big Badlands of South Dakota. *Fieldiana, Geology Memoirs 4*.
Early consideration of "taphic factors" and how they influence paleoecological interpretations of fossil mammal assemblages.
Dodson, P. 1971. Sedimentology and taphonomy of the Oldman Formation (Campanian), Dinosaur Provincial Park, Alberta (Canada). *Palaeogeography, Palaeoclimatology, Palaeoecology 10*:21–74.
Early treatment of formation-scale vertebrate taphonomy.
———, Behrensmeyer, A. K., Bakker, R. T., and McIntosh, J. S. 1980. Taphonomy and paleoecology of the dinosaur beds of the Jurassic Morrison Formation. *Paleobiology 6*: 208–232.
Comprehensive taphonomy of the dinosaur-bearing Morrison Formation. Methods of large-scale taphonomic data collection.
Fiorillo, A. R. 1988. Taphonomy of Hazard Homestead Quarry (Ogalla Group), Hitchcock County, Nebraska. *Contributions to Geology* (University of Wyoming) *26*:57–97.

Taphonomy of a Miocene mammal bone bed with much emphasis on bone orientation data.

———. 1991. Taphonomy and depositional setting of Careless Creek Quarry (Judith River Formation), Wheatland County, Montana, U.S.A. *Palaeogeography, Palaeoclimatology, Palaeoecology 81*:281–311.
Taphonomy of a Late Cretaceous dinosaur bone bed. General methods.

Frison, G. C., and Todd, L. C. 1986. *The Colby Mammoth Site: Taphonomy and archaeology of a Clovis kill in northern Wyoming.* Albuquerque: University of New Mexico Press.
Detailed taphonomy of a mammoth site associated with human artifacts, with actualistic experimentation on stream-sorting of elephant bones.

Gnidovec, D. M. 1978. Taphonomy of the Powder Wash Vertebrate Quarry, Green River Formation (Eocene), Uintah County, Utah. Unpublished master's thesis, Fort Hays State University, Kansas.
Taphonomy of a diverse lacustrine microvertebrate locality interpreted as a coprocoenosis, with bone flotation and settling experiments.

Graham, R. W., Holman, J. A., and Parmalee, P. W. 1983. Taphonomy and paleoecology of the Christensen Bog Mastodon Bone Bed, Hancock County, Indiana. *Reports of Investigations No. 38.* Illinois State Museum.
Taphonomy of a taxonomically diverse bog deposit.

Hendey, Q. B. 1981. Palaeoecology of the Late Tertiary fossil occurrences in "E" Quarry, Langebaanweg, South Africa, and a reinterpretation of their geological context. *Annals of the South African Museum 84*:1–104.
Good review of types of bone damage.

Hill, A., and Walker, A. 1972. Procedures in vertebrate taphonomy; notes on a Uganda Miocene fossil locality. *Journal of the Geological Society of London 128*:399–406.
Review of the types of bone assemblage data collected from the Bukwa II fossil locality (Miocene).

Hook, R. W., and Hower, J. C. 1988. Petrography and taphonomic significance of the vertebrate-bearing cannel coal of Linton, Ohio (Westphalian D, Upper Carboniferous). *Journal of Sedimentary Petrology 58*:72–80.
Petrographic analysis of bone-bearing sediment samples.

Hunt, A. P. 1991. Integrated vertebrate, invertebrate, and plant taphonomy of the Fossil forest area (Fruitland and Kirtland formations: Late Cretaceous), San Juan County, New Mexico, USA. *Palaeogeography, Palaeoclimatology, Palaeoecology 88*:85–107.
Survey of vertebrate, invertebrate, and plant remains in Upper Cretaceous alluvial strata.

Hunt, R. M., Jr. 1978. Depositional setting of a Miocene mammal assemblage, Sioux County, Nebraska (U.S.A.). *Palaeogeography, Palaeoclimatology, Palaeoecology 24*:1–52.
Detailed treatment of quarry mapping.

———. 1990. Taphonomy and sedimentology of Arikaree (lower Miocene) fluvial, eolian, and lacustrine paleoenvironments, Nebraska and Wyoming; A paleobiota entombed in fine-grained volcaniclastic rocks. *Geological Society of America Special Paper 244*:69–112.
Geologically oriented taphonomy of mammal-bearing nonmarine facies.

Kusmer, K. D. 1990. Taphonomy of owl pellet deposition. *Journal of Paleontology 64*:629–637.
Sampling, preparation, and analysis of modern owl pellets.

Lawton, R. 1977. Taphonomy of the dinosaur quarry, Dinosaur National Monument. *Contributions to Geology* (University of Wyoming) *15*:119–126.
Taphonomy of dinosaur bones preserved in channel deposits with a focus on sorting and orientation.

Lehman, T. M. 1982. A ceratopsian bone bed from the Aguja Formation (Upper Cretaceous), Big Bend National Park, Texas. Unpublished master's thesis, University of Texas, Austin.
Taphonomy of three dinosaur bone beds.

Maas, M. C. 1985. Taphonomy of a late Eocene microvertebrate locality, Wind River Basin, Wyoming (U.S.A.). *Palaeogeography, Palaeoclimatology, Palaeoecology 52*:123–142.
Thorough treatment of the taphonomy of a vertebrate microsite.

Martill, D. M. 1985. The preservation of marine vertebrates in the Lower Oxford Clay (Jurassic) of central England. *Philosophical Transactions of the Royal Society of London B311*: 155–165.

Taphonomic overview of a Jurassic marine setting.
———. 1987. A taphonomic and diagenetic case study of a partially articulated ichthyosaur. *Palaeontology 30*:543–555.
Detailed methods section with a unique mapping technique.
McGrew, P. O. 1975. Taphonomy of Eocene fish from Fossil Basin, Wyoming. *Fieldiana, Geology Memoirs 33*:257–270.
Taphonomy of classic lacustrine beds of Green River Formation. X-ray analysis of fish-bearing shales.
Potts, R., Shipman, P., and Ingall, E. 1988. Taphonomy, paleoecology, and hominids of Lainyamok, Kenya. *Journal of Human Evolution 17*:597–614.
Taphonomic evaluation of a Middle Pleistocene hominid/artifact site.
Pratt, A. E. 1989. Taphonomy of the microvertebrate fauna from the early Miocene Thomas Farm locality, Florida (U.S.A.). *Palaeogeography, Palaeoclimatology, Palaeoecology 76*:125–151.
Detailed methods for sampling and processing microvertebrate-bearing sediments.
Rogers, R. R. 1990. Taphonomy of three dinosaur bone beds in the Upper Cretaceous Two Medicine Formation of northwestern Montana: Evidence for drought-related mortality. *Palaios 5*:394–413.
Dinosaur taphonomy in floodplain sediments, general methods.
Sander, P. M. 1987. Taphonomy of the Lower Permian Geraldine Bonebed in Archer County, Texas. *Palaeogeography, Palaeoclimatology, Palaeoecology 61*:221–236.
Taphonomy of a mixed vertebrate-plant assemblage in fluvial deposits.
———. 1992. The Norian Plateosaurus Bonebeds of central Europe and their taphonomy. *Palaeogeography, Palaeoclimatology, Palaeoecology 93*:255–299.
Comparative taphonomy of plateosaur bonebeds.
Saunders, J. J. 1977. Late Pleistocene vertebrates of the Western Ozark Highland, Missouri. *Reports of Investigations No. 33*. Illinois State Museum.
Thorough taphonomic investigation of a Pleistocene spring site.
Shoshani, J., Fisher, D. C., Zawiskie, J. M., Thurlow, S. J., Shoshani, S. L., Benninghoff, W. S., and Zoch, F. H. 1989. The Shelton Mastadon Site: Multidisciplinary study of a late Pleistocene (Twocreekan) locality in southeastern Michigan. *Contributions from the Museum of Paleontology, The University of Michigan 27*(14):393–436.
Detailed taphonomic and paleoecologic study of a diverse Pleistocene locality preserved in a lake.
Trewin, N. H. 1986. Palaeoecology and sedimentology of the Achanarras Fish Bed of the Middle Old Red Sandstone, Scotland. *Transactions of the Royal Society of Edinburgh: Earth Sciences 77*:21–46
Freshwater fish taphonomy with emphasis upon lacustrine stratigraphy and sedimentology.
Turnbull, W. D., and Martill, D. M. 1988. Taphonomy and preservation of a monospecific titanothere assemblage from the Washakie Formation (Late Eocene), southern Wyoming. An ecological accident in the fossil record. *Palaeogeography, Palaeoclimatology, Palaeoecology 63*:91–108.
Taphonomic and sedimentologic study of a monospecific bone bed.
Visser, J. 1986. Sedimentology and taphonomy of a *Styracosaurus* bonebed in the Late Cretaceous Judith River Formation, Dinosaur Provincial Park, Alberta. Unpublished master's thesis, University of Calgary, Alberta.
Detailed taphonomic and sedimentologic study of a dinosaur bone bed preserved in a meandering fluvial system.
Voorhies, M. R. 1969. Taphonomy and population dynamics of an early Pliocene vertebrate fauna, Knox County, Nebraska. *Contributions to Geology Special Paper No. 1*. University of Wyoming.
Classic study in vertebrate taphonomy with a detailed methods section.
Wilson, M. V. H. 1980. Eocene lake environments: Depth and distance- from-shore variation in fish, insect, and plant assemblages. *Palaeogeography, Palaeoecology, Palaeoclimatology 32*:21–44.
Comparative nearshore/offshore taphonomy of freshwater fish assemblages.
Wood, J. M., Thomas, R. G., and Visser, J. 1988. Fluvial processes and vertebrate taphon-

omy: The Upper Cretaceous Judith River Formation, south-central Dinosaur Provincial Park, Alberta, Canada. *Palaeogeography, Palaeoclimatology, Palaeoecology* 66:127–143.
Process-oriented taphonomic survey of dinosaur-rich fluvial deposits.

Zangerl, R., and Richardson, E. S., Jr. 1965. The paleoecological history of two Pennsylvanian black shales. *Fieldiana, Geology Memoirs* 4.
Classic early treatment of vertebrate taphonomy and paleoecology with details methods and great illustrations.

Acknowledgments

I am grateful to S. M. Kidwell, C. May, and J. Wilson for editing the manuscript and C. Abraczinskas for drafting Figure 3.3.

4

Macrovertebrate collecting

Field organization and specimen collecting

Patrick Leiggi, Charles R. Schaff, and Peter May

Collecting fossil vertebrates and corresponding geologic data is the major focus of any paleontological field expedition – and the vertebrate paleontologist's favorite outdoor activity. Preservation of fossilized vertebrates varies enormously, and collectors should always approach collecting as a delicate exercise by taking certain precautions before they begin. The field paleontologist can save a great deal of time if careful planning is done well in advance. Fossil collectors must think beyond simple extraction of fossil specimens and consider safety during transportation from the field to the laboratory, reassociation and restoration of the fragmented material collected, and methods to be used in their final preparation. Preplanning will ensure proper organizational procedures, as well as the availability of the right equipment and supplies thus making for a successful expedition. In addition, a thorough reconnaissance of the research area and nearby service towns is vitally important before one begins.

Personnel

In any field expedition careful consideration in choosing experienced personnel will help determine the ultimate viability of all field information needed by the principal investigator. Because most fieldwork is performed in very remote areas, it is desirable to have a staff that will form a cohesive working unit and perform the tasks required to survive in both living and working environments in isolated areas that can often be dangerous. It is essential that all staff be required to complete CPR (cardiopulmonary resusitation) and first-aid training offered by local Red Cross instructors or similar agencies.

A normal progression of field personnel hierarchy and their responsibilities is as follows:

Principal investigator. The principal investigator provides the focus and direction for recovering paleontological and geologic field data for research purposes and often is the primary person who is responsible for the expedition.

Crew chief. The crew chief organizes and directs the field staff in how data are to be collected and assigns duties on site and in camp. The crew chief is usually the person who has planned the project well in advance and has obtained permission for the expedition, determined the size of crew needed, budgeted and created supply accounts for the season, and organizes the setting up of the base camp and the transportation of crew, equipment and supplies, and specimens.

Geologist. The primary function of the geologist is to measure and record stratigraphic data that will be useful to the principal investigator in determining the location of fossil horizons in stratigraphic sections. Proper sampling of geologic specimens helps determine taphonomic implications.

Field assistants. Field staff are required to follow the directives of the crew chief and, in cases of multiple site localities, may be required to be site leaders directing smaller crews in recovering field data. Field assistants may be personnel who are already under the supervision of the principal investigator, graduate students and well-trained volunteers.

In cases where volunteers are recruited for fieldwork, a letter should be sent to them well in advance with the following information:

- Living and working environments, including temperature and weather conditions and suggestions of type of clothing needed
- Housing available or suggestions of the type of tent that best suits the living environment
- Personal tools and equipment required to perform work duties
- General responsibilities in camp and on site
- Typical daily schedules
- Information about the rigorous nature of fieldwork
- Field address, name, and telephone number of field crew chief or other contact person
- Request for information on any health problems and special medications used
- A required notification of the volunteer's insurance carrier and waiver of any liability for individuals or institutions managing and directing field operations

Overstaffing a field expedition can only result in wasted time and will prematurely exhaust a field budget already constrained by funding limitations. Therefore it is important for all members of the field party to report on a daily basis to the crew chief so that the expedition's progress can be monitored for a unified effort.

Permission first

Once a geologic formation has been selected and researched by the principal investigator, the areas to be investigated for paleontological survey are provided to the crew chief by the principal investigator. The crew chief then determines landownership from the geologic, topographical, and section maps. These various maps indicate the ownership of private lands, knowledge of which is essential before you begin. Formal permission needs to be obtained from all landowners.

In cases of public lands that are administered and protected by federal, state, or provincial governments, applications for field research must be submitted well in advance with information about the research to be conducted and specific areas of study. The permitting agency often provides the qualified institution with a general permit that allows the field party to search and locate fossil remains. If you anticipate major sediment disturbance or large-scale excavation, it is usually required that you also obtain a special use permit. It is important that the crew chief keep the agency informed throughout the project so that the agency has a better understanding of the project's impact on the environment; in turn, the agency can notify and instruct the field crew of any required limitations to the operation. It is also important to remember that most public lands are leased for grazing or farming purposes and it is just as important to obtain the lessee's permission so that there will be no interference with the lessee's operation. Courtesy should always be exercised.

If you are requesting permission to work on private land, a visit with the landowner is necessary. It would be helpful to explain the type of research you are conducting and why. Many landowners are very familiar with the geology in their area and can provide additional data for your records. In cases where landowners are reluctant to offer permission, do not press the issue. Instead, ask if you can drop by and visit another time. Perhaps if they get to know you better, there is a possibility they will let you on their land in the future. Always remember that we are the guests and they are our hosts. Any negative interaction may result in serious consequences to your institution and the profession. Always ask and obtain permission first.

Transportation

The efficiency of any field operation depends to a great extent on the availability of transportation. Transportation is crucial for the crew's accessibility to fossil localities, delivering equipment and supplies, and the safe transport of specimens. Four-wheel-drive vehicles that are properly equipped are absolutely necessary in isolated, rugged terrain. A field camp should never be left without enough transportation for all staff, equipment, and supplies, especially because of emergencies. Suburban-type vehicles are best for transporting large crews to a site. Half ton to one ton trucks are better for delivering equipment, supplies, and specimens. In extreme cases, where there is absolutely no vehicular access and traveling on foot is out of the question, helicopters may be used. This is, of course, very expensive and should only be considered if absolutely necessary.

Probably the most difficult type of transport in the field is specimen transport. All specimens must be stabilized and securely wrapped when collected and made ready for transport, even when four-wheel-drive vehicles and other heavy equipment cannot find or obtain access to a site. It is vitally important that no matter what type of transportation is being used, site leaders should be well trained in tying knots, so that specimens can be securely confined to their settled position. Another transportation problem can be the size and weight of field specimens. The all-terrain-vehicle (ATV) may be a viable way to transport specimens or deliver equipment and supplies to a remote site. When ATVs are not available, or allowed, a game cart/dino wheel (see Figures 4.1a and 4.1b), is a useful solution to the transport problem.

Trailers, often used to carry equipment, supplies, and specimens, also have extra space for personal gear. Whenever a trailer is loaded for travel on the road, tarps should be securely placed over the top of a load, to keep materials from flying out and protect other motorists on the road. If specimens are to be transported in a trailer from the field to the museum, careful loading is essential because a trailer is more likely to bounce and not absorb shock as well as a motor vehicle. Usually the largest and heaviest field casts are loaded near the front of the trailer or over the wheels, the same as in any four-wheel-drive truck bed. Overloading of field casts will only weaken and fracture specimens by continuous knocking and vibration within the plaster-burlap wrap. Very fragile specimens should be properly packed and crated before loading for transport. Make an itemized list of specimens while loading is taking place so that specimens can be properly off-loaded at the museum.

Base camp construction

A skilled construction specialist – either yourself or another person – is an important member of the field crew. Considerations for choosing a base camp site include road accessibility to fossiliferous outcrops; protection from inclement weather; and easy access to water, for supply to site and for cleanup. In most cases crew members will have their own tents or tepees, and an area must be specified for their living quarters. A kitchen/meeting place should be constructed in a central location of the camp. This structure (see Figure 4.2) provides a dry environment for cooking, eating meals, and a shady meeting place at the end of a long, hot day in the sun. It also helps in waiting out the rainstorms that so frustrate field paleontologists. Construction of the structure is not difficult if you know basic principles of wood construction or framing. With help from the crew, an entire camp can be constructed in one day.

Truck tarps that have "D" rings sewn along the borders make a very durable, weather-proof roof. By using the dimensions of the tarp or tarps for the roof and one side of the structure, it is easy to design a simple box-style structure. In windy conditions it is a good idea to bury vertical posts that have a 1- to 2-foot pad secured to the bottom side of each support, which will prevent the post from pulling up during high winds. Cross-bracing of the elevated structure and plywood sides provides additional support.

A list of equipment for use in the base camp structure follows:

- 6- to 8-foot countertop (or compact folding table) for preparing meals
- Propane refrigerator with regulator hookup for perishables (can be purchased at an RV [recreational vehicle] center)
- 2 propane stoves with regulator hookups

Figure 4.1A

Figure 4.1A. The late Robert Makela posing with the dino wheel. The dino wheel is used for transporting specimens, equipment, and supplies to remote areas. (Leiggi photo.)

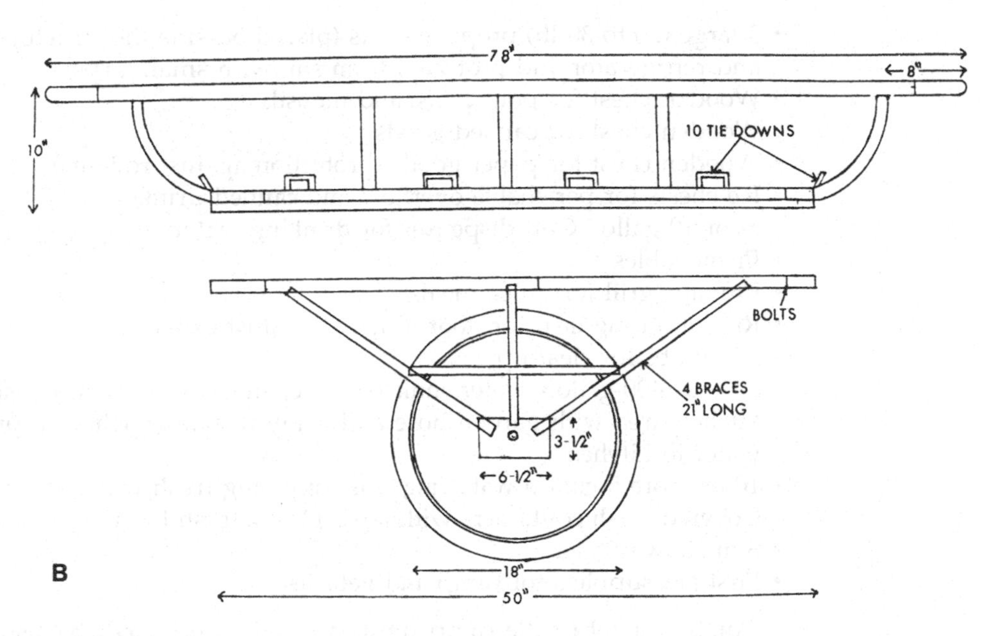

Figure 4.1B. Detailed drawing of the dino wheel. (Museum of the Rockies drawing and photo.)

Figure 4.2. Camp base structure provides a place for cooking and eating meals, a meeting place for field crew, and a shady spot after a long, hard day in the sun. (Museum of the Rockies photo.)

- 3 large (20 to 30 lb) propane tanks (placed outside the structure behind stoves and refrigerator and protected from sun with small tarps)
- Wooden chest for pots, pans, and utensils
- Wooden chest for canned goods
- Wooden chest for paper goods (protection against rodents)
- Ice chests for perishable overflow and canned drinks
- 5- or 10-gallon Gott dispenser for drinking water
- Picnic tables
- Propane grill for quick meals
- Rug or strong tarp for floor to keep the dust down
- Washtubs for cleanup
- 300- to 400-gallon water tank (size depends on crew size) elevated outside kitchen area with garden hose and spray nozzle attached to provide running water to kitchen
- 10 or more 5-gallon water jugs for supplying fresh water to tank
- Covered trash containers with large plastic trash bags
- Sun shower bags
- First-aid supplies for camp and vehicles

Another asset to the camp site is a small wood shed. By using three sheets of plywood and some two by fours, in an hour, you can build a structure that protects generators and other equipment and supplies. Also, the shed provides a central location for supplies and equipment and makes inventory much easier.

Figure 4.3. A backpack (or fanny pack) contains the fossil collector's tools and equipment. In addition, genuine army surplus add-ons aid in the utility of the prospecting kit with convenient attachment clips for both storage and canteen.

Equipment, supplies, and tools

Field equipment falls into two categories: collecting and excavating. Fossil vertebrate remains occur in rock of varying matrices or as surface fragments that have weathered from the parent rock. The preservation features of the bone and the induration of the rock can vary significantly. These criteria will determine the types of tools and the variety of supplies and equipment needed to collect specimens. Careful preparation and planning for all contingencies are essential when working in remote areas where supplies and equipment are not otherwise available.

Choosing proper collecting tools is an essential requirement for reliable collecting techniques. Field paleontologists usually prefer a lightweight, compact prospecting kit with multifunctional items (see Figure 4.3). The following list of tools, equipment, and supplies reflects the variety and scope of the paraphernalia used in collecting fossil vertebrates (see Figures 4.4a, 4.4b, and 4.4c) and are part of a crew member's everyday personal gear or stored in camp. You will not need everything on the list all of the time, but having it on hand or knowing where to obtain it when needed can save a lot of time.

Personal equipment

- Backpack with outside pockets
- Insulated canteen
- 10× magnification hand lens
- Marsh pick or U.S. Army pick-mattock
- Rock hammers (pointed and chisel shape)
- Whisk broom
- Work gloves
- Good quality paintbrushes (camel hair – 1", 2", and 3") and air blower

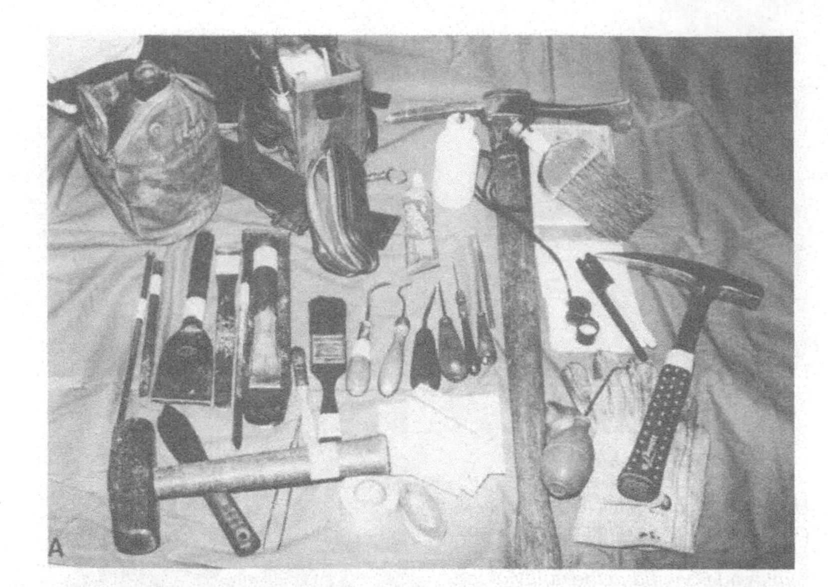

Figure 4.4A. A variety of equipment and supplies that can be carried in a fossil collector's prospecting kit.

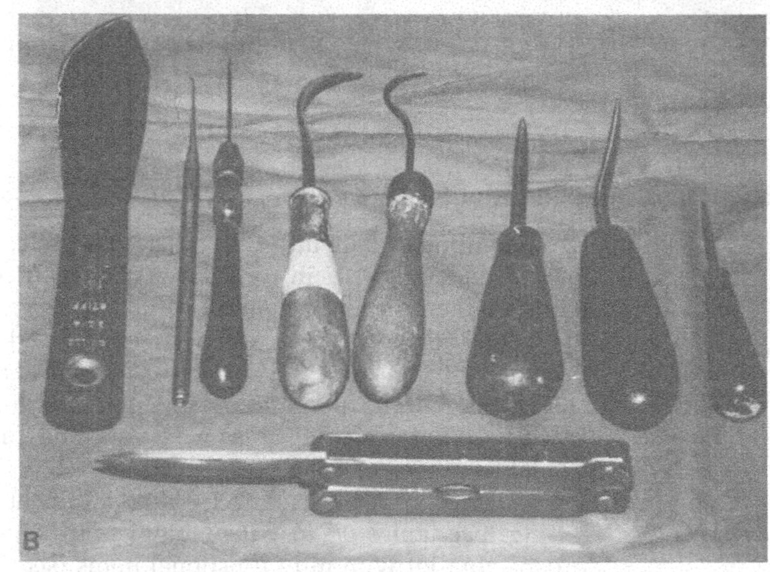

Figure 4.4B. The sharp pointed probes represent the different styles of hand collecting tools. From left to right are a modified spatula, dental probe, pin vise, two sizes of curved awls, straight awl, miniature awl, and a multipurpose utility tool with a knife blade. These tools are most useful in the initial stages of exposing fossil bone.

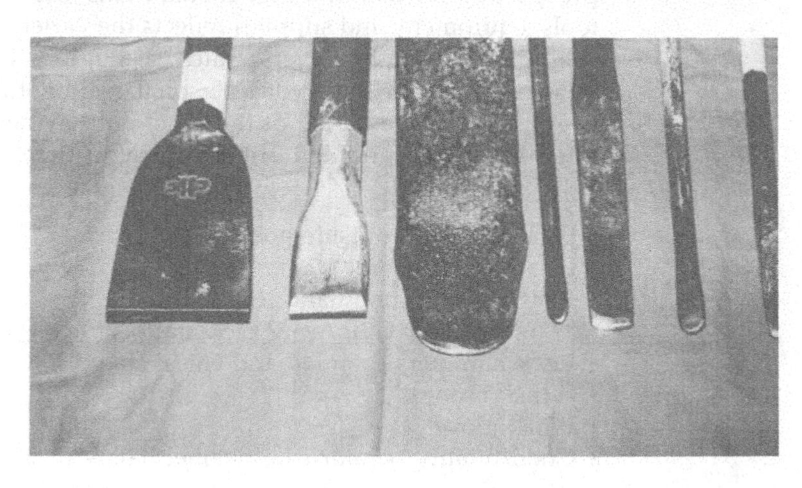

Figure 4.4C. The excavation of fossil vertebrates can be accomplished with different shaped chisel blades. The chisels, from left to right: mason chisel, heavy duty 1¼-inch cold chisel, a flat wide chisel (made from a car leaf spring), and the last four are chisels with rounded blades of various lengths.

- Pin vise, awl, dental probes, and ice pick
- Swiss army knife
- Tweezers
- Cold (steel) chisels (½", 1", 1 ½")
- Professional quality wood chisels (¾", 1")
- Scissors
- Measuring tape (metric and standard)
- 8- to 10-ounce polyethylene bottle for acetate consolidant
- 20- to 30-cc plastic syringe
- Tube of epoxy adhesive
- Cloth collecting bags with attached specimen labels or Ziploc bags of various sizes
- Specimen labels for labeling inside specimen bags and plaster jackets
- Johnson & Johnson Specialist plaster bandages
- Tissue paper and masking tape for wrapping specimens
- Specimen vials and containers
- Camera
- Field notebook with waterproof pen

Camp equipment and supplies

- Large picks
- Shovels
- Brooms
- Two 2½-gallon polyethylene containers for mixing 10% acetate consolidants
- Wooden chest for wrapping materials such as paper towels, tissue paper, aluminum foil, and field labels
- Gasoline containers that are always kept filled
- Gasoline-powered generator
- Gasoline-powered jack hammer with chisels and star bits (size depends on type of excavation)
- Gasoline-powered compressor
- Air guns with chisels and extension air hoses with quick release connectors
- Electric extension cords
- Tool chest with common items, such as saws, hammers, level, screwdrivers, pliers, vice grips, hand drill, noise protectors, safety glasses, small sledge hammer, etc.
- Assortment of nails, screws, and staples
- Bale wire
- Diamond sharpening stone
- Water jugs for on-site casting
- Pry bars
- Screen wash boxes
- Pourable buckets and basins for mixing casting plaster
- Leaf springs cut at 2 ½-foot lengths and sharpened for breaking pedestals underneath plaster jackets
- Tarps for covering sites and specimens
- Casting plaster
- Acetone

- 5-gallon covered container of VINAC B-15 polyvinyl acetate beads
- Burlap bags or plastic-coated rolls of burlap
- 5- and 3-gallon water dispensers for drinking water on site (as many as needed)
- "Come-along" and chain
- Topographical and other geologic maps/aerial photographs
- Global Positioning System (an electronic hand-held navigator)
- Packing boxes
- Bridge nails, 8"–12"
- Solid braid nylon rope, 1/4", 1/2" spools, and assorted nylon ropes
- Portable radio transmitters
- Modified game cart/dino wheel

A checklist of these aforementioned materials will enable the crew chief quickly to assemble the tools, equipment, and supplies needed for any field expedition. Other crew members would also find this checklist useful, especially in regard to personal equipment.

Field specimen collecting

Fossilization varies greatly, as not all fossil bone is perfectly fossilized. Furthermore, because most specimens are extremely delicate (and may already be fragmented), it is of the utmost importance that proper procedures, using appropriate equipment and supplies, be implemented in order to ensure maximum information is collected from the fossils.

Vertebrate paleontologists employ several basic collecting procedures:

1. Prospecting/surface collecting-gathering of "float" or fragments from the surface that have eroded from their place of deposition and may or may not indicate their original horizon
2. In situ collecting: fossil remains articulated or associated that are preserved in the burial place with minimal specimen transport
3. Quarry collecting: generally involves multiple concentrations of vertebrate remains that are considerably mixed on a distinctive bedding plane with obvious specimen transport
4. Taphonomic collecting: similar to quarry collecting except all specimens are carefully mapped and measured for specimen orientation before final removal (see Chapter 3)
5. Underwater screen collecting: washing fossil bearing sediments from, for example, an ant hill, fissure fill, or lag deposit (see Chapter 5)

In addition, sediments that contain low distribution of vertebrates may be washed through a series of different size screens to concentrate the bone elements and teeth.

Accordingly, the primary responsibility of the members of the field party is to recognize and collect fragments of fossil bone. Sometimes the tantalizing bone fragments eroding out of a hill or isolated bones known as "float" are the only vital clues that the fossil collector has to locate more of the specimen. The collector should fit these fragments together so that no pieces are overlooked. In addition, the surface area can be dry-screened for any remaining fragments. Small specimens usually can be collected in a couple of hours or less; medium to large articulated skeletons and bone assemblages may require a few days or weeks to remove. The excavation of a large sauropod dinosaur may even require the work of several persons for an entire field season or longer. Today, even with new kinds of sophisticated equipment available, fossil collectors must still rely on their professional experience and training when recovering vertebrate fossils from the field. There are many different ways of collecting fossils, and each can be adapted to the requirements of specific situations.

It is essential that anyone who is about to go into the field is properly equipped, keeping in mind the type of data that need to be collected. The principal investigator and an experienced field crew can recover an enormous amount of vertebrate and associated data in a short period of time. The proper direction and training of amateurs can only enhance the field experience.

Prospecting and surface collection

Prospecting is one of the most rewarding aspects of field paleontology. The moment of discovery creates a tremendous amount of excitement for the discoverer and crew. Whenever a field party is about to set out prospecting, it is essential that each member be equipped with the right tools for the job. Field prospecting is often done alone but can overlap with that of other field party members in the same vicinity. It is extremely advisable to locate oneself frequently in regard to landmarks. Stop occasionally and study the landscape from where you left a vehicle or camp. Do this by looking back from where you started. This is the same view you will be seeing when returning. Sometimes it is wise to team up with an experienced field member.

A geologic formation has been identified and researched by the principal investigator, and often, a list of possible faunal groups has been provided to the crew. Researching geologic/topographical maps, section maps that show representations of public and private lands, and county maps that show individual ownership of private lands is essential before you begin. After becoming familiar with these maps, it becomes easier to figure out the type of area and terrain that need to be surveyed and who you must obtain permission from before you begin.

Even after you become familiar with the maps and the survey area, experience is still important. For example, an experienced prospector can tell you that badlands are very deceptive, and anyone can become confused and get lost. It is also important that crew members show the principal investigator or crew chief on a map what areas they intend to prospect. A meeting place and time should also be considered. If a crew member does not show up at an announced time and place, there is a good possibility that he or she is hurt or lost.

Before prospecting, the principal investigator and geologist should explain the sedimentology of the area to be surveyed and inform the crew of the most likely sediments where fossil vertebrates weather out and where fossil bone can be found. How do you recognize fossil bone? Fossil bone can be heavy, dark, light, mottled, soft, hard, crumbly, shattered, or encrusted with different colored and textured sediments obscuring its true shape. Fossil bone tends to crack at right angles to the grain, whereas recent bone splinters along the long axis. There are no substitutes for hands on experience in recognizing the different colors, shapes, and textures that characterize fossil bone and teeth.

The most common mistake of any amateur prospector is to go directly to the source or sediments where fossils lay buried. Instead, the base of any outcrop requires a slow, careful, systematic search first. Some sedimentary rock is loose or soft, and by walking on the hill, you can create a slide and quickly bury the clues of fragmented material that have eroded out of the hillside. These important pieces of the specimen then would remain unnoticed or not collected. By starting from the bottom of the hillside you can see the last eroded fragment and follow the trail of fragments to their original source. Avoid gluing each fragment together, as this requires a lot of time. It is, however, important to see what fragments fit together so that you are sure you have all of the specimen. Clean or fresh breaks in fossil bone indicate that pieces still need to be found.

First uncover the specimen with small hand tools, such as pin vises, awls, or ice picks, and brush away the loosened matrix with a good paintbrush. Never use your fingers because the pressure can damage the specimen. Determine whether or not the specimen needs consolidation. If the specimen is rock hard, the fragments can be removed. If the bone is soft or unstable, apply a small amount of polyvinyl acetate (VINAC B-15 solution, 10–15%) to the specimen or the amount needed with a squeeze bottle or with an eyedropper to stabilize the specimen. Overgluing surface bone can result in the specimen's falling apart or make cleanup in the laboratory very difficult. Let common sense be your guide.

Another important step in uncovering bone is determining how much bone is present. Once stabilized, carefully wrap each fragment in tissue, keeping associated pieces together as much as possible; use a little (not a lot) masking tape to keep the wrap together; and store all wrapped fragments in the same cloth specimen bag. If additional bags are needed for the same specimen, mark them – for example, 1/5, 2/5, 3/5, and so on. Indicate the locality (quadrangle, section, township, range), primary description, field number, formation remarks that are useful to the preparator or researcher, name of the collector, and date on the attached tag of the bag. A duplicate field label should be placed inside the bag in case the outside label is torn or lost during transport. Ziploc bags can work well when cloth bags are not available but not advisable. Paper bags should never be used because they tear easily and specimens can be damaged beyond repair.

Recording field data

Locality information should be plotted directly on a topographical map with the corresponding field numbers and an entry of all information should be made in the field notes. If photographs are taken, be sure to record in the field notes a description and sequence of each photograph with the roll, frame number, and date.

United States Geological Survey Quadrangle 7.5 minute 1:24000 scale topographical maps are the most commonly used in the field because they are easy to read. Most of these maps can be obtained on loan from a university library or purchased directly from a USGS office or a public lands district office, such as the Bureau of Land Management. A copy of a particular state's "Index to Topographic and Other Map Coverage" catalog issued by the USGS is useful to have on hand when requesting maps. A variety of these and other scale maps are becoming available for office and lap top computers. With this software, the principal investigator has the option of using a laptop computer in the field and can quickly plot field localities with their corresponding field numbers directly on the map. Furthermore, the use of modems can advance the same information with field notes to the principal investigator's main computer, thus saving time when returning from the field. Computer systems and software can differ so it is recommended that you consult a computer expert on these systems. The following is an example of how to further enhance locality coordinates.

GPS Ensign, the Global Positioning System, available from Trimble Navigation, is a compact, hand-held, portable, waterproof, navigator/receiver that tracks and uses several GPS satellites simultaneously and updates position data each second. The GPS receives signals from 24 satellites orbiting the earth 12 miles high and plots latitude and longitude or Mercator grid position when used anywhere in the world. Another advantage of the GPS is its three-dimensional position fix, giving not just latitude and longitude but also altitude using a signal from a fourth satellite. The Ensign model can store up to 100 way points (coordinates) and mark points in its memory (operated by four AA batteries). The GPS is regulated by the U.S. Department of Defense, and therefore all GPS systems are subject to accuracy positions within 100 meters. The remarkable Ensign model is moderately expensive and establishes an effective triangulation system for precise locality and site coordinates. Its application for both paleontological and geologic field studies dramatically enhances retrieving locality information, as some GPS models have input for downloading into a computer.

Removing overburden with hand tools

In any field exercise the rule among paleontologists is always to show up prepared for anything. If fragmented or washed specimens can be followed to their parent rock or bone horizon, the prospector inevitably must start digging. In most cases uncovering a meter or two of overburden should give the paleontologist a good idea of what lies buried in the ground.

Probably the most difficult task in the field is to remove tons of sediment in the approach to uncovering macrovertebrate information. Most importantly, before you begin, careful thought to the type of sediment and other information should be considered (see Chapter 3). In many cases people tend to attack the hill with picks and shovels, without recognizing plant fossils or invertebrates that can aid in taphonomy studies. The person in charge should map out an area to be removed, determined by making trenches along the hill to figure the extent of the fossil horizon.

However, never uncover more overburden than you can handle. In other words, if you cannot complete the removal of a 20-meter area of the fossil horizon, try 10 meters first. Undisturbed sediment is best for protecting fossils and you can always return to the site another time.

By systematically removing sediment (layer by layer) with picks and shovels, and having spotters (dirt pickers) take a close look at the loosened dirt,. most or all of the data surrounding the fossil horizon can be recovered. In most cases water-worn bone or "float" can be found. It is important that the geologist or taphonomist record this information and be able to tie in the information stratigraphically. Remember, all fossils, whether they are vertebrates, invertebrates, plants, or unidentifiable curiosities, are important to any taphonomist who is trying to determine the event or events that took place millions of years ago. Collect and make a record of everything.

Removing overburden with heavy equipment

For the most part, field collecting usually requires only hand tool excavation. The most powerful tool you will most likely ever use is a large jackhammer. Jackhammers are either electricity or air driven. Electric generators are less bulky to transport than an air compressor which is why most field crews prefer electric jackhammers. The most common size hammer is 65 pounds; it is also beneficial to have a lighter model approximately 20 to 25 pounds. Heavier jackhammers are necessary to remove large portions of dense matrix, such as a well cemented sandstone or a calcium carbonate. The lighter jackhammers are useful for lighter matrices and trimming up or squaring off the work area. Bosch manufactures an excellent 65-pound electric jackhammer, and many chisels and points are available for a variety of purposes. Many companies manufacture lighter jackhammers, like Black & Decker and Stanley.

When heavy types of equipment are used, it is important to remember safety as required by the manufacturer and common sense in regard to the site. Use spotters to look for any unexpected fossils that may show up in the area you are trying to remove. Never be in a hurry or try to force the jackhammer beyond its capabilities. A tired worker on a jackhammer can do more damage than good to a site and will most likely get hurt in the process. Large rocks should be inspected for fossils before they are broken into smaller, more manageable pieces. Pry bars are especially helpful during the jackhammer process of overburden removal.

On rare occasions heavier equipment, such as backhoes, bulldozers, and graders, is needed to remove large areas of matrix for site accessibility. Operators of large machinery are not trained to be careful of fossiliferous sites. Again, it is extremely important to have spotters on site and to communicate effective commands between the operator and spotters. Thus, it is extremely important that the spotter or site leader explain to the operator that a limited amount of surface is to be removed at one time. There have been instances when an excited crew and a heavy equipment operator removed too much sediment at one time and ended up destroying an unexpected skeleton that was above the targeted fossil horizon. Working slowly and methodically layer by layer will prevent this type of disaster or at least limit the damage.

No matter what type of heavy equipment is used, never use it to remove sediment all the way to the fossil horizon. It is best to stop a half or full meter above the horizon. From this point on, pick and shovel work should be carefully executed.

Bone beds and articulated or associated skeletons

Any specimen in its parent rock, whether isolated, in a bone assemblage, or articulated/associated material, should always be approached by removing sediment from above and working down to the specimen to its complete uncovering. Specimens should never be uncovered from the side or undercut unless you are ready to remove the specimen from the parent rock completely. One of the biggest mistakes we have seen in the field during the uncovering of bone material is an attempt to fully prepare the specimen on the spot. This process should be left to qualified preparators or technicians in the laboratory. Another mistake is to try and uncover more bone than time permits you to carefully remove at the end of the field experience. These two scenarios ultimately lead to damage of the specimen and the data will then quickly become almost useless to the principal investigator and other associated researchers. Crews should always be reminded that they are only to collect and record the data needed for research. Nothing more, nothing less.

Properly uncovering fossil bone requires a great deal of patience and concentration. It is easy to hurry this process, especially when uninitiated because they are eager to see the entire bone and learn its identity. A site leader needs to pay strict attention to the crew during this process to ensure that each specimen is carefully developed. Implements generally needed are awls, ice picks, and dental tools for scraping away the surrounding sediment. In harder matrices, such as limestones and densely cemented sandstones, a hammer and various size cold chisels are needed. Professional quality wood chisels with beveled edges and gasket scrapers are excellent for medium to coarse grained sandstones. Proper safety precautions should be exercised when using any of these tools: Protect your eyes with safety goggles or glasses. Instead of your fingers, use good quality paintbrushes to brush away loose sediment. Accumulated sediment should be examined for any trace fossils and then discarded by using a whisk broom, dustpan, and bucket.

Consolidation of uncovered bone should always be kept at a minimum if possible. Too much consolidant used in the field can make it difficult for the lab technician when preparing the specimen later on. The most common rule of consolidation of specimens in the field is to use the same or similar materials that will eventually be used on the specimen in the laboratory. If the lab technician will not be joining you in the field, it is important that you consult before any consolidants are used. Each project may require some changes from the ordinary. We recommend using a thinned polyvinyl acetate solution, preferably VINAC B-15 (see Chapters 1 and 2).

By following proper data collection procedures (see Chapter 3), the field crew can commence removal of bone from the parent rock. This procedure is not difficult, but, again, patience and concentration are important. When working with isolated specimens or specimens in a bone assemblage that are disarticulated, it is easy to systematically map out a course of action that will ensure safe collection of each specimen. In this situation the crew will number each bone individually from the map drawn and plotted by the site leader. In most cases each bone will be removed individually.

Figure 4.5. A field crew working in a dinosaur bone bed. Note prepared field casts in the rear. (Robert Harmon photo.)

Before trenching around the specimen it is a good idea to have the materials ready for plaster casting/jacketing. Cut burlap strips 6 to 8 inches wide and 3 to 4 feet long (depending on the size of bone to be cast), should be rolled and dampened. A bucket with some water for the plaster mixture should also be nearby, along with damp paper towels or aluminum foil to be used as a separator. The fieldworker then begins removing sediment from around the bone but not by undercutting the specimen. By removing the sediment in a horizontal fashion a vertical support wall or pedestal is created. The pedestal should be far enough below the in situ specimen. If the specimen is stable, it does not hurt to make a slight lip at the bottom of the trench at the base of the pedestal as long as the sediment is also stable (see Figure 4.5).

Now it is time to cast the specimen:

1. On a piece of paper write the site name, field map number, locality number, and any other important data and place in a well-sealed Ziploc bag over the top of the bone.
2. Place the damp paper towel or aluminum foil directly over the exposed areas of bone and the protected field data. This measure will prevent plaster from adhering to the bone.
3. A medium consistency plaster mixture should then be prepared. A too-thin mixture will not set properly and the cast will end up being too weak. (Adding plaster to water until islands appear on the surface of the water is usually how a plaster-to-water ratio is measured.)
4. Immerse dampened rolls of burlap in plaster mixture and roll plaster into burlap.
5. Unravel plaster-soaked rolls of burlap around and over the top of the entire bone and pedestal. Be sure that all sections of burlap overlap one another for better strength. This action creates the "top cast/jacket."
6. In cases of larger specimens, extra layers of plaster-soaked burlap will be required, so this process will need to be repeated.
7. Undercut the sediment under the cast and finish casting. This action will prevent the bones from slipping out of the cast when it is turned over.
8. Before the cast is set, etch the site name/number and corresponding field map number into the plaster.
9. Larger bones should always have extra support in the cast. This is accomplished by placing wood (two by fours) or other rigid materials in the cast during the casting procedure. Make sure that the ends of any rigid materials used in casts are wrapped completely so that they will not slip out during the transport process.
10. When the plaster cast has fully set, take a flat piece of steel, such as a section of a car's leaf spring, that has been tooled with a beveled edge and place the edge at the base of the pedestal. Tap the steel with a 3-pound sledgehammer to "pop" the cast.
11. When the cast is popped and turned over, protect the exposed sediment (bottom of the cast) by covering it with plaster-soaked burlap. Etch numbers in the cast, as described in step 8.

Articulated or associated skeletons must be approached a little differently than the removal of individual skeletal elements. In cases of smaller skeletons it is best to figure on casting the entire animal at one time. This will ensure the skeleton is in situ and can be more carefully uncovered and mapped in the laboratory. In cases of large skeletons, the skeleton will have to be "blocked out" by removing some vertebrae in key locations. Planning and engineering the entire removal of the skeleton have to be done before any trenching or pedestaling (see Figure 4.6). The procedure for casting the specimens is the same as previously described but on a much larger scale. Unfortunately, with this type of situation, the field crew cannot make simple pedestals. Tunneling under the large blocks is necessary in almost every instance, and this practice must always be done with extreme caution.

Tunneling is not a time to have inexperienced personnel around. This procedure can be very dangerous, and rumor has it that a student lost his life some

Figure 4.6. Tyrannosaurus rex *(MOR 555) just before blocking and field casting. After casting, the entire animal weighed more than seven tons. (Photo by Bruce Selyem, Museum of the Rockies photo.)*

years ago while tunneling under a large specimen block for casting. The best procedure to follow is not ever to put either your head or body in an uncompromising situation. This means that you would actually be able to reach the other side of the block without having to endanger yourself. In any event, tunneling is a slow process because you want to be sure that the specimen block is supported at all times. If you have never practiced a tunneling procedure before, we do not recommend your trying this on your own without the help of a very experienced fieldperson. Sediments vary and circumstances are different from site to site.

In general, two crew members work toward the middle of the block after a top cast has been set in place over the entire block. The top cast should already have rigid supports made of wood or steel already in the cast. Excavation of the tunnel should be carried out slowly and systematically. During every couple of inches of sediment removal from the underside of the block, another layer or two of plaster-soaked burlap should be added to the cast. Eventually when the tunnel is completed there is an entire bandage around the top and bottom of the cast. Always have a good supply of large wood blocks for additional support of the specimen block. Other key areas that need to be tunneled are part of the engineering portion of this type of procedure. In essence, it is always necessary to understand how the large block will stay supported by a number of pedestals or blocks of wood.

No matter what you do in the field, safety for yourself and others is of prime importance. There is no fossil that is worth trying to collect if it is too dangerous to do so.

Acknowledgments

The authors wish to thank William W. Amaral, Robert Harmon, and Allison Gentry for their help and suggestions.

Bibliography

Camp, C. L., and E. D. Hanna. 1937. *Methods in paleontology.* Berkeley: University of California Press.

Clemens, W. A., 1963. Fossil mammals of the Type Lance Formation, Wyoming: Part 1. Introduction and multitubericulata. *University of California Publ. Geol. Sci. 48*:105.

Converse, H. H. 1984, rev. 1989. *Handbook of paleo-preparation techniques.* Florida Paleont. Soc.

Hibbard, C. W. 1948. Techniques of collecting microvertebrate fossils. *Contributions of the Museum Paleontogy, University of Michigan 8*(2):7–19.

Kummel, B., and D. Raup. 1965. *Handbook of paleontological techniques.* San Francisco: Freeman.

Makela, R., and P. Leiggi. 1989. Reinventing the wheel. *Curator 32/2*:123–130.

Rixon, A. E. 1976. *Fossil animal remains: Their preparation and conservation.* London: Athlone.

Collecting in caves

Frederick Grady

Caves are repositories for a large percentage of Quaternary vertebrate fossils, and some contain much older material. Because of their special nature, caves put certain constraints on the collection of vertebrate fossils. Such obstacles such as vertical pitches, tight squeezes, and water filled passages must be negotiated occasionally in order to get to localities. Logistics of collecting differ from those of surface collecting according to the difficulty of the cave. Personnel need to be trained in the use of the sophisticated equipment required for given cave situations.

Vertebrate fossils in caves vary considerably in their state of preservation. In rare instances soft tissue is preserved, but all too frequently bones and teeth are extremely crumbly. High humidity in caves can preclude the use of conventional hardeners. Substances such as polyethylene glycol (Carbowax) can be used to collect fragile specimens safely. Specimens must be packed securely for the often difficult trip out of the cave.

Discussion

Caves, particularly limestone caves, are one of the best repositories of Quaternary vertebrate fossils. Virtually every state in the United States contains at least a few caves, though not always limestone. The physical conditions in caves place certain constraints on the collection of fossil vertebrates. Paleontologists often hear about fossil vertebrates from amateur cave explorers. Because caves often have vertical pitches, tight crawl ways, and water passages, advance planning is necessary before attempting to collect such fossils.

All supplies for collecting in caves must be packed in easy-to-carry containers, such as small backpacks or ammunition cans that have a shoulder strap attached so that both hands can be kept free for negotiating obstacles in the cave. Hard hats with attached lights are also necessary. Usually lights are powered by batteries or acetylene generated by calcium carbide. Food and liquid for the trip must be packed as well. Carbohydrates are usually preferred over fats and proteins because they provide quicker energy. The best liquids for cave use are water and fruit juices. Drinks that contain caffeine should be avoided because they tend to dehydrate the body. Water found in caves should not be drunk, as it is often polluted. Clothing should be appropriate for the conditions in the cave. Extra dry clothes packed in a plastic bag are often welcome additions. One should never attempt a project far inside a cave without the help of experienced cavers.

Vertebrate remains found in caves are as varied as those found outside, but collection techniques used outside cannot always be used in a cave. When well preserved bones and teeth are found, they can be wrapped in several layers of soft tissue, such as toilet paper or paper towels. The specimens should be carefully labeled and packed in a hard container such as an ammunition can (see Figure 4.7) for removal from the cave. Even in a cave where walking is easy, it is desirable to pack specimens in hard containers, for a dropped pack or a fall can destroy even a well preserved bone.

Typical hardeners such as those dissolved in acetone usually cannot be used in caves because of the high humidity. Well preserved though fragile larger specimens can be jacketed in plaster bandages. Before applying plaster bandages, wrap the specimen in several layers of soft paper separator. Packaged surgical bandages are ideal for cave work, as they save time and leave less mess in the cave (see Figure 4.8). Jacketed specimens are best placed in hard containers for removal from the cave, for they are awkward to carry and a blow on a jacketed specimen can damage the bone within. Plastic fishing tackle boxes with the inner trays removed are large enough for plaster jacketed peccary skulls and fossils of similar size. Foam padding can be used to keep the jacketed bones from moving around in the packing containers.

Sometimes bones and teeth are firmly cemented in limestone, and such remains can be virtually impossible to collect without destroying them. However, if the connection between the specimens and the surrounding rock is reasonably thin, careful use of a hammer and chisel can free such specimens. Exposed parts should be protected before starting to hammer. Occasionally a saw, preferably one with tungsten carbide teeth, can be used to cut away the connecting limestone. Limestone is softer than both steel and tungsten carbide, so with patience and ingenuity, such as removing the blade from the saw handle to work around tight areas, some specimens in limestone can be collected without subjecting them to the potentially shattering effects of a hammer and chisel.

Frequently vertebrate fossils found in caves are so fragile as to seem all but impossible to collect. On the basis of work over the past decade, it is clear that in many cases such specimens can be collected or at least partially salvaged.

Figure 4.7 (left). Packaging field-jacketed specimens in ammunition cans and fishing tackle boxes.

Figure 4.8 (right) Jacketing peccary bones in plaster bandages.

Polyethylene glycol (PEG; molecular weight 3350), that is sold under the name of Carbowax, is a partially water soluble compound that melts at about 58°C. (Chaney 1988). PEG can be melted in a small pot over a heat source such as jellied naptha, generally sold as Sterno. A piece of hardware cloth can be made into a stove. Isolate the specimen from the surrounding matrix as much as possible and either paint or pour the liquid PEG over it until the specimen is completely covered. When the PEG has hardened, it can be reinforced by a plaster jacket, as appropriate for larger specimens. The specimen can then be undermined and turned over. Small specimens that are not plaster jacketed can be turned by sliding a putty knife or trowel a centimeter or so beneath the underlying matrix and carefully lifting. After removing excess matrix from the other side, put additional PEG on the exposed specimen and jacket the second side if appropriate. Depending on the humidity, PEG may remain fairly sticky inside a cave. Small specimens should be wrapped in tissue and removed from the cave in hard containers.

Once PEG has been used on a specimen, other hardeners and glues will probably not stick to it. Excess PEG can be removed from specimens by the use of a hot air torch or by putting the specimens on a metal tray covered with absorbent paper and heating in an oven at low temperature. Such specimens will always have to be handled with extreme care. In the past several years several rare field specimens from West Virginia caves have been collected using this method (Grady 1981a, 1981b).

Small bones and teeth are often found in well-preserved concentrations in caves. The usual practice is to collect the bone-bearing matrix in sacks. If there is no apparent stratification of a site, arbitrary intervals 10 to 25 cm thick are designated. Each bag should contain some sort of waterproof label such as a plastic vial containing a paper label or a piece of flagging tape marked with an indelible marker. Glass vials should be avoided in cave work for obvious reasons. The size of the cave usually dictates the size of the bags used. Five to ten kg is reasonable for most purposes and can be either hand-carried or put into small backpacks. In large walking caves, when sufficient help is available, bags can be passed to the entrance by a line of several people. The line may have to be moved forward several times. In this manner 50 bags can be moved several hundred meters in a couple of hours.

Bags of matrix occasionally have to be stored for a considerable amount of time before processing. They should be put in a dry place and their contents suitably labeled, spread out to dry. Bags should not be stacked one on top of another, for they have a tendency to rot, thus spilling and mixing their contents.

Processing cave matrix is best done by wet screening (Guilday, Martin, and McCrady 1964). This method recirculates well for water for washing matrix through screens on waist-high racks. The finest screen used was fly screen with 1.0–1.5 mm openings. More recently the use of mosquito net bags for washing matrix has been shown to have certain advantages. The fine netting retains small teeth that go through fly screen and the net bags are extremely portable and can be used anywhere there is a water source (Grady 1978).

Footprints and scratches are occasionally found in caves, sometimes a considerable distance from a known entrance (Youngsteadt and Youngsteadt 1980). Prints and scratches should be photographed and sketched before attempting to make casts. Casting can be done in plaster, silicone, paraffin wax, or Alginate dental compound. Tracks in soft sediment are likely to be damaged or destroyed in the casting process, which, therefore, should be done selectively.

Acknowledgments

For more than ten years, amateur cavers too numerous to name have assisted me in the collection of fossil vertebrates from caves and have hauled out hundreds of bags of bone-bearing matrix. Dr. Nicholas Hotton III critically read this manuscript and made many useful suggestions.

References

Chaney, D. S. 1988. Techniques used in collecting fossil vertebrates on the Antarctic Peninsula. *GSA Memoir 169*:21–24.

Grady, F. V. 1978. Some new approaches to microvertebrate collecting and processing. *Chisler 1*(1):1–4.

———. 1981a. The Hamilton Cave jaguar. *Potomac Caver 24*(4):3.

———. 1981b. Saber toothed cats from West Virginia caves. *D. C. Speleograph 37*(9):9.

Guilday, J. E., P. S. Martin, and A. D. McCrady. 1964. New Paris No. 4. A Late Pleistocene Cave deposit in Bedford County, Pennsylvania. *Bull. Natl. Speleo. Soc. 26(4)*:121–194.
Youngsteadt, N. W., and J. O. Youngsteadt. 1980. Prehistoric bear signs and black bear (*Ursus americanus*) utilization of Hurricane River Cave, Arkansas. *Bull. Natl. Speleo. Soc.* 42(1):3–7.

Site documentation

Darren Tanke

In the spring of 1935, paleontologists William A. Parks of the Royal Ontario Museum (Toronto) and Charles "Charlie" M. Sternberg of the Geological Survey of Canada (Ottawa) each thought it would be a useful project to relocate, map, and mark previously excavated dinosaur quarries in the badlands of southern Alberta, Canada, and publish the results in map form. It was originally intended to conduct this project in both the Steveville district (now Dinosaur Provincial Park) and in the Drumheller Valley, but the latter did not happen in earnest until 1986, when the Royal Tyrrell Museum of Palaeontology assumed this responsibility. Prior to 1935, Charlie's brother Levi had marked a number of quarries he had excavated for the Royal Ontario Museum, plus others collected by other institutions that he was aware of, but this effort was not synthesized in any form. Memos (May 1935) from Charlie to E. M. Kindle (chief of the Palaeontology Division, Geological Survey of Canada) expressed a desire and interest to map both areas, with the Steveville area having first priority if only one region could be surveyed.

During the summer of 1935, F. P. DuVernet of the Topographical Survey mapped about 90 staked and unstaked quarries in the Steveville area. In 1936, Charlie, now assisted by his son Ray, returned to the Steveville region. All of the quarries were examined and identified. Some new ones were located, and Ray pinpointed these on maps by plane table. Although Charlie's 1936 field notes are not specific, he and Ray also staked at least 73 of these quarries (Sternberg 1936a) and a preliminary map sheet was produced (Sternberg 1936b). By 1950, Levi and Charlie had marked 112 quarries in the Steveville district and a final topographical map of the staked dinosaur quarries was published (Sternberg 1950). In 1954, Levi Sternberg and crew staked 6 Royal Ontario Museum and 5 National Museum of Canada quarries in Dinosaur Provincial Park, bringing the total number of staked quarries to 123 (Colwell-Danis 1988). The same numbering sequence was then carried on in the Drumheller Valley, when, in 1956, two sites (one, 124, from 1923, and the other, 125, from 1955) were staked by Wann Langston, Jr., and other staff of the National Museum of Canada. This continuance of the same numbering sequence in two different areas was not recorded and caused much confusion in later years, as it was widely believed that additional quarry staking of at least 123 quarries had already been conducted in the Drumheller Valley. However, field surveys for the missing quarry stakes, and requests for amateur fossil collectors to report any quarry stakes seen in their travels in the Drumheller region (Graham 1987), produced no additional results, suggesting only that two quarries were actually staked. This was later determined to be the case.

Subsequently, in 1988, it was decided to keep the quarry numbering system in Dinosaur Park continuous and unto itself and have separate quarry numbering systems for Dinosaur Provincial Park and Drumheller Valley. To avoid confusion, the 124 and 135 Drumheller Valley quarries have retained their original number designations, but all other sites currently marked by the Royal Tyrrell Museum in this area are numbered chronologically, starting with the designation L1500 (L = locality). Some of these are also prefixed by the map sheet number. As a result there are two quarries bearing a number 124 stake and two quarries numbered 125.

A quarry map sheet for the Drumheller Valley, similar to that generated for Dinosaur Provincial Park (Sternberg 1950), will be produced in the future, when enough sites have been staked to make such an endeavor feasible. An updated version of the 1950 Sternberg map of Dinosaur Provincial Park is also planned.

As part of the University of Alberta's paleontological fieldwork in 1966 and 1967, Jane Colwell put in ten temporary metal stakes to mark mammal-rich localities in Dinosaur Provincial Park and surrounding area, and the region north of the Dominion Experimental Range Station (One four) east of Manyberries, Alberta. These markers consisted of a 10-inch (25-cm) long steel bolt pounded into the ground, and the preface "JMC" and a locality number were stamped into the 1-inch (2.25-cm) square head. The plan was for these markers to be replaced a year or two later by more permanent stakes set in concrete, but this did not see fruition. Despite these shortcomings, several of the markers are still in place, the numbers are quite legible, and the number designations are currently used by Royal Tyrrell staff for site reference.

Since the fall of 1981, the author has been in charge of the Royal Tyrrell Museum's quarry staking project in Dinosaur Provincial Park (Tanke 1981, 1982, 1986, 1988). As of 1986, he has also been heading fieldwork relocating and staking Drumheller Valley quarries (Tanke 1986, 1988, 1990) with Maurice Stefanuk (Drumheller) and, more recently, a continuation of the same project with support from Loris S. Russell (Royal Ontario Museum, Toronto) and his wife Grace (Russell 1987–1989, in prepation). Future quarry staking plans will eventually cover the rest of the province, although the numbering sequence to be used for these stakes has yet to be determined. In both Dinosaur Provincial Park and Drumheller Valley, the Royal Tyrrell Museum has permanently marked more than 70 excavations (mostly dinosaurian, some dating back to 1910), by placement of an information-bearing bronze-capped iron pipe or stake set in concrete. Although such a technique has been used in Alberta and, more recently, in Montana (S. Sampson, personal communication 1989), it would appear that no standard technique has been established, and that each institution has had to approach the problem of quarry staking in its own way with varying degrees of success.

Therefore it would seem prudent to outline here the proven quarry staking technique utilized by the Royal Tyrrell Museum of Palaeontology as a guide to other institutions wishing to mark their excavated sites. Whereas the Royal Tyrrell Museum uses this technique almost exclusively to mark excavated fossil vertebrate localities, the same procedure could be used to mark permanently archaeological sites, microfossil/bone bed assemblages, geologic horizons, type sections, formational contacts (Quarry 109 in Dinosaur Provincial Park marks the Judith River/Bearpaw formation contact), and a wide range of other outdoor locations. At the Devil's Coulee hadrosaur nest/egg site in southern Alberta, quarry stakes (each with a small hole drilled through the middle of the bronze head and set only inches above ground level) serve as exact reference points for mapping and surveying work when using a transit.

Why stake?

There is ample justification for staking an excavated paleontological quarry. With the passage of time, erosion, fading memories, and attrition of the excavators, a specific locality's whereabouts can be confused or lost entirely. Occasionally it is desirable to return to a specific site many years after the specimen was secured: It may be to examine the area for additional skeletal material that may have eroded out in the interim, or, more frequently, someone may want to study the lithology of the quarry and to collect geologicals samples for biostratigraphic (Russell and Chamney 1967; Russell, in preparation), stratigraphic (Russell 1986), taphonomic (Dodson 1971), palynological, sedimentologic (Brinkman 1990; Eberth 1990; Eberth, Braman, and Tokaryk 1990), or other studies. The faunal list accompanying the Steveville map sheet (Sternberg 1950) has served as a ready information source of specimens collected for various aspects of dinosaur faunal analyses (Russell 1967; Bakker 1972; Farlow 1976; Beland and Russell 1978, 1979, 1980). In his study of dinosaur quarry lithologies and dinosaur taphonomy in Dinosaur Provincial Park, Dodson (1971) relied heavily on old photographs taken during excavation and quarry stakes for relocation and final confirmation of his study areas (Dodson 1968; personal communication 1984). Some quarries were never photographed making relocation impossible. Such appears to be the case for some of the Lawrence Lambe and Barnum Brown quarries that were not relocated and staked by C. M. Sternberg back in 1936. Site data for these specimens, including types (AMNH 5356 *Dromaeosaurus*), unfortunately appear to be lost forever. Once quarry stakes are in place, they also become handy points of reference, especially for confirming locations on topographical maps or air photos, and as reference points for future specimen finds (i.e., specimen found 35 ft/10.7 m NNE and 8.2 ft/2.5 m below base of Quarry 7 stake). Such utilization of a quarry stake is most useful in pinpointing a locality when used in conjunction with a legal land description of UTM coordinates.

Once a specimen is excavated, the resulting hole is abandoned and largely ignored. When one considers the high rate of erosion in most soft sediments (annual erosion rates in Dinosaur Provincial Park are about 0.75 inches/1 cm according to Bayrock and Broscoe 1972; see also Campbell 1970, 1974, and Bryan and Campbell 1980, and citations therein), it is not surprising that all traces of even a large excavation can be lost in a few years, especially if the skeleton was entombed in soft marine shales or bentonitic clays (Russell and Chamney 1967). Russell and Chamney (1967) noted that "3–4 inches (7.5–10.5 cm) of hard white sandstone (were) lost to erosion in one quarry made in 1916." Relocation of old quarries is difficult, and it is best not to trust memory or even field notes. Ideally a quarry should be staked soon after specimen removal.

What to stake?

A wide variety of fossil localities could warrant quarry stakes, depending on budget and research interests. In the highly fossiliferous Dinosaur Provincial Park, nearly 200 excavations have been staked to date, and many more quarries worthy of staking are worked by thc Royal Tyrell Museum of Palaeontology annually (in 1982, 12 excavations were worth marking). In several areas, quarry stakes are only 20 feet (6 m) apart, and in others up to five stakes can plainly be seen from a high vantage point. Although the majority of staked quarries yielded a complete or fairly complete skeleton or skull, some sites consisted of isolated bones, such as a mammal jaw (Quarry 130) or pterosaur bone (Quarries 139, 147, and 158). In Dinosaur Provincial Park it would probably be useful to stake all bone beds and microsites, but the sheer numbers of these would make such an endeavor cost and time prohibitive and would distract from the natural beauty of the park. Therefore it is the Royal Tyrrell Museum's current policy to quarry-stake the following sites:

- All type specimens
- Partial to complete dinosaur skeletons
- Good skeletons of turtles, crocodiles, champsosaurs, and the like
- Isolated skulls, such as those of crocodiles or dinosaurs
- Bone beds of significant research interest or historical importance
- Dinosaur eggshell sites
- Footprint localities
- Unusual or extremely rare occurrences – for example, pterosaur bones
- Quarries previously excavated but not staked by other institutions
- Restaking loose or stolen stakes placed by other institutions

Quarry stake construction

The quarry stake (see Figure 4.9) is a simple, inexpensive unit consisting of two or three components depending on the construction method adopted. These are a bronze head and a length of pipe, joined either by a lock pin or by braze welding. The completed quarry stakes currently used by the Royal Tyrrell Museum cost about $20 (Canadian) each.

The head (Figure 4.10a–c) of the quarry stake carries the site data on its face (Figure 4.10a) and therefore is constructed of a hard nonrusting metal. The heads of Royal Tyrrell Museum quarry stakes are made of bronze (B metal bronze – 85% copper and 5% each of zinc, lead, and tin; Society of Automobile Engineers metal no. 40). The head is sand cast at a foundry. It is more economical to order large numbers of quarry stake heads, enough to last several years. A 1982 order of 54 large bronze heads cost the Royal Tyrrell Museum less than $15 (Canadian) for each. A 1990 price quote from the same foundry was $16 (Canadian) each for a similar-sized order.

The shape and size of the head can be designed to suit the institution's re-

Figure 4.9. Assembled Bronze/Iron quarry stake viewed from the side, overall length 42–55 inches (107–140 cm).

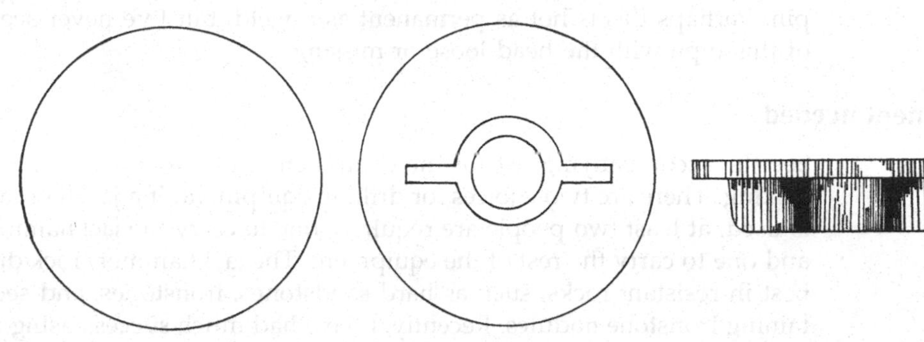

Figure 4.10. Bronze quarry stake head of the type installed by the Royal Tyrrell Museum in Dinosaur Provincial Park and Drumheller Valley in (left to right): top (script omitted from face), bottom and side views, showing method of construction. Actual size = maximum 4 inches (10 cm) in diameter. On Royal Tyrrell Museum bronze quarry heads, the institution name runs along inside curve of top half of round face and the place-name runs along inside curve of bottom half of face. "Quarry No. ________" and "Year ________" script are centered in between.

quirements or budget. Different shapes and sizes have been used in Alberta in the past. All are either round or square in outline when viewed from above. Square-shaped heads invariably become damaged, with the corners bent downward, presumably from hammer blows delivered by vandals. Round-shaped heads appear more resistant to damage. Size can vary too, as long as the face is large enough to carry the information you wish to convey. Some information can be incorporated into the quarry head at the time of casting, such as the name/logo and place-name of the institution, Quarry No. _____ (leave area blank for numbers to be stamped in later), year (19_____), and land description (i.e., Lsd. _____, S._____, T._____, R._____, W._____). Lettering can be either raised or indented on the bronze castings. Indented lettering is preferred because raised lettering is susceptible to damage or destruction by a vandal with a hammer. It is also easier to level the stake at the time of installation using a bubble level if the surface is flat. On the bottom of the bronze casting there needs to be a socket with an appropriate-sized hole to receive the pipe. For added support, a pair of flanges on opposite sides of the socket are built into the prototype (Figure 4.10b and c).

Any diameter of pipe can be used as long as it fits snugly within the socket beneath the bronze head. Quarry 61 in Dinosaur Provincial Park was restaked by us, and it appears that the pipe used in this old quarry stake was from salvaged galvanized plumbing pipe. The Royal Tyrrell Museum has used galvanized or black rolled pipe 7/8 to 1 inch in diameter cut into 41-inch (104-cm) or 54-inch (137-cm) lengths, respectively. The pipe should be cut exactly at right angles to the long axis so the bronze head sits at right angles to the pipe. Pieces of metal (such as bolts) could be welded onto the sides of the middle section of the pipe to give anchor points within the concrete. Alternately, grooves can be cut into the pipe's lower surface using a metal saw. If highly visible quarry stakes are desired, the pipe can be spray-painted a fluorescent color.

The best method of joining the bronze head and iron pipe is simply to braze weld them. Some of the older quarry stakes were assembled by drilling through the pipe and the base of the bronze head, and joining the two with a locking pin. Perhaps this is not as permanent as a weld, but I've never seen any stakes of this type with the head loose or missing.

Equipment needed

See the accompanying list for the equipment and materials required for quarry staking. There are two choices for drilling equipment. If a jackhammer/rock drill is used, at least two people are required, one to carry the jackhammer/rock drill and one to carry the rest of the equipment. The jackhammer/rock drill will work best in resistant rocks, such as hard sandstones, ironstones, and sediments containing ironstone nodules. Recently, I have had much success using hand augers. Not only are these lighter, but they are far superior to the jackhammer/rock drill when drilling into softer, friable rock types, such as siltstones or bentonitic clays. Earlier quarry staking expeditions also made use of hand augers. In 1954 movie footage of the Royal Ontario Museum expedition in Dinosaur Provincial Park includes a short quarry staking sequence showing an awkward drilling procedure using a 7-foot (2.2-m) long dynamite drill with chest brace. One man held the chest brace above head level while another cranked the auger at midshaft. While conducting fieldwork in Dinosaur Provincial Park in 1983, Clive Coy (Royal Tyrrell Museum) found a damaged homemade version of a hand auger suitable for quarry staking. It differs little from the type we now use. The hand augers used by the Royal Tyrrell Museum were originally designed for log home construction (Scotch eye augers). The blades wear out fairly quickly (approximately 15 to 20 drilling jobs in rock of soft to medium hardness) but are restored by adding layer upon layer of welded beads that are filed down to a sharp edge. We use two diameters, 1 ½ inches (3.75 cm) and 2 inches (5 cm). If the augers do not come with handles, these can easily be fashioned out of wood doweling turned on a lathe. The shaft of the smaller model was cut in half and a 17-inch (43-cm) section of pipe fitted between the ends to increase its overall length. By using hand augers, the weight and number of pieces of equipment are greatly reduced, and only one person is required to install quarry stakes.

Equipment, materials, and supplies needed for quarry staking

- Portable gas-operated jackhammer with drilling capabilities (changeable chuck)/ or hand augers
- 36-inch (91-cm) long drill rod with 1 ⅜ inch (35-mm) diameter bit (carbide tipped bits are available in most sizes for use in very hard rock)
- 17-inches (43-cm) long drill rod with 3 ½ inch (90-mm) diameter bit
- Gasoline/oil mixture for jackhammer/rock drill
- Water (1 gallon/4.5 L)
- Hacksaw with spare blades
- Tape measure
- Small folding shovel

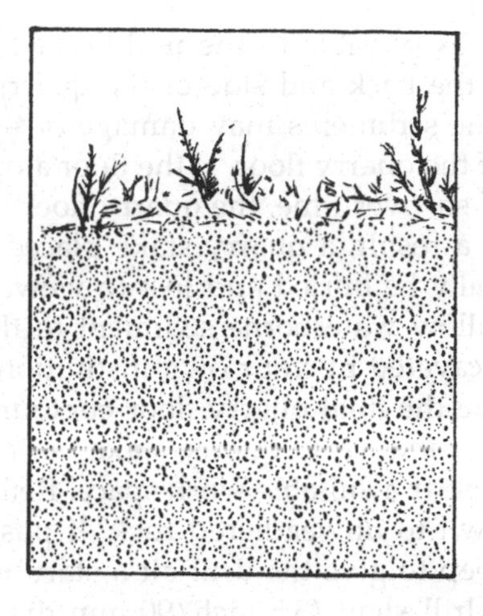
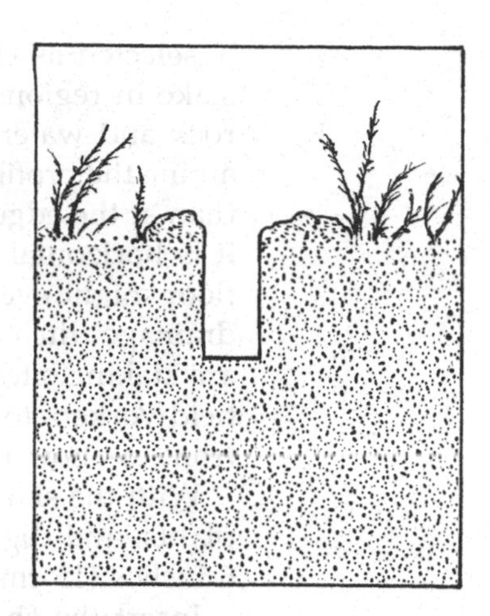
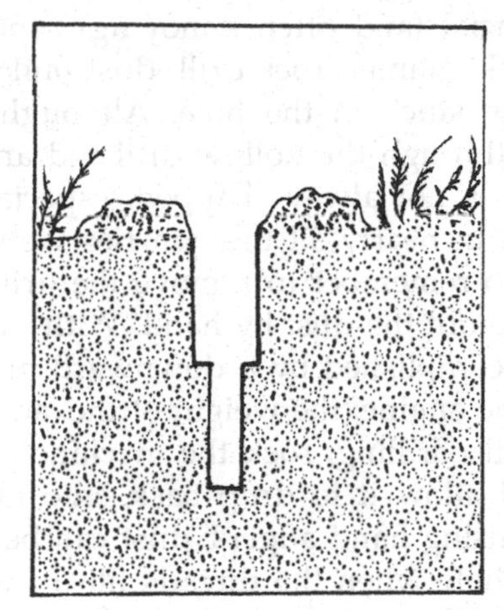
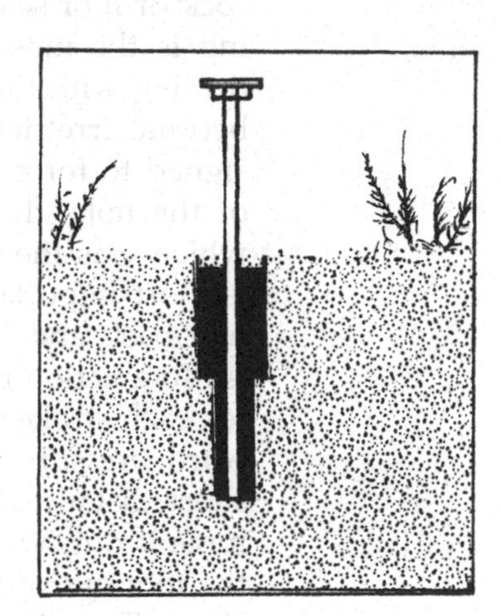

Figure 4.11. Cross section of quarry floor showing various stages of the quarry staking procedure. (A) Before drilling. (B) After drilling the large diameter hole. (C) After drlling the small diameter hole. (D) Finished quarry with stake in place, concrete poured, and hole filled.

- Bubble level
- Funnel
- Mixing bucket
- Redimix concrete ("just add water" variety)
- 3/4-inch (2-cm) diameter wooden dowel stirring stick, 3 feet (1 m) long
- Small squares of plastic sheet
- Elastic bands or wire twist ties
- Awl or narrow-bladed knife
- Crack hammer and steel die number stamp set (optional)
- Bronze/iron pipe quarry stake
- Backpacks for jackhammer/rock drill and other equipment

How to stake

The larger hole is done first, and then a smaller diameter hole is drilled through the floor of the first (Figure 4.11c). Upon relocating the site to be staked, an area is selected as close as possible to the middle of the quarry. Avoid placing the stake in regions at the back and side of the quarry near walls where slumping rock and waterborne sediments may damage or eventually bury the stake. Examine the profile of the quarry floor. Is the floor along the back wall much deeper than at the edge? If so, over time, the quarry floor will fill in with sediment until it is horizontal. As a result, a quarry stake placed in a low spot on the quarry floor might eventually be buried and lost to view. It may be necessary to dig a drainage ditch to allow waterborne sediment to flow out of the quarry. If you are staking sites excavated by your own institution, it would be wise to instruct fieldworkers to have the final quarry floor contours higher at the back than the front to allow runoff and avoid quarry stake burial.

With the shovel, clear a suitable work space. Remove all loose rock debris and plants, and dig down to unweathered rock. If possible, dig a depression about 6 inches (15 cm) deep to give the drill bit a starting point.

Insert the short drill shaft (3.5-inch/90-mm diameter) into the jackhammer/rock drill or select the large hand auger. Always start the larger hole first. Drilling is the most crucial (and often annoying) aspect of quarry staking. When drilling with the jackhammer/rock drill, dust buildup can cause the drill bit to become irretrievably stuck in the hole. Although most jackhammers are designed to force air through the hollow drill rod and to blow the rock dust out of the hole, this does not always happen, especially when the rock is damp, which was the case in nearly all the quarries we staked. With this in mind, it is most important to remember to remove the drill bit from the hole on a regular basis and scoop out the dust by hand. With the jackhammer/rock drill, the drilling action is accomplished by a downward push and a 1/4 turn of the bit for each stroke of the engine; the weight of the jackhammer provides the downward pressure. While drilling, rock the machine back and forth, and lift it up short distances and allow it to drop back onto the accumulated rock dust to force it out. When using hand augers, dust can be removed by carefully lifting the tool out of the hole when the threads fill up with sediment and shaking it clean. Depending on the rock type, the drilling procedure usually takes 30 to 45 minutes, but can take up to 3 hours or as little as 5 minutes. In my experience, a hand auger with a sharp blade is usually faster than a jackhammer/rock drill when drilling friable rock.

Once you have drilled the first hole as deep as possible (Figure 4.11b), switch drill bits or the hand auger. Center this new, smaller drill on the floor of the bigger hole. Make sure that the drill rod or hand auger stands vertically during drilling, or a leaning quarry stake will result. If using the jackhammer/rock drill at this stage, it is important to ensure that the drill bit does not get stuck, as it will be nearly impossible to retrieve. Most of our drilling problems occurred at this stage. Lift the jackhammer/rock drill 4 to 8 inches (10 to 20 cm) every 15 to 20 seconds and allow it to drop down onto the accumulated rock dust. Listen to the pitch of the engine. If it suddenly gets higher, this means the bit is getting stuck and the machine is experiencing difficulties in turning the bit. Drill this hole as deep as necessary (Figure 4.11c). If the hole cannot be drilled deeper because of dust or rocks, the quarry stake pipe can always be cut shorter with a hacksaw. If the hole is accidently drilled too deeply, small rocks dropped into the hole can be used to raise the quarry stake to the desired level. After removing the drill rod, the remaining dust can be scooped out of the bigger hole by hand, and dust in the smaller hole can be removed by twisting and pressing the pipe end of the quarry stake into the hole. The pipe is then carefully withdrawn, and the plug of dust inside removed by banging it against a rock or dug out with an awl or knife. By doing this several times, the dust in the smaller hole can be removed (Figure 4.11c). Drilling the smaller diameter hole usually takes less time, anywhere from 5 to 60 minutes depending on rock type.

Using the mixing bucket, water, Redimix concrete, and stirring stick, mix the concrete according to the instructions on the bag. Make sure the mixture is well stirred. Invert the quarry stake, insert the funnel, and fill the pipe with concrete. Upon hardening, this will give the pipe more strength and will thwart those who try to saw the bronze head off. Liquid concrete can be retained in the pipe by a small piece of plastic sheet covering the opening and held in place by a piece of wire or an elastic band. Place the quarry stake into the hole, and turn it so the writing faces the front of the quarry. The height of the bronze head above the quarry floor can vary. Most Royal Tyrrell Museum stakes sit with the bronze head about 28 inches (71 cm) above the quarry floor. It might be more desirable to have the head sitting at the stratigraphic level of the excavated specimen, if this is known. The stake should sit neither too high (where it could be bent by people or large animals such as cattle) nor too low (where it risks being overlooked or buried by sediment washing into the site). Pour the rest of the concrete mixture into the hole. Pack the concrete mixture with up and down motions of the stirring stick. The gravel in the mixture will settle to the bottom, leaving liquid concrete at the top, so add more small hard rocks that have been collected locally.

Now, using the metal stamp set and crack hammer, stamp the quarry number, year and other relevant data into the bronze head. This step is optional if the numbers were stamped before going to the quarry.

Use the bubble level on the pipe and top of the bronze head to check vertical and horizontal alignment, and make adjustments if necessary. Do a visual check of the stake from several angles and distances to confirm this.

Using the shovel, refill the work area with loose rubble to restore the quarry floor to its original contours. Tamp the debris down with your feet, especially at the base of quarry stake (Figure 4.11d). Sprinkle leftover water onto rock debris around base of pipe. Wetting helps the concrete set properly by reducing water absorption by debris. Clean the quarry stake of any concrete that may have slopped onto it.

Photograph the staked quarry from various angles and distances to help relocate it in the future. Ensure that some of your photographs include the skyline or local prominent topographical features. Keep a file of these photographs at your institution. Was the site full of plants upon your arrival? This should be noted in your file. In some recent quarries in Dinosaur Provincial Park, disturbance of the bentonitic clay through excavation activities has invited the growth of tall weeds. These were removed at time of staking but have since returned to hide the stake. Such an accumulation of weeds might be a feature to look for when future researchers wish to relocate the quarry.

Now would be a good time to accurately plot your staked quarry onto topographical maps and air photos, if this has not already been done.

Do not touch the quarry stake for a couple of days. Once the concrete has set, a firm upward tug on the bronze head should confirm the success of your operation.

Quarry stake survivability

If properly installed, a quarry stake should last a long time. Of the quarry stakes installed in Dinosaur Provincial Park by Charlie and Levi Sternberg (Sternberg 1936a), most are still firmly in place. Erosion has apparently buried several, but perhaps these can be relocated with a metal detector. The stake in Quarry 125 (Drumheller Valley) is still in the rock but lies on its side and only the head and a short length of pipe are still visible, apparently a victim of quarry wall and overburden slumping. In Quarry 164, marked in the fall of 1981, the exposed stake was half as high a year later. The cause of this is uncertain. It appears that the stake is sinking into the rock, but it is more likely the result of the "puffing-up" action of bentonitic clay as moisture penetrates the siltstone. Water erosion and frost around the base of some quarry stakes have exposed the concrete plugs and weakened the surrounding rock, loosening them. Most such affected stakes were installed on sloping surfaces where erosion rates are higher than level areas. One stake was bent and completely out of the rock, possibly the result of large animal (cattle) activity. Most fossil localities are in inaccessible, hot areas with tortuous geographic features and therefore receive little human traffic. Unless you know the quarry stakes are there and are looking for them, they can go unnoticed by even experienced field staff. Quarries 59 and 60 in Dinosaur Provincial Park, probably staked in 1936, are highly visible and have always been accessible to countless numbers of the general public, yet they are still in excellent condition. On the other hand, some quarry stakes have been lost entirely. Quarry 124 (Drumheller Valley), near the Bleriot Ferry was staked in 1956, and remained there until about 1985 (Stefanuk 1986, personal communication), when someone succeeded in slipping the pipe out of its concrete plug. The site has since been restaked. In all the sites staked by the Royal Tyrrell Museum, using the technique described above, the stakes are still firmly anchored.

By making the bronze head as simple as possible (i.e., omission of institutional logo), they are less desirable to souvenir hunters. Perhaps a "penalty for removal" warning should be included on the castings, although this would then require backup legislation. The Royal Tyrrell Museum is currently exploring the feasibility of such an addition on its bronze heads. Quarry stakes should be inspected and reported whenever relocated. At that time they should be checked to ensure they are still firmly anchored. Loose quarry stakes should be restaked.

Our quarry staking method provides a quick, inexpensive way of permanently marking the locations of historically and scientifically valuable sites for future generations. Some may suggest such an effort to be wasteful, but this is not true. If C. M. Sternberg (1936b, 1950) had not carried out his staking operations in Dinosaur Provincial Park, several subsequent studies (i.e., Dodson 1971) would not have been possible. Future generations of geologists and paleontologists will thank you for staking your quarries today.

Acknowledgments

I would like to thank the following individuals for their assistance in the Royal Tyrrell Museum of Palaeontology quarry staking project and their help in making this chapter possible: Dr. Phillip J. Currie (Royal Tyrrell Museum), and Dr. Loris S. Russell (Royal Ontario Museum, Toronto) and his wife Grace. Jane Danis and Fred Orosz (Royal Tyrrell Museum) have provided logistical and/or field support. Field assistance was provided by Clive Coy, Pat Harrop, Moira Hirst, Mike Klassen, Paul Nielson, Micheal Ryan, Scott Sampson, Maurice Stefanuk, John Walper, and Linda Watson. Maurice Stefanuk (Drumheller, Alberta) has been helpful in relocating many of the older dinosaur quarries in the Drumheller Valley, and I am grateful to him for his valuable contributions. Dr. Philip J. Currie, Brooks Britt, Clive Coy, Ken Kucher, and Jim McCabe (Royal Tyrrell Museum) read the manuscript and provided useful suggestions. Richard G. Day (National Museum of Natural Sciences, Ottawa) and Jane Danis have provided historical information on the earlier quarry staking efforts. The figures were drawn by Sandra Urschel.

References

Bakker, R. T. 1972. Anatomical and ecological evidence of endothermy in dinosaurs. *Nature 238*:81–85.

Bayrock, L. A., and A. J. Broscoe. 1972. A quantitative study of badland erosion, Dinosaur Provincial Park, Alberta. In: *24th International Geological Congress Guidebook*, Field Excursion C-22, pp. 95–101.

Beland, P., and D. A. Russell. 1978. Paleoecology of Dinosaur Provincial Park (Cretaceous) Alberta, interpreted from the distribution of articulated remains. *Canadian Journal of Earth Sciences 15*:1012–1024.

———. 1979. Ectothermy in dinosaurs: paleoecology evidence from Dinosaur Provincial Park, Alberta. *Canadian Journal of Earth Science 16*:250–255.

———. 1980. Dinosaur metabolism and predator/prey ratios in the fossil record (pp. 85–102). In: R. D. K. Thomas and E. C. Olsen (Eds.), *A cold look at the warm-blooded dinosaurs*. AAAS Selected Symposium 28. Boulder, Colo.: Westview Press.

Brinkman, D. 1990. Paleoecology of the Judith River Formation (Campanian) of Dinosaur Provincial Park, Alberta, Canada: Evidence of vertebrate microfossil localities. *Palaeogeography, Palaeoclimatology, Palaeoecology 78*:37–54.

Bryan, R. B. and I. A. Campbell. 1980. Sediment entrainment and transport during local rainstorms in the Steveville Badlands, Alberta. *Catena* 7:51–65.

Campbell, I. A. 1970. Erosion rates in the Steveville Badlands, Alberta. *Canadian Geographer 14*(3):202–216.

———. 1974. Measurements of erosion on badland surfaces. *Zeitschrift fuer Geomorphologie, Neue Forschungen*, Supplement Band *21*:122–137.

Colwell-Danis, J. 1988. Quarry staking at Dinosaur Provincial Park, Alberta: A unique type of locality data conservation (abstract). Society for the Preservation of Natural History Collections, Programs and Abstracts, Carnegie Museum of Natural History, Pittsburgh, p. 16.

Dodson, P. 1968. Unpublished field notes, Dinosaur Provincial Park, Alberta. On file in Royal Tyrrell Museum of Palaeontology library.

———. 1971. Sedimentology and taphonomy of the Oldman Formation, Dinosaur Provincial Park, Alberta, Canada. *Palaeogeography, Palaeoclimatology, Palaeoecology* 10:21–74.

Eberth, D. 1990. Stratigraphy and sedimentology of vertebrate microfossil sites in the uppermost Judith River Formation (Campanian), Dinosaur Provincial Park, Alberta, Canada. *Palaeogeography, Palaeoclimatology, Palaeoecology 78*:1–36.

———. D. Braman, and T. Tokaryk. 1990. Stratigraphy, sedimentology and vertebrate paleontology of the Judith River Formation (Campanian) near Muddy Lake, west-central Saskatchewan, and comparisons with selected sites in Alberta and Montana. *Bulletin of Canadian Petrology and Geology 38*(4):387–406.

Farlow, J. O. 1976. A consideration of the trophic dynamics of a Late Cretaceous large-dinosaur community (Oldman Formation). *Ecology 57*:841–857.

Graham, D. 1987. Dinosaur site information requested. *Fossil Trails 20*(3):6.

Russell, D. A. 1967. A census of dinosaur specimens collected in western Canada. *National Museum of Canada, Natural History Papers No. 36.*

———. and T. P. Chamney. 1967. Notes on the biostratigraphy of dinosaurian and microfossil faunas in the Edmonton Formation (Cretaceous), Alberta. *National Museum of Canada, Natural History Papers, No. 35.*

Russell, L. S. 1986. Exploring a great dinosaur graveyard. *Rotunda 19*(2):20–29.

———. 1987–1989. Report on field work, for the Royal Tyrrell Museum of Palaeontology. Annual reports on file in Royal Tyrrell Museum of Palaeontology library.

———. in preparation. Biostratigraphy of the Horseshoe Canyon Formation.

Sternberg, C. M. 1936a. Unpublished field notes, Steveville, Alberta. On file in Royal Tyrrell Museum of Palaeontology library.

———. 1936b. Preliminary map 969a, Steveville sheet, Alberta. *Geological Survey of Canada Paper 36–18.*

———. 1950. Steveville west of the 4th meridian, with notes on fossil localities. *Geological Survey of Canada Map 969A.* Scale 1:31,680 (1 inch to 1/2 mile).

Tanke, D. H. 1981, 1982, 1986, 1988, 1990. Unpublished quarry staking field notes, Dinosaur Provincial Park and Drumheller Valley. On file in Royal Tyrrell Museum of Palaeontology library.

5

Microvertebrate collecting: Large-scale wet sieving for fossil microvertebrates in the field

Malcolm C. McKenna, Ann R. Bleefeld, and James S. Mellett

Under the right circumstances, large-scale wet sieving (washing, wet screening, etc.) in the field at a semi permanent site can result in great increases in efficiency in exploration for microvertebrate specimens, especially in low-grade fossiliferous matrix. Our experience has resulted in a shift in emphasis regarding the role of equipment versus personnel, away from Hibbard's (1949) original ideas and toward the use of many screen-bottomed boxes in which concentrate is dried while still in the box. We do not deal here with laboratory methods, the supplemental use of chemicals, or small-scale wet sieving.

This chapter focuses on wet sieving (washing, wet screening) methods applied on a large scale in the field in order to obtain microvertebrate specimens from low-grade but nevertheless quarriable fossiliferous sites. These field methods increase efficiency of collection of useful samples of small microvertebrate fossils from large volumes of weakly consolidated sedimentary rock, but are not the only options available to field paleontologists. Each fossiliferous site is different, so collection methods must be chosen with care. Large-scale wet-sieving methods should be used only after it has been determined that little or no significant damage to delicate specimens will occur and after appropriate testing on a small scale has suggested their efficacy.

We are not concerned here with simple dry screening or with wet sieving by means of various types of "washing machines" that utilize pump-circulated water (Henkel 1966; Clemens 1973:162; Ward 1974, 1981, 1984; Freudenthal 1976; Orth 1983). Nor do we deal with separation of fossils from matrix under laboratory conditions utilizing heavy liquids, magnetic separation, jigging, or ultraviolet fluorescence. We also sidestep the use of acids, hydrogen peroxide, kerosene, or detergents like sodium metaphosphate (Calgon, e.g.). Such chemical enhancements are not economically feasible at the initial large-scale exploratory sieving level, but are more attractive for small-scale efforts or later on when concentrate can be reprocessed under laboratory conditions. Processing with Calgon costs upward of $40 per ton of matrix (J. D. Bryant, personal communication). For excellent reviews of laboratory methods and combinations of small-scale field methods with laboratory methods, see Green (1970) and Freeman (1982).

Some years ago one of us published descriptions of the large-scale wet-sieving methods used by various field parties of the University of California and the American Museum of Natural History to recover small fossil bones, teeth, and other significant specimens from marginally fossiliferous sedimentary deposits of Cretaceous and early Tertiary age (McKenna 1960:3–5, 1962, 1964, 1965). Similar reports based on experience at other institutions were published by Patterson (1956:3–5), Clemens, (1965), Sahni (1969), Lillegraven (1969:13–15), Rigby (1980: 1–4), and Archibald and Clemens (1982:3). These methods were directly based on those invented by Claude W. Hibbard in 1928 and developed by Hibbard and Henry Jacob in the period from 1936 to 1949 (Hibbard 1949, 1975). The relatively unconsolidated late Cenozoic sediments processed by Hibbard broke down more easily in water than when subjected to simple dry screening. Hibbard found that he could process large quantities of fossiliferous matrix efficiently by wet sieving. The relatively abundant fossils contained in the resulting concentrate sustained only minimal damage as a result of being sieved in water. But when we (among others) applied Hibbard's techniques to recover fossils from generally less fossiliferous and also somewhat more lithified late Mesozoic and early Cenozoic sediments, we began to modify the procedure in order to deal with new challenges.

In this chapter we present an historical account of the evolution to the present of our own elaboration of Hibbard's wet-sieving methods, particularly as we have applied them to somewhat consolidated sediments on a large scale. Many other workers have solved the same problems somewhat differently, so in a few cases we compare approaches.

Early work

Wet sieving for microvertebrate fossils dates back at least to Theodore Plieninger's work in Mesozoic rocks in Europe in 1847 (see Kulhne 1968:13; see also Owen 1871:3). In the British Isles, at his residence in Bath, in 1858, Charles Moore wet-sieved about 3 cubic yards of Rhaeto-Liassic coarse, friable sand transported there from a fissure at Holwell. Moore sorted concentrate over a period of two years, which yielded 29 haramiyid teeth and a very large number of other vertebrate remains (Owen 1871:3; Savage 1960 and personal communication to McKenna 1983). Moore (1869) described his methods in an obscure paper on fissures in Carboniferous rocks.

In North America, the first recorded wet sieving for fossils was done by J. L. Wortman in 1891 for the American Museum of Natural History. However, Wortman's use of wet sieving merely enabled him to recover and clean the scattered fragments of the type specimen of *Palaeonictis occidentalis* at one site and an individual representing *Oxyaena lupina* from another site, found in early Eocene (Wasatchian) sediments in the Bighorn Basin, Wyoming (Wortman 1892:145–146). Wortman was, of course, in each case collecting as many fragmentary parts as he could of a single individual. Presumably, Wortman never employed wet sieving as a primary means of obtaining a large number of specimens from a quarriable concentration as Plieninger and Moore had done in Europe. Wortman carried sacks of matrix on horseback and also by wagon to water, where it was carefully sieved, "after the manner of the placer miner" (1892; 1899:141). Possibly, Wortman cleaned and concentrated the fragments of his two Eocene individual animals by swishing matrix-filled cloth sacks (such as feed sacks) about in a body of water until the fines had departed through the mesh, or he may have possessed a screen sieve. However, the American Museum's annual report for 1891 was written entirely by H. F. Osborn and yields no clues. Wortman's work was a step up from simple surface collecting, but his technique cannot be considered to have been "mass production."

By 1906, in wet sieving to recover whatever unseen specimens there might be in promising matrix, Barnum Brown of the American Museum of Natural History segregated Late Cretaceous microvertebrate remains from sediments from the Hell Creek Formation of Montana.

> A few fragments of mammal limb bones and a tooth were found on the surface, after which two wagon loads of pay dirt were hauled to a spring and washed. After the clay and sand was washed out and the material dried many interesting specimens were recovered besides parts of several mammal limbs and fourteen teeth. This material gives evidence of a much greater Laramie fauna than has been heretofore described. (Brown 1906 and letter to H. F. Osborn, July 26, 1906; Sloan 1976:146)

We do not know what kind of mesh Brown used, but he was following more in the footsteps of Plieninger and Moore than of Wortman, although he was probably quite familiar with the details of Wortman's fieldwork. However Brown did it, his was the first truly exploratory wet sieving to be done in North America. It was also the first such effort in Lancian Cretaceous sediments, 50 years before the University of California's work at Lance Creek, Wyoming, in 1956 (see Clemens 1963, 1965:140; Estes 1964).

Burlap bags such as feed bags, seed bags, or any other appropriately woven sack can be partially filled with dry matrix and placed directly in water and agitated until the fines pass through the mesh. This method of wet sieving without use of wire screen has proven effective in some circumstances, especially when dealing with dense claystones, but it nevertheless has some disadvantages. If breakup of the matrix is incomplete in one soaking cycle, then the wet bags and their contents will take a long time to dry before they can be placed in the water again for further processing. Moreover, roots in the matrix will not be removable from the concentrate during wet sieving and will interfere with drying and sorting. Either the concentrate and any remaining matrix must be dumped onto a tarpaulin or one must have plenty of sacks, all without holes, and adequate time. Dumping concentrate from bags in order to dry it may damage contained fossils. Lumps of concentrate, whether in or outside a bag, will dry slowly, especially if roots are still present. Nevertheless, this method of wet sieving sometimes works well on a small scale. It is particularly adaptable to situations in which the sieving equipment must be lightweight in order to be transported easily by backpack or aircraft, or for exploration in areas distant from home base. For the most mobile situations, sacks can even be constructed from mosquito netting, readily available in several mesh sizes (Grady 1979; Chaney 1988).

Wet sieving with screen

It is generally agreed that handmade wire mesh first appeared in Europe in the middle of the seventeenth century. Its use in the United States began early in the eighteenth century, mainly for the protection of food (Lynes 1963:130). Such handmade wire screen was expensive and hard to obtain until 1859, when "machinery for weaving wire screens" was patented by J. M. Schuyler and W. Zern of Pottsville, Pennsylvania (G. Terry Sharrer 1984, personal communication [letter from G. Terry Sharrer, curator of Agriculture, Smithsonian Institution, in reply to query from Ellen R. Riedel, reference librarian, Black Gold Library Information Center, Santa Barbara, California]). The product was apparently available for military use during the Civil War, and its mass production appears to have been a by-product of the war (Lynes 1963; Boorstin 1976). Steel replaced iron in the manufacture of wire screen, and by 1885 "galvanized wire cloth was on the market" (Lynes 1963:131).

In any case, window screen made of wire mesh became widely available in the United States during the 1870s. We do not know if Wortman used wire screen in 1891, but by 1896 J. B. Hatcher had evidently tried screen because he warned, "Another way to secure these small teeth is to transport the material to a small stream and there wash it in a large sieve in the water, the finer material being washed away, but this treatment is too harsh to give the best results, what few jaws there are always being broken to bits" (1896:119).

Modern wet sieving derives directly from methods employed by Claude W. Hibbard, who independently invented this route to mass production in June 1928, while employed for the summer as camp cook for a University of Kansas expedition working at Edson Quarry in the late Cenozoic of western Kansas (Hibbard 1975). Hibbard's initial effort was to construct several window-screen baskets, with which he wet-sieved matrix from Edson Quarry in water in a buffalo wallow about 2 miles from the quarry. Hibbard was unaware of the efforts of his predecessors. Moreover, his employers had little interest in "unimportant" small fossils. Beginning in 1936, now on the staff of the University of Kansas and able once more to devote attention to vertebrate microfossils, Hibbard emphasized wet sieving in his now professionally sanctioned but undersupported fieldwork. In 1949, by then at the University of Michigan but still doing fieldwork in Kansas, Hibbard was employing about a dozen wire-mesh screens worked by a small crew of students or volunteers rather than a single screen manipulated entirely by one person (Hibbard 1949; McKenna 1950, personal observations). Moreover, he had by that time incorporated the screens in high wooden boxes so that the floor and the lower parts of two opposite sides of the box were made of screen (see also McKenna 1962, fig. 6; Clemens 1965, fig. 1). The high sides permitted Hibbard to fill the boxes to any desired level of matrix. The purpose of incorporating the two partial screen sides was to let water current, if any, pass through the sides of the box near the bottom, helping to carry away the fines. Like Plieninger and Moore long before him, Hibbard was interested in the efficiencies of wet sieving for exploration, but he worked on a larger scale and was more than a thousand miles from his laboratory in Michigan. Because of the availability of motor vehicles, he was better able to solve the problems of transport of raw matrix to local water. The work was more efficiently performed in the field relatively close to the collecting sites than it would have been if Hibbard had tried to transport a large amount of raw matrix to the laboratory. However, because field equipment was at that time difficult to fund, student workers, rather than equipment, were still heavily emphasized. Thus there were few screen-bottomed boxes per person, the boxes were deep, and burlap bags were each loaded with as much as 150 pounds of matrix in order to use fewer sacks.

In Hibbard's work, the matrix was often sufficiently moist when collected so that predrying was believed necessary before the wet-sieving process could begin. Raw matrix was extensively predried on coarse-screened racks covered with canvas before it was loaded into the deep screen-bottomed boxes for processing in water. The intended purpose of this predrying was to reduce lumping of contained clay during subsequent wet sieving. However, the predrying operation meant that delicate specimens were subject to additional opportunities for breakage when raw matrix was dumped from overfilled sacks onto drying racks or when later moved from there to the wet-sieving operation.

A special work site was used for soaking and shaking the deep screen-bottomed boxes. They were placed either in standing water or, preferably, running water. Generally, one person would do most of the work in the water, while the others carried the boxes, filled and dumped them, and performed other chores.

In 1950, McKenna visited Hibbard's camp in Meade County, Kansas. Hibbard was then using about a dozen deep wooden boxes made on two opposite sides from 1" × 12" lumber and from 1" × 8" lumber on the other two (McKenna 1962, fig. 6). The boxes were about 18 inches square and were kept from deforming or falling apart by a diagonal handle made from 1" × 2" lumber lacking knots. Lath skids and cleats protected the screen, which was nailed to the side pieces and overlapped by the end pieces. After the matrix was processed, concentrate remaining in the boxes was dumped immediately onto towels borrowed from the University of Michigan's Athletic Department. This freed the boxes for a succession of new and generally copious loads of predried matrix.

Hibbard was thus able to use relatively few boxes many times per day, but the ground near the washing site throughout the day was a sea of spread towels covered with drying concentrate, each anchored by six stones because the towels were vulnerable to high winds and passing livestock (see Figure 5.1). Hibbard's method emphasized towels and drying racks rather than long residence of matrix or concentrate in a large number of boxes used for both wet sieving and drying. As late as the early 1980s, field crews from various institutions still used Hibbard's method with few changes (see, e.g., Archibald and Clemens 1982, fig. 1).

Beyond personnel

Over the years we have frequently experimented with Hibbard's wet-sieving technique, and have gradually modified it to fit the requirements of processing ever larger quantities of different types of fossiliferous matrix, ranging from dense clays to loose channel sands. In general, we have extended our field ex-

Figure 5.1. Claude W. Hibbard at Meade County, Kansas, about 1950. Towels in the foregrond courtesy of the Universty of Michigan Athletic Department. The wind is calm.

ploration to ever more diffuse "concentrations" of bone-bearing matrix, even profitably wet sieving large quantities of weakly consolidated rock from Mesozoic sites that produced as little as about one mammal tooth per ton of matrix (Lillegraven and McKenna 1986:3).

Although our field crews have remained about the same size since the 1950s, the ratio of equipment to people has changed drastically and the use of mechanical equipment has become more important. Each successive summer's fieldwork has produced significant changes in the field methods used, partly through general improvement and partly through adapting to the unique characteristics of various sites. Occasionally, presoaking in kerosene or detergents has been tried on a small scale, generally for test purposes, but this enhancement has not been used on a large scale because of expense and for reasons of environmental protection. Chemical treatment is more appropriate for small-scale operations or for laboratory work.

The following description of the 1983 version of the process used at a Late Cretaceous site (Hatcher 1896; Clemens 1960, 1963; Estes 1964) at Lance Creek, Wyoming, is therefore a generally valid model of our current large-scale wet sieving, although the peculiarities of any particular site such as the one we were excavating in 1983 will not be duplicated exactly at other sites. Readers will wish to modify these procedures to suit their particular needs.

By 1983, many years of experience in a variety of situations had increased the number of screen-bottomed boxes in use by a crew at any one time from the dozen used by Hibbard to about 300. This approximately thirtyfold increase allowed small batches of matrix to remain completely through the drying phase in the boxes within which they were originally wet-sieved. For several reasons, screen-bottomed boxes had become shallower than the heavy, deep ones used by Hibbard and then by various crews from other institutions in the 1950s and 1960s. Although the price of lumber had risen dramatically, the cost of boxes was kept at the lowest possible level (about $3 at that time) by using shallower ends and sides made of rough-cut lumber. Clear, planed lumber was necessary

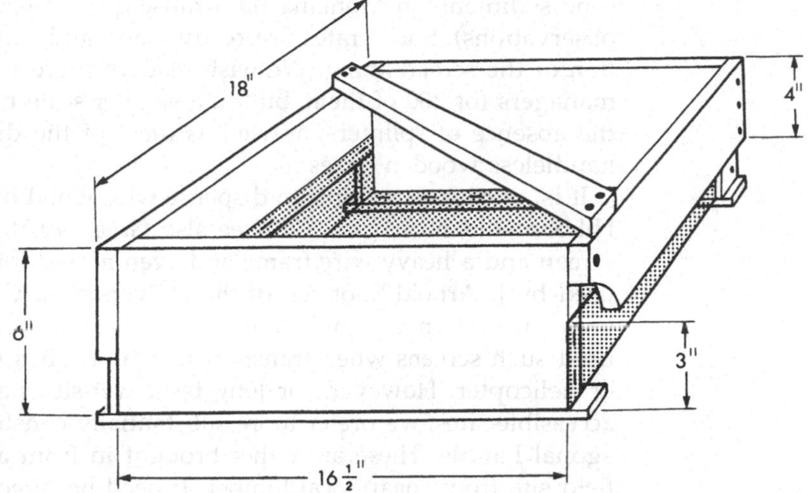

Figure 5.2. Suggested design of a wet sieve (washing box), evolved from Hibbard's (1949) pioneering design. Such boxes are useful in large-scale operations at semipermanent camps. Many variants are possible, depending on the availability and costs of materials, need for portability, nesting, ease of storage, type of screen used, and so on.

only for the handles of the boxes, for strength and to spare crew members from splinters. Standard screen-bottomed boxes constructed by us in 1982 and 1983 utilized 1" × 6" × 16.5" sides and 1" × 4" × 18" ends (see Figure 5.2). Screen was stapled with a staple gun to the sides before being protected by the nailed ends, skids, and cleats. Resin-coated nails were generally used. Each box weighed about 4 pounds when dry and empty. So long as the matrix doesn't overflow the box at the beginning of the sieving cycle, shallower depth has no effect on optimum potential matrix content of a box. Shallower boxes cost less to build and, importantly, if the boxes have less volume than those constructed by Hibbard, more boxes can be stacked at a given spot for soaking in the water, storage, or transport. Most important of all, light boxes are simply easier to work with.

Some workers prefer to construct two or more sizes of boxes without handles, so that the various sizes can be nested (J. K. Rigby, Jr., and R. E. Sloan, personal communication; Bleefeld, personal observations). The innermost box can, of course, have a handle. Nesting reduces storage bulk and can be important in transportation, but lack of a handle weakens box construction and also forces crew members to hold each box with both hands simultaneously while stooped or otherwise bent over. One cannot change hands for relief, nor can one easily remove objects like roots or stones without placing the box on the ground or in the water. In order to rotate and otherwise manipulate boxes during the wet-sieving process in still water, more fatiguing body motion is required than is the case with boxes with handles. Nevertheless, for work at certain localities – for example, Bug Creek Anthills in the uppermost Hell Creek Formation of McCone County, Montana, where the matrix is both unconsolidated and rich – little stress to a box need occur, so the advantages of handleless boxes can outweigh the disadvantages in certain cases. However, we do not recommend such boxes for general use over many years under a variety of conditions. They are not best for most applications.

Crews from the State University of New York at Stony Brook have substituted durable plastic milk crates lined with screen for boxes when wet-sieving Paleocene sediments in Montana (D. Krause, personal communication and personal observations). Such crates are ready-made and modular, needing only installation of the screen lining. We wish readers success when they ask grocery store managers for 300 of them, but on a smaller scale they offer advantages, such as the absence of splinters, as well as most of the disadvantages associated with handleless wooden boxes.

It is, of course, possible to dispense with wood or other support altogether, as Hibbard did at first in 1928 (see also Green 1970, fig. 5). Boxes constructed of screen and a heavy wire frame and even nested for ease of transport were once used by J. Arnold Shotwell of the University of Oregon in an exploration program in eastern Oregon's late Cenozoic (Shotwell 1968, fig. 6). We, too, have used such screens when transport to remote sites was necessary by packhorses or helicopter. However, for long term wet-sieving programs at single, readily accessible sites, we prefer more substantially constructed boxes, braced by a diagonal handle. These are either brought in from a storage site or made at the field site from cheap local lumber. If need be, precut lumber and screen can be transported more economically than finished boxes. For a truly remote field site, large numbers of lightweight nested screens or mosquito netting bags can be constructed beforehand and shipped to and from the field.

Although a few brass screens of smaller mesh size (24 openings per inch, about 1 mm each) are used for test purposes in our fieldwork, standard metal window screen, readily available at lumber yards and hardware and feed stores, remains our screen of preference for use in wooden boxes for large-scale sieving operations. Aluminum or galvanized window screen, never plastic screen, is used: After initial use, plastic screen sags in the center of the bottom of the box and thus prevents rapid drying of concentrate clumped in the sagged area. Concentrate should be evenly and thinly spread on a flat screen bottom, so that it will dry rapidly. Metal screen is much better than plastic screen for this purpose, because it stays relatively flat. However, if plastic screen is all that is available, it can be backed up by large-mesh wire "hardware cloth."

Mesh size of such commonly used metal window (fly) screen (approximately 1/16 inch/1.4–1.6 mm) does permit a number of small isolated teeth and bones, as well as small snails, charophytes, and other organic matter of interest, to pass through the screen as the matrix is wet-sieved. If these small specimens are not already available in adequate numbers, or if the total fossil content of the sediment is to be collected for some special reason, then smaller meshed brass screen or two sets of screen of different mesh sizes can be employed. For instance, Lillegraven (1972) used brass screen with 40 openings per inch in small-scale wet sieving of matrix presoaked in kerosene, using pumped water. But if a large amount of matrix is available to be wet-sieved, an adequate sample of the same kinds of small teeth will probably be retained in jaws. This is often the case with multituberculate anterior premolars and posterior molars. For obvious reasons, such jaws are more valuable than isolated teeth. Finer meshed screen unquestionably retains smaller items of interest in the concentrate, but the amount of concentrate that one must look through to find them rapidly increases to the point where very large amounts of crew time must be spent in sorting (picking) the concentrate. The limit of resolution of the human eye at a distance of 18 inches is about 0.5 mm, so unless some optical aids such as magnifiers or microscopes are available, many small specimens that are trapped by the small mesh will be swept away or lost anyway during sorting. Drying time of fine-grained concentrate also increases drastically if the thickness of the concentrate layer on the screen bottom is too great. Sometimes several days per box are required for drying of concentrate if the layer is too thick, which ties up equipment and may idle the crew. When standard window screen is used, fines pass completely out of the box, allowing the concentrate to dry in several hours in normal weather.

The following account of daily activity is based on our efforts to wet-sieve large amounts of raw matrix from University of California Museum of Paleontology (UCMP) locality V-5711, a Late Cretaceous vertebrate site in Wyoming known commonly as Bushy Tailed Blowout. This Lancian site has been worked occasionally by many institutions since its discovery by the University of California Lance Creek Expedition of 1957. Most recently, the American Museum of Natural History expeditions of 1981–1985 have wet-sieved about 300 tons of channel sand and ox-bow fill from this site. In the 1890s, Hatcher had tried to wet-sieve matrix in the intermittent water of Lance Creek from nearby localities, but had abandoned the effort in favor of using a flour sifter because he believed that wet screening produced excessive breakage to jaws of small mammals (Hatcher 1896:119). However, even a gross of Hatcher's famous flour sifters would never have been effective in processing large quantities of raw matrix. It is virtually certain that he only used the device to sift anthills constructed from sand grains and fossils by the ant genus *Pogonomyrmex* (Lull 1915; Turnbull 1959; Clark 1966; Adams 1984).

At present, when we are thus occupied, our wet-sieving operations in the western United States employ about 300 operable boxes concurrently. This generally works out to be about 50 boxes per crew member per cycle. Attrition usually necessitates building or rebuilding about 50 boxes per year, generally best done during the first few days of the field season. In contrast to Hibbard's version of the process, concentrate is dried in the same box in which it is wet-sieved. Weak, wet fossils are thus not dumped out at the worst possible time, and the drying power of the wind and sun is used to the greatest possible extent under safe conditions. Although we have not encountered substantial breakage of the type mentioned by Hatcher (1896:119), drying of matrix in the same box in which it has been wet-sieved has the advantage that fragments of broken specimens are kept together. If seen at this stage, damage can be repaired before fragments become dissociated by further processing.

Storage of the 300 boxes in the off-season or transporting them to another locality is seldom an insurmountable problem. Total volume of that number of boxes is about 400 to 500 cubic feet, and cooperative ranchers generally have barns or outbuildings with available space. Transport by trailer is easily accomplished if the distance is not great; otherwise a rental truck may save time and expense.

A day's work

On typical days serving as models for this account, 280 of the full complement of 300 screen-bottomed boxes were in use on the bank of a small reservoir near Lance Creek, Wyoming. In good weather, boxes were cycled twice a day.

Through recycling, about 520 boxloads of raw matrix from UCMP locality V-5711 were processed per day by six people, most of whom also had other duties. About 20 boxes were devoted to a second wet sieving of previously acquired concentrate.

Between 10 and 20 pounds of raw matrix were placed in each box, which required a ready supply of about 2600 to 5200 pounds of matrix for each cycle. Usually, at times of peak production, about 5 tons of raw matrix per working day were processed by the crew, but the average production was almost always somewhat less, because of various logistic factors and unpredicted adverse weather conditions.

Hibbard had loaded matrix into burlap feed sacks for transport to his field camp, where wet sieving was conducted. At the quarry he generally loaded sacks rather heavily, as the survivors from his crews attest. These sacks then had to be lifted, carried some distance, and then loaded and unloaded from a truck that transported the matrix to the processing site. Unless a trailer was used, workers had to bend over when loading or unloading roofed vehicles other than open pickups.

Although sacks do keep matrix and broken specimens generally associated while in transit from the quarry, we have found that, except for tests, sacks can be dispensed with altogether at this stage if the quarry site is readily accessible, as is the case with UCMP locality V-5711. If the proper equipment is available, a single individual can now be sent to an accessible quarry, driving a pickup and towing a four-wheel flatbed trailer. After arrival at the quarry, that individual can also load the trailer single-handedly, using a mechanical front-end loader kept at the quarry for that purpose.

The use of mechanized excavating equipment (e.g., Clark Company's Bobcat) as a routine part of the operation has contributed in a very positive way to successful large-scale sieving. Most importantly, it has helped to reduce strained backs and exhaustion among crew members. Further, the front bucket excavator makes one cut about 4 feet wide and picks up about 5 cubic feet of material at a time. Digging that amount by hand in somewhat indurated matrix would require scores of shovel thrusts, each one of which has the potential to slice through or pulverize teeth, jaws, and other fossil specimens. In our opinion, lifting out a larger amount of matrix in one scoop is somewhat easier on the fossils contained in it. Finally, a backhoe attachment, in addition to the front-end loading capability, permits frequent sample cuts to be made in the walls and floor of the excavation. Thus, if need be, one can easily trace fossiliferous horizons or make taphonomic reconstructions of the quarry.

Approximately six to ten scoops with a small front-end loader (Bobcat or equivalent) are all that are required to load the trailer. Normally, no sacking is required when loading the trailer. Generally, however, we also load the pickup with various test sacks as well (partly for traction in order to tow the trailer and partly because the space is available). If many sacks are loaded, however, a second person may be required at the quarry because it is inefficient in a given interval for one person alone to attempt to fill, carry, and load 30 or 40 sacks of matrix.

Ideally, trips to the quarry in hot weather should be undertaken early in the morning and again late in the day in order to avoid midday heat. However, the full crew is more urgently needed at the processing site early in the day in order to set the drying process in motion early, so that the first of the day's two drying cycles can be finished by the lunch break. As a result, trips to the quarry are generally made near the ends of cycles rather than at the beginning. The afternoon trip is especially crucial in that enough raw matrix to supply a cycle must be on hand on the trailer at the beginning of the next morning's cycle if that work is to be completed in time to begin the second cycle of the day. Therefore, a load of matrix collected in the afternoon is generally stored on the trailer overnight, away from any danger of destruction by flash flooding. The matrix should be protected from the elements by a tarpaulin, so that it will be dry when placed in water the next morning.

At the beginning of a day's operations the trailer is repositioned close to the water (see Figure 5.3). Raw matrix piled on the trailer is shoveled into boxes or poured into boxes from sacked matrix. The correct amount of matrix per box is a function of a large number of factors, to be determined by experience, and may vary substantially from day to day. The fraction of clay present is generally the most important factor. In each screen-bottomed box one should aim to produce a layer of concentrate thin enough to dry readily.

Boxes filled with an appropriate amount of matrix are placed in the water at the shore or double-stacked in deeper water if water depth is adequate. They are allowed to soak until such time as each individual box is processed. Filling and carrying the boxes into the water may take an hour or more. In order to keep the wet-sieving process working as smoothly as possible, a small inventory of matrix must be maintained at the processing site. Such stored matrix is generally kept in bags rather than on the trailer. This frees the trailer for further trips to the quarry. It is not efficient to transfer large quantities of matrix from the trailer to temporary storage, but if this must be done, it is best to pile it on tarpaulins in a dry place.

Usually, as regular boxes with raw matrix are being placed in the water to soak, a second wet sieving of pooled concentrate from a previous day is done at a special location nearby in the same water body, often with special, shallow boxes having large areas of screen. If this is done early enough, these special boxes will dry in time to provide sortable "super concentrate" late in the cycle.

Once the regular boxes filled with an appropriate amount of raw matrix are in the water and have had time to soak sufficiently so as to break up dry clay as much as it will without further attention, the main activity of a wet-sieving cycle can begin. This work can be a long, boring affair, in that each person will have perhaps 50 boxes to process per cycle. The contents of each box are gently agitated by raising and lowering the box and by gently manipulating lumps of sand or clay by hand. While lowering the box, it should also be rotated slightly, to shift matrix on the screen. The boxes are lifted entirely from the water often enough to allow inspection of the contents. Unwanted stones and lumps of clay can be tossed out at this time. All roots, leaves, rabbit and sheep dung, and other light floating material should periodically be allowed to float completely out of each box by tilting it, because such materials will impede drying and interfere with efficient sorting if they are allowed to remain with the concentrate.

Generally, it is easier on the backs of crew members working in the water if freshly filled boxes are carried a short distance out to deep water (if available) and doubly stacked there until processed, in order that workers can stand up straighter while working. Tennis shoes are a must in order to avoid hidden glass, barbed wire, and various other sharp objects on the bottom.

On completion of wet sieving, boxes are usually carried to shore two at a time (see Figure 5.4) and then to a slope where large numbers of boxes can be stacked domino-style in triangular arrays facing the point where the sun will be at the half way point in the drying process (see Figure 5.5). Usually it is impractical to stack more than about 100 boxes in such an array. Arrays should not be set out to dry on any pathways used by livestock. Boxes should be leaned against one another so that screens are kept off the ground, allowing for thorough air circulation. If enough space is available, boxes requiring about the same amount of time to dry should be set in the same array. If one person stacks the boxes once they reach the shore, he or she can expect to walk up to 10 kilometers per day, half of the time carrying two boxes.

When the concentrate in the boxes is completely dry, it can be either sorted or reprocessed if any clay is still present. As Hibbard was well aware, clay will not break up further unless it is first completely dry before being placed in the water again. In the case of UCMP locality V-5711, clay balls are a prominent component of the matrix; they require several sievings in order for them to break down to a small enough particle size to pass completely through the screens. The first sieving generally reduces the volume of matrix to about 2% of the original amount; later sieving is never so effective, but a second sieving can reduce the volume of concentrate to about 1% of the original amount in many cases, and more than that in some. For that reason resieving is nearly always worth the effort, provided the action is gentle and breakage to fossils is not excessive. However, every quarry is unique in the quality of its matrix and fossils, so methods will have to be adapted to circumstances. Obviously, the amount of final concentrate should be kept to the most practical minimum, either for sorting purposes or for economical further processing, such as with heavy liquids.

Stacked boxes afford shade to small animals of various kinds during the heat of the day. In 1983 we had at least one resident rabbit, several toads, and occasional small rodents. It should be remembered that not only people but also rattlesnakes find such wildlife interesting. Sometimes the routine of washing can be punctuated by a surprise.

The choice of an appropriate body of water is important and should be made by weighing many factors. The location can be anywhere, even in a university or museum courtyard, but we have found it preferable to do large-scale work as close to the quarry as practical in order to minimize heavy wear on mechanical equipment. For safety reasons as well as convenience, camp should also be close to the processing site. In case of injury this can be quite significant, and in case of flash flooding rapid action at night may be required in order to save equipment from a quick trip to the nearest ocean. Running water is preferable, but in arid country little-used stock tanks or larger reservoirs are often the main source of dependable water for processing. In the case of all water use, permission from owners and/or appropriate managers must be obtained beforehand and mitigation of any possible damage must be arranged. It is necessary to ascertain what the effects of a wet-sieving operation will be on the environment nearby or downstream. However, damage is generally quite minimal and, in the case of streams, often less than would occur in a single storm

Figure 5.3. Arrival of matrix at the screening site, a reservoir near Lance Creek, Wyoming, June 1983. Note stacked boxes from previous day's second cycle. Matrix distributed to boxes from nonsacked pile on the trailer is being put in water to soak.

Figure 5.4. Sieved concentrate being carried to drying array. Generally this occupies one person continuously – in this case, Dr. Zofia Kielan-Jaworowska, 1983.

Figure 5.5. Triangular arrays of boxes in which concentrate is drying; near Lance Creek, Wyoming, 1983. The arrays should face the sun and not shade one another.

under natural conditions. However, in the case of small stock tanks, minor filling with sediment can occur.

Large-scale wet sieving can be conducted in still water, moving water in streams or rivers, and even in ocean or lake surf if carefully supervised. For both rivers and streams, levels and water velocities may rise suddenly and unexpectedly, especially at night, so some sort of tethering is especially necessary, as is a deflection system for dealing with large flotsam such as floating logs. J. A. Lillegraven and T. H. Rich used anchored floating wooden frames to constrain floating wooden boxes in the Red Deer River in Alberta, Canada (Lillegraven 1969, fig. 3, 4), thus circumventing the effects of rising river level on bankside boxes.

Needless to say, tetanus immunization is also necessary, especially in that cows and other livestock, both dead and alive, may be contributing to the contents of the water being used.

Sorting

Sorting (picking) of concentrate is done at times naturally arising from lulls in the all-important drying cycle. Because of weather conditions and other factors, there is a varying schedule of slack time before lunch and late in the day. Either all or part of the crew can sort concentrate when not otherwise occupied, and most crew members enjoy this activity for the fun of discovery. Rotation of duties is useful for morale purposes. We give sorting a secondary priority in our work, but we do not store all the concentrate for later laboratory sorting. Rather, sorting is carried out at least in part on the spot in order to keep track of productivity in the quarry. This is especially important when a low-grade but large fossiliferous site is being investigated. In general, however, concentrate builds up at a much faster rate than it can be sorted, so that after some weeks of work as many as 50 bags of doubly sieved concentrate may be on hand. At that point activities can be moved to a more pleasant locale and the danger of mutiny partially averted! Concentrate still unsorted at the end of a field season can either be sorted under laboratory conditions or stored over the winter. However, in our experience, unless volunteers can be recruited there is little time during the academic year to sort large amounts of concentrate under laboratory conditions. Also, for sorting purposes bright sunlight is far superior to most indoor lighting. For these reasons, much concentrate is sorted in the field at the end of the summer at a pleasant locale, but a small inventory of concentrate is maintained over the winter at the box storage site in mouseproof canisters in order to have enough on hand at the beginning of the next season to occupy weather-generated slack time if such occurs.

At the end of a second, sometimes a third, wet sieving, concentrate is temporarily stored in doubled burlap bags, woven plastic bags, or polyethylene sacks of high quality. We emphasize that new, solid polyethylene, woven plastic, or burlap bags of high quality, with tight mesh, should be used for storage of concentrate in a dry place. Used bags are often available at bargain prices, but they do not have the tensile strength and seam durability of new bags. In addition, used bags may contain residual stored grain that may have been treated with mercuric fumigating agents. These compounds are highly soluble in water

Figure 5.6. Drs. Zofia Kielan-Jaworowska and Malcolm C. McKenna at a sorting table with concentrate from a morning cycle, Lance Creek, Wyoming, June 1983.

and toxic to wildlife, livestock, and crew members, so we strongly advise against the use of bags that lack a known history.

We recommend that concentrate be completely dry before it is stored in bags or transported. This will prevent concentrate from being "packed in" when specimens are wet and therefore weak. Moreover, we have found that storage of wet concentrate causes any remaining clay particles to clump and also to adhere to fossils. This makes recognition of specimens during sorting more difficult and adds extra work when the specimens are prepared in the laboratory. Dry, dust-free, and clay-free concentrate is necessary for the best results. Moreover, if burlap or other cloth bags are used, they will soon rot if not stored in a dry place.

Each bag is tied and marked as to its source and date of collection. Both an internally placed note and an externally placed tag are used for each bag. Such notes should be written in either pencil or indelible ink on durable paper and tags.

When sorting is undertaken (see Figure 5.6), a small quantity of concentrate is poured onto a canvas-covered wooden table or several such tables constructed from crude lumber. Long tables (e.g., 8' × 3') are ideal. Crew members doing the sorting should sit *facing* the sun for best results. Shadows should be avoided and polarized sunglasses should *not* be used because such glasses prevent seeing the glint from wear facets on teeth. Hats and ultraviolet-blocking creams are thus advised. Sorting early in the morning or during the late afternoon is inadvisable because of low sun angles. Sorting in poor light should be avoided completely.

From the central pile, concentrate should be spread in a layer one particle thick in front of each sorter. This is easily done by using the hand as a brush. There is no need for special equipment of any kind other than simple hand lenses for those who require them. On a small scale, fussy gadgetry (Kuihne, 1971; Hovestadt and Hovestadt 1982) may work for some people, but we have found that humans, looking through large quantities of concentrate particles in single layers in broad daylight, are in the long run more efficient than powerful magnets, conveyor belts, ultraviolet lights, photoelectric separation, and various more exotic techniques. Though they might work on a small scale, these methods simply are not effective on a large scale or as reliable under varying field conditions as the human observer. We do not even routinely use microscopes in the field while sorting concentrate processed with window screen, but for concentrate obtained with finer meshed screen a microscope is essential. The use of a microscope at the field sorting table will slow sorting significantly and also will rapidly result in a rather dirty microscope.

The single layer of particles on the sorting table should be searched completely with the eye in some form of raster pleasing to the sorter and then, after items of interest are picked up and various particles are turned over in order to see all sides, the particles remaining should be swept permanently from the table. For picking up small items, a wet brush is sufficient; tweezers should be avoided because of the damage they can cause to delicate specimens.

Inexperienced sorters tend to waste much time by not spreading the particles in a layer thin enough to permit all particles to be scrutinized in an organized and efficient way. They tend to paw through the same pile several times, seeing some particles many times and missing others completely. Ideally, particles should be scanned once thoroughly and then permanently discarded. Discarded particles should not be returned to the table because of the danger of confusion with previously discarded particles that might come from a different source. Sorters should learn a comprehensive search pattern and conduct it decisively and rapidly. True, some items of interest will be lost, but experience has shown that the overall efficiency of rapid sorting results in more and better fossils in the long run.

A supply of small plastic boxes, vials and corks, gelatin capsules, tissue paper, various solvents and glues, labels, and other curatorial paraphernalia should be on hand to treat and store valuable specimens, but some of this work can be postponed until night or the relative inactivity of a rainy day. At the time the specimens are first seen, an initial division of the recovered fossils into major taxonomic categories is useful but not essential. However, a proliferation of receptacles on the sorting table is often cumbersome and can interfere with other types of cans or bottles in use. For this reason we generally divide the fossils into "interesting" and "not so interesting" categories and save the taxonomic work for the laboratory.

Conclusions

Concerning wet sieving there is a major philosophical difference between ourselves and many other workers in the field. Hibbard and his associates used about a dozen deep boxes with standard window-screen bottoms and partial screen sides. These were used over and over again during the day's work because concentrate was not dried in the same box in which it was wet-sieved. The process used a minimum of permanent equipment. Our modification of Hibbard's process involves about 30 times as many boxes, but is somewhat easier on the specimens. More importantly, we process a large amount of often relatively low-grade matrix through standard-sized window screen mesh in shallower boxes, knowingly losing some tiny isolated teeth and other minute objects of possible interest in order to process enough matrix to make likely the acquisition of the same kinds of teeth still in place in jaws of the same taxa. Moreover, while diaphyses of long bones may slip through screens or remain unrecognized during sorting, we have found that epiphyses of humeri and tibiae and complete scapulae and tarsal and carpal elements are preserved using this method. Such work requires a quarry with a large amount of proven reserves. Other workers tend to use less than 50 boxes, generally with finer mesh in order to catch more nearly the total of the contained tiny teeth. However, they inevitably process far less matrix in a season and thus must expect to find fewer teeth in jaws. Such methods are, of course, appropriate for very small concentrations of fossils at a poor site, in order to collect as many as possible of the limited number of fossils present or for certain kinds of taphonomic studies (see Korth 1979), but we do not recommend them for routine wet sieving of large but low-grade concentrations at a good site if the primary purpose of the work is to build a large collection of informative specimens.

Empirically, the ratio of jaw fragments to isolated teeth of the same taxon in our work varies widely, but jaws are always rarer. Shaw (1964) analyzed the question of the probability of finding at least one specimen of a particular species in a sample. By analogy, the probability of collecting at least one jaw of a taxon increases toward certainty as more and more matrix is processed. In our experience, most other crews process much less matrix per day than we do; if the fraction were one tenth, it would take them ten times as long to encounter the same number of jaws, which are intrinsically much more valuable than isolated teeth of the same animals. Moreover, the unit cost per specimen is much lower in a large-scale wet sieving operation, and in this real and unpredictable world, we believe it may not always be possible to get in another nine seasons of work on a particular quarry. Our prevailing philosophy has been to attempt to collect the largest possible sample of fossil mammals and other vertebrates within any allotted field season. Running a fully mechanized, large-throughput, economy-of-scale operation allows us to do just that.

Acknowledgments

We thank the various unsung and long-suffering associates who have helped Hibbard's techniques to evolve. Figure 5.1 was obtained through the courtesy of the University of Michigan. Lorraine Mecker of the American Museum of Natural History drew Figure 5.2. Chester Tarka wrestled with the perennial problem of converting our own bad photographs into usable illustrations. A. Serio of the Smithsonian Institution kindly supplied the letter by G. T. Sharrer to Reference Librarian Ellen R. Riedel and also provided other references regarding early use and patenting of wire screen. Many helpful people have supplied information about Hibbard's work and about their own modifications of it.

References

Adams, D. 1984. A fossil hunter's best friend is an ant called "Pogo." *Smithsonian 15*:99–104.

Archibald, J. D., and W. A. Clemens. 1982. Late Cretaceous extinctions. *American Scientist 70*:377–385.

Boorstin, Daniel J. (Ed.). 1976. America in two centuries: An inventory. In *Subject matter index of patents for inventions issued by the United States Patent Office from 1790 to 1873, inclusive,* Vol. III. New York: Arno Press.

Brown, B. 1906 [MS]. Report of expeditions of 1906 to the Montana Laramie. In 1906 Annual Report of the Department of Vertebrate Paleontology, American Museum Natural History.

Chaney, D. S. 1988. Techniques used in collecting fossil vertebrates on the Antarctic Peninsula. *Geological Society of America Memoir 169*:21–24.

Clark, J. 1966. Go to the ant. *Bulletin of the Field Museum Natural History 37*(6):1.

Clemens, W. A. 1960. Stratigraphy of the type Lance Formation. In *Report of the 25th International Geological Congress,* Pt. 5, pp. 7–13.

———. 1963. Fossil mammals of the type Lance Formation, Wyoming. Part I. Introduction and multituberculata. *University of California Publications in Geologic Science 48*:1–105.

———. 1965. Collecting late Cretaceous mammals in Alberta. *Alberta Society Petroleum Geology, 15th Ann. Field Conference Guidebook.* Part I. Cypress Hills Plateau, pp. 137–141.

———. 1973. The roles of fossil vertebrates in interpretation of Late Cretaceous stratigraphy of the San Juan Basin, New Mexico (pp. 154–167). In J. E. Fassett (Ed.), *Cretaceous and Tertiary rocks of the southern Colorado Plateau.* Memoir Four Corners Geological Society.

Estes, R. 1964. Fossil vertebrates from the Late Cretaceous Lance Formation, eastern Wyoming. *University of California Publications Geologic Science 49*:1–187. In 1983 an emended and annotated version of this important paper was privately printed and distributed by Estes.

Freeman, E. F. 1982. Fossil bone recovery from sediment residues by the "interfacial method." *Palaeontology 25*(3):471–484.

Freudenthal, M. 1976. Introduction. In M. Freudenthal, T. Meijr, and A. J. van der Meulen, Preliminary report on a field campaign in the continental Pleistocene of Tegelen (the Netherlands). *Scripta Geol. 34*:1–24. Leiden.

Grady, F. vH. 1979. Some new approaches to microvertebrate collecting and processing. *The Chisler 1*(1):1978. Reprinted in *Newsletter of the Geological Curators Group 2*(7):439–442.

Green, M. 1970. Recovering microvertebrates with acetic acid. *South Dakota Geological Survey Circular 40*:1–11.

Hatcher, J. B. 1896. Some localities for Laramie mammals and horned dinosaurs. American Naturalist, February 1, pp. 112–120.

Henkel, S. 1966. Methoden zur Prospektion und Gewinnung kleiner Wirbeltierfossilien. *Neues Jb. Geol. Paläont.* Mh. 3:178–184.

Hibbard, C. W. 1949. Techniques of collecting microvertebrate fossils. *Contributions of the Museum of Paleotology, University of Michigan 3*(2):7–19.

———. 1975. [Edited] Letter from C. W. Hibbard to W. G. Kühne. Claude W. Hibbard Memorial Vol. 3. *Papers on Paleontology No. 12.* Museum of Paleontology, University of Michigan, pp. 135–138.

Hovestadt, D. G., & M. Hovestadt. 1982. An endless belt for scanning fossil residues under the microscope. *Tertiary Research 4*:9–11.

Korth, W. W. 1979. Taphonomy of microvertebrate fossil assemblages. Ann. Carnegie Museum, *48*(15):235–285.

Kuehne, W. G. 1968. Contribuição para a fauna do Kimeridgiano da mina de lignito Guimarota (Leiria, Portugal). 1 partie. History of discovery, report on the work performed, procedure, technique and generalities. Mem. (N.S.) Serv. Geol. Portugal, no. 14, pp. 9–20.

———. 1971. Photoelectric separation of microfossils from gangue. *Proceedings of the Geological Society of London 1970* (1664):221–222.

———. 1971. Collecting vertebrate fossils by the Henkel process. *Curator 14*(3):175–179.

Lillegraven, J. A. 1969. Latest Cretaceous mammals of upper part of Edmonton Formation of Alberta, Canada, and review of marsupial-placental dichotomy in mammalian evolution. *University of Kansas Paleontology Contributions,* art. 50 (Vertebrata 12).

———. 1972. Preliminary report on Late Cretaceous mammals from the El Gallo Formation, Baja California del Norte, Mexico. Natural History Museum, Los Angeles County, *Contributions in Science* no. 232.

———. and M. C. McKenna. 1986. Fossil mammals from the "Mesaverde" Formation (Late Cretaceous, Judithian) of the Bighorn and Wind River basins, Wyoming, with definitions of Late Cretaceous North American land-mammal "ages." *American Museum Novitates*, no. 2840.

Lull, R. S. 1915. Ant-hill fossils. *Popular Science Monthly*, September, pp. 236–243.

Lynes, Russell 1963. *The domesticated Americans*. New York: Harper & Row.

McKenna, M. C. 1960. Fossil mammalia from the early Wasatchian Four Mile Fauna, Eocene of northwest Colorado. *University of California Publications in Geologic Science 37*.

———. 1962. Collecting small fossils by washing and screening. *Curator V*(3):221–235.

———. 1964. Mining for fossils in Wyoming. *Nature and Science*, May 15, pp. 10–11.

———. 1965. Collecting microvertebrate fossils by washing and screening (pp. 193–203). In B. Kummel and D. Raup (Eds.), *Handbook of paleontological techniques*. San Francisco: Freeman. [Abridgment of McKenna 1962.]

Moore, C. 1869. Report on mineral veins in Carboniferous limestone and their organic contents. *Report of the British Association for the Advancement of Science*, Trans. of Sections, 1868, pp. 428–429. [Not seen; Section (e.g., Bristol, Exeter, etc.) not known to us.]

Orth, Bernard. 1983. Recherche appliquée a la concentration d'un sédiment fossilifère mise au point d'un traitement complexe. Thèsis. Ministère de l'Éducation Nationale, École Pratique des Hautes Études, 3 eme Section, Sciences de la Vie et de la Terre.

Owen, R. 1871. *Monograph of the fossil Mammalia of the Mesozoic formations*. London: Palaeontographical Society, Volume for 1870.

Patterson, B. 1956. Early Cretaceous mammals and the evolution of mammalian molar teeth. *Fieldiana: Geology 13*(1):1–105.

Rigby, J. K., Jr. 1980. Swain Quarry of the Fort Union Formation, middle Paleocene (Torrejonian), Carbon County, Wyoming: Geologic setting and mammalian fauna. *Evol. Monographs, 3*.

Sahni, Ashok. 1969. Techniques in prospecting for terrestrial microvertebrates. *Journal of the Palaeontology Society*, (India) *13*:38–43.

Savage, R. J. G. 1960. Cenozoic mammals in North America. *Nature 188*:200.

Shaw, A. B. 1964. *Time in stratigraphy*. New York: McGraw Hill.

Shotwell, J. A. 1968. Miocene mammals of southeast Oregon. *Bulletin of the Museum of Natural History, University of Oregon*, no. 14.

Sloan, R. E. 1976. The ecology of dinosaur extinction. In C. S. Churcher (Ed.), *Essays on Palaeontology in honour of Loris Shano Russell. Miscellaneous Publications of the Royal Ontario Museum*, Life Sciences, pp. 134–154.

Turnbull, W. D. 1959. Ant colony assists fossil collectors in Wyoming. Chicago *Natural History Museum Bulletin 30*:6–7.

Ward, D. J. 1974. Sieve technology. *Tertiary Research, Colour Supplement no. 2*, pp. 3–4.

———. 1981. A simple machine for bulk processing of clays and silts. *Tertiary Research* 3(3):121–124.

———. 1984. Collecting isolated microvertebrate fossils. *Zoology Journal of the Linnaean Society 82*:245–259.

Wortman, J. L. 1892. Narrative of expedition of 1891. VI. In H. F. Osborn and J. L. Wortman. 1892. Fossil mammals of the Wahsatch and Wind river beds. Collection of 1891. *Bulletin of the American Museum of Natural History* 4(1):81–147.

———. 1899. Restoration of *Oxyaena lupina* Cope, with descriptions of certain new species of Eocene creodonts. *Bulletin of the American Museum of Natural History 12*:139–148.

6

Laboratory preparation

Macrovertebrate preparation

Peter May, Peter Reser, and Patrick Leiggi

Once the specimens have been collected and shipped to the home institution, their preparation begins. The laboratory in which the work is done has to be well equipped to process the specimens properly, and the preparator has to be fully versed in the correct methods and procedures. Most institutions with a paleontology department have a history of preparation techniques that have been passed on from preparator to preparator. The techniques used in these different institutions probably do not differ radically; the goals are always the same, the methods similar, tools identical, and in most cases the quality of work is excellent. Modern technology has introduced some new tools, and adhesives and consolidants have become, for the most part, trustworthy and consistent. Still, the end result of a beautifully prepared specimen rests on the skills of the preparator.

The laboratory

Every paleontology department has an area set aside for the removal of the matrix from fossils found during the field season. Some of these areas are large and elaborate with many separate rooms set aside for different uses; others are small, functional areas, usually stretched to their limit. Both of these facilities can produce beautifully prepared specimens, depending on the equipment available and the skill of the preparator. The area of concentration of most vertebrate paleontological laboratories usually follows the area of expertise of the curators doing the research. The work can vary immensely in scale and scope: It may include collecting, preparing, casting, and mounting dinosaur skeletons for displays in a museum or extensive preparation of one type of fossil animal for many years in a university research laboratory. Preparators today have to be familiar with every aspect of their work so that they can fulfill the jobs required of them in any situation.

The future of most paleontological collections has been in the hands of fossil preparators since the first vertebrate fossils were discovered. In the early days the materials used to conserve the specimens were whatever was at hand. Over the years conservation materials have developed to the point where a complete chapter is devoted to the various adhesives, consolidants, and hardeners available to the modern preparator. The preparator's job is to ensure that all specimens are treated with the utmost care and conserved for future generations with the highest quality of materials that will have a minimum of degeneration, thus ensuring the specimens' place in a collection and for research forever.

Another aspect of quality conservation is the capability to reproduce the prepared fossils by taking molds of the specimens and casting them for research comparisons and study or for mounting complete specimens for display purposes. Most modern vertebrate paleontological technicians must be accomplished not only in preparation and conservation but also in mold making. They must be capable of producing complex molds and making casts from these molds in exact detail. As if this isn't enough, the mounting of skeletons can also be added to the list requiring the preparator to have a thorough knowledge of skeletal anatomy and familiarity with welding techniques. These techniques are discussed in detail in various chapters of this volume.

Laboratory function

When one considers the various jobs a preparator has to do, the design of a paleontological laboratory can become a complicated proposition. Foremost in any design is the health and safety of the employees working in the lab. Usually any health concerns can be resolved by installing an adequate ventilation system capable of handling the noxious residue produced by most of the tasks performed. Every workbench, workstation, and specialty room should have enough ventilation so that odors and particles produced do not carry farther than the immediate task area. Storage of all chemicals should be in approved cabinets and a serious attempt should be made to keep track of their use and expiration dates. Labeling should be prominent. Every country has explicit laws that govern these procedures, and all information is available at governmental occupational health and safety offices. For the guidelines that are in force in your state, province, or country, contact the highly trained individuals in these offices before proceeding to design or work in a paleo lab.

First, an office must be available so that the preparator has somewhere quiet to fill out specimen sheets and prepare for the tasks at hand, as well as space for files for suppliers, field notes, and research articles, (usually copies, so that the originals don't become damaged) and a general workbench for examining specimens before work begins.

A preparation area is next on the list (see Figures 6.1a, 6.1b, and 6.1c). In the preparation area there should be a fine air abrasive unit, small rotary tools (dental turbine, e.g.) hand tools, picks, and so on. The area should be well-ventilated and connected to a dust collection system. A binocular microscope is mandatory so that progress on the finer preparation can be monitored. A rough preparation area is needed for removal of plaster jackets and the use of chipping hammers, Airscribes, and hammers and chisels, ideally on mobile worktables, with access to a dust collection system to keep control of airborne silica (a designated substance). This room will end up being the dustiest and will require constant housekeeping to keep the particles under control.

For molding and casting, the ideal situation is to have a separate room with a self-contained exhaust system or a well-ventilated worktable. The size of the casting area depends on the scale of the casting program. A very large area is required for full scale skeletal casting, whereas if limited to small specimen casting, fume hoods will be sufficient.

The mounting area is usually a part of the main lab. Here a workshop should be located capable of maintaining large-scale skeletons. Basic requirements for mounting are an arc welder or other form of electric welding machine, some oxyacetylene tanks and gauges for cutting and bending, a chop saw, bench grinder, drill press, a large vise, and a multitude of hand tools. This area can also serve as the field tool maintenance shop.

These are the basic requirements that the lab should be equipped for and capable of accomplishing. In today's world of specialization it is rare to find people accomplished in all of these tasks, and the majority of technicians have learned their skills through trial and error on the job. The occupation of the vertebrate paleontological preparator is probably one of the few jobs left where a general knowledge is required of so many technical procedures and one is often called upon to apply them.

Mechanical preparation

Tools

Tools to have on hand are cast cutter (see Figure 6.2), various saws and knives, chipping hammers, chipping guns, pneumatic engravers, electric engravers, angle grinder, small rotary grinder, chisels, awls, picks, dental picks, hammers, and any other tool that will aid in removing the matrix. Also be sure to have adhesives, consolidants, and the appropriate thinners around the workstation. Safety equipment should consist of hearing protection, eye protection, dust masks, appropriate ventilation, and an eyewash station.

Pneumatic engravers

At least three manufacturers in the United States make tiny reciprocating air hammers or engravers: the Chicago Pneumatic Airscribe, the Ingersoll Rand EP50 Air Engraving Pen, and the AERO. All of these use a carbide stylus ground to a conical point that reciprocates at a very high rate, averaging 38,000 moves per minute. The force of this impact is adjustable and the air exits from the working end of the tool, clearing debris from around the stylus as it is used. Special modifications can be made to the Chicago Pneumatic tools to make them more suitable for preparation work.

Electric engravers

A large number of engraving tools are on the market. Of these, the Burless Vibrograver and the Dremel engraver are the two most applicable because they have a variable adjustment mechanism, and the bit (stylus) can be easily changed.

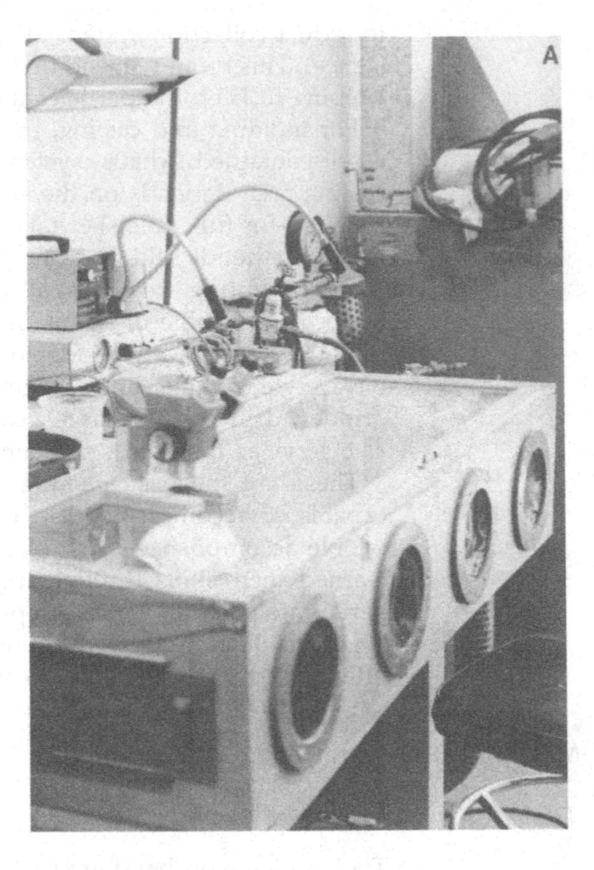

Figure 6.1. Equipment and tools in a vertebrate paleontology laboratory. Air abrasive unit and abrasives, dental moter, work unit with dust collector, microscope, fiber optic lamp, 10 power head piece magnifier.

Figure 6.1A.

These tools are also reciprocal, but the frequency and power of the strokes are not as great as with pneumatic engravers, and they have no built-in chip blower.

Sandblasters

The principal instruments used in micropreparation are the various models of the SS White air abrasive machine. These machines deliver an airstream of variable velocity carrying abrasive particles of varying hardness through a wide variety of nozzles. By adjusting these three variables (airflow, type of abrasive, and nozzle configuration), it is possible to arrive at the right combination needed for the job at hand. However, these machines are expensive and require both a blast cabinet and a dust recovery system. Sandblasters are also most likely to cause damage to specimens, other equipment, and personnel through careless use.

Rotary tools

Rotary tools are available in great variety and come in three basic designs: flexible shaft driven by an electric motor, hand-held electric motor, and hand-held

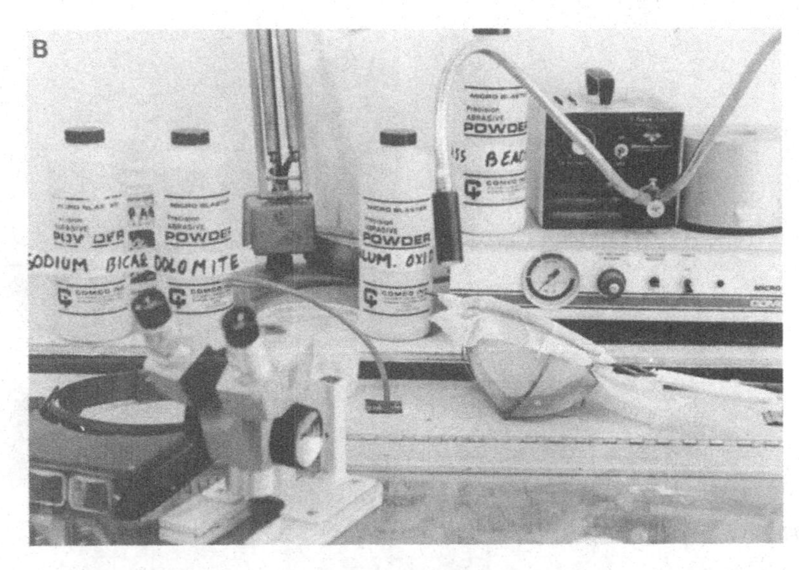

Figure 6.1B. (Caption on facing page)

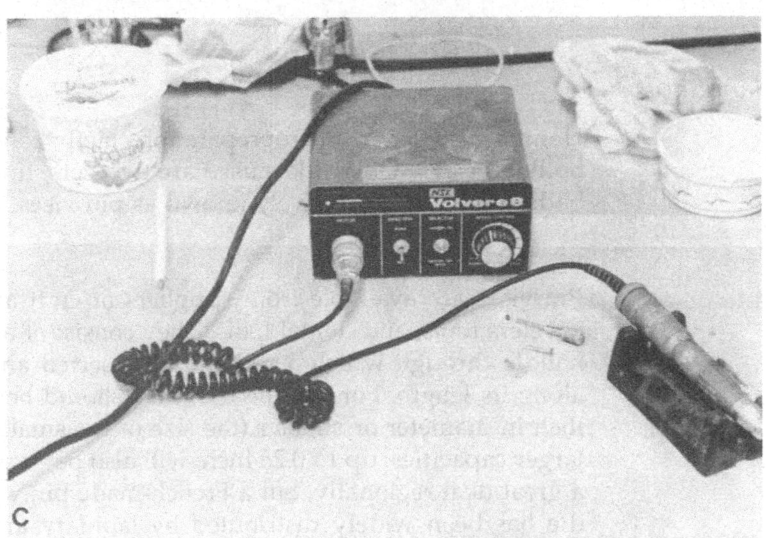

Figure 6.1C. (Caption on facing page)

pneumatic turbine. All are characterized by a small hand piece containing a collet, keyed chuck, or snap-in holder to secure the shaft of small burrs, bits, or grinding stones. A dental drill is a familiar example of this sort of unit, and a typical preparation laboratory example can be seen in any of the several models of flexible shaft tools made by the Foredom Company.

Grinders

A small, stationary bench grinder is needed to sharpen the needles employed in hand preparation. The Foredom Company makes a very useful miniature grinder that can be fitted with a stone on one end and a rubber polishing wheel on the other. A slightly larger unit, able to mount ''greenstones'' capable of grinding carbide, will also be needed.

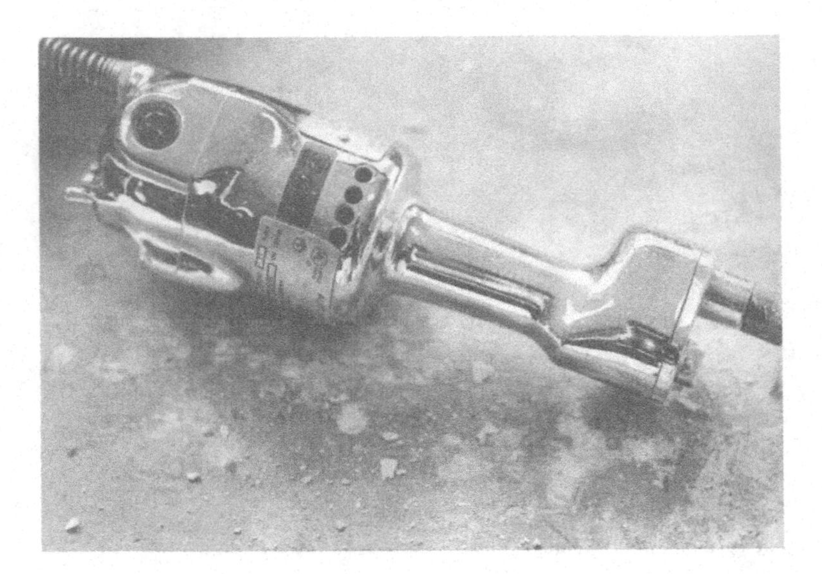

Figure 6.2. Cast cutter.

Hand tools

Hand tools used in micropreparation chiefly consist of infinite variations on the bodkin or awl. Other tools used are tweezers, tiny brushes, and various types of bodkin handles collectively known as pin vises.

Pin vises

Pin vises are available from suppliers of craft and hobby tools, machine tools, jewelers tools, and dental tools. They consist of a collet chuck joined to a hollow handle through which a rod can be inserted and gripped tightly at any point along its length. For routine use, they should be able to tightly grip a rod 0.022 inch in diameter or smaller (the size of the smallest sewing needles). Some with larger capacities, up to 0.25 inch, will also be occasionally required. Sources vary a great deal regionally, but a French-made pin vise with a striated wooden handle has been widely distributed by lapidary and jewelry supply houses. It is inexpensive and well suited for micropreparation.

Needles

The term refers to any thin metallic rod held by a pin vise and used to remove matrix. Carbide rod for this purpose is available from machinist supply houses and is relatively expensive. Various grades of small diameter drill rod are available from the same sources but are less expensive. The most practical, least expensive, and most thoroughly usable micropreparation needles are sewing needles in all sizes, which can be had in any variety store.

Dental picks

A quick look through the pages of a dental supply catalog will show an immense selection of picks. These can be considered needles with their own handles. A representative sampling of the short, stubby types is useful, but the long, thin, springy ones should be avoided because they are very difficult to control. The quality of top-of-the-line dental hand tools made before World War II is unexcelled. Should they become available, all reasonable effort should be made to obtain these.

Tweezers

Tweezers (forceps) are an indispensable part of the tool kit. Fine-pointed steel tweezers are available from medical or jewelry suppliers, hobby and craft stores, and laboratory supply companies. Those made of nonferrous metals or plastic should be avoided because they lack precision and durability.

Brushes

Obtain several artist brushes of the smallest pattern, with the smallest diameter bristles possible. These are best purchased at an artist supply store where the required type is sure to be found.

Miscellaneous tools

Some of these tools may seem unconnected to preparation, but they do have application in specific circumstances. A soldering jig, or "third hand," consisting of a weighted base and one or more alligator clips can be found at electronic or jewelry supply houses. A Water Pik dental hygiene appliance is available from drugstores, and a child's toy water pistol can be had at any variety store. Old mimeograph styluses are also handy. Ordinary erasers, corks, fish aquarium air-pump tubing, and feathers are some odds and ends that have proved useful.

Optics

Magnifying devices consist of three kinds: magnifying glasses in a frame (frequently incorporating a lamp), loupes of various types (sharing the common feature of being worn on the user's head), and microscopes. Of the three, the use of loupes is most limited because at higher magnifications it is difficult to keep the head constantly at the exact distance needed to bring things into focus. However, they are very useful when one needs to lean over a large specimen to examine a particular area in detail. Stationary magnifying glasses with a large surface area see a great deal of use when hand-held specimens are being prepared with power tools because they exaggerate the differences in surface height, making it easier to work carefully, and they serve as a shield from flying debris. People vary greatly in their preference for these low-level magnifiers and some are unable to use one or the other, so access to a variety of types is desirable.

Microscopes

A microscope is the main optical tool in this line of work. A binocular model of the sort known as a dissecting or industrial is preferred rather than the medical type because the area between the objective lens(es) and the base must be free of such obstructions as slide stages. Fossils of various sizes must be able to fit into this space. Accordingly, the gap between the objective lens and the specimen, known as the working distance, should be as great as possible. Tools, fingers, lights, and the fossil itself all have to be maneuvered in this small arena.

The working distance should be at least 3 inches. The problem of cramped working room (not to be confused with working distance) is mainly solved by various sorts of boom stands that allow the microscope to be moved horizontally 30 inches or more and vertically 15 inches or so to clear large blocks of matrix. A certain loss of stability is the price that must be paid for the added mobility; and consequently the boom stand is not suitable for very delicate work performed at more than 25×. However, several major manufacturers offer microscope optical assemblies that can be conveniently moved between stands. This allows a single microscope to be used in several different situations.

Quite a few types of microscope optical systems are now on the market. The sort best suited to micropreparation features a continuously focusable zoom. The most important optical quality is depth of focus. Because the usable range of magnification is 10× to 60× or slightly beyond, systems of decent quality would normally require two sets of eyepieces (ocular lenses) to cover this range of magnification. Most zoom systems focus both ocular lenses through a single objective lens. The system that provides the optimum three-dimensional image uses individual eye tubes, each with its own objective lens. The more expensive and durable microscopes provide a glass lens cap to seal the objective lens end from dust and dirt. At this writing, the average price of a suitable microscope is $1000.

Microscope maintenance

As a rule, the paleontology preparation lab is a hostile environment for microscopes. Abrasive dust that may scratch lenses or cause premature failure of the adjusting mechanism is produced close by. Dust should be removed with a soft brush and blown off the lenses with an aspirator. Never use compressed air because it will force grit inside the microscope and scratch lenses with incidental particles. The work area should be vacuumed at regular intervals so that dust buildup is kept to a minimum. Also, other preparation activities should be separated from the micropreparation area.

It is possible to clean the microscopes too often in this situation. Given the presence of ambient grit, every cleaning brings abrasive particles in forceful contact with the instrument, so cleaning should occur only when dirt is noticeable on the lenses. After all dust is brushed from the glass, use lens paper moistened with distilled water or alcohol to thoroughly clean the lens. These suggestions apply particularly to microscopes used with air abrasive machines. With a little common sense, a good microscope will remain functional for decades.

Microscope illumination

The lights needed for micropreparation are purchased as separate units. There are two useful types. One consists of a transformer and two high-intensity, focusable lamps on adjustable goosenecks. This arrangement affords control of brightness and intensity, and accommodates the constant reaiming necessary to eliminate shadows from fingers, hands, and projections of the specimen itself. The second type of light is a ring illuminator consisting of a fluorescent bulb completely circling the objective lens. It produces a very even light, eliminating shadows from fingers and tools. Together, these provide the optimum combination. If forced to choose one over the other, the dual gooseneck lamp is preferable because it can be adjusted for low-angle aiming. Both configurations are available in fiber optic models, which are noticeably more comfortable to work with because they don't produce heat. This is a significant factor in working on delicate fossils and justifies the higher cost of fiber optics.

Initial assessment

Upon their arrival in the laboratory, most paleontological specimens are well wrapped in plaster jackets. These specimens could have been very recently collected on the latest field trip or may have come from storage, having been collected many years ago. The first thing the preparator must do is gather as much information as possible about the project on hand. Usually the place to start is the field notes (see Figure 6.3a, 6.3b, 6.3c, and 6.3d) made when the specimen was collected. Field notes can be a wealth of information – or they may not exist at all. If you are fortunate enough to have a record of when, where, and how a specimen was collected, it will add immensely to the information you have before preparation begins. If the locality is given, you will have a good idea of the type of matrix you have to deal with. If the notes are thorough, they will also tell you if the matrix is solid or very friable. Field notes may also include a drawing of the specimen and how it was collected, with all blocks being numbered. Thus the block to be opened is well described, possibly down to the individual elements contained in this particular block.

Procedure

Until now, the process has been concerned with assessment and preparation for the job at hand. Once the backside of the jacket is off and the block is stabilized, the specimen is now ready for the rigors of preparation. After you research the available information about the contents of the plaster block, it is time to remove the protective plaster jacket. When a specimen is collected in the field, the procedure usually ends up with a block of matrix wrapped thoroughly in plaster bandages. The last plastering to be done is usually a cap that encases the bottom of the block; its edges are visible all around the side of the jacket. This will be the first section of plaster and burlap to be removed.

To remove the cap, work the edge of the plaster/burlap to see if it is removable by hand or needs a strong plaster knife or drywall saw. If it is difficult, a cast cutting saw (see Figure 6.2) is the ideal tool for the job: It does not create a lot of dust and is very gentle, as the blades do not rotate at high rpm but instead reciprocate quickly through the plaster. Cast cutting saws can be equipped with diamond blades, and because they originally were a medical tool, a dust collection system can be easily adapted. If the work is being done on an unfamiliar jacket – for example, an old one from the warehouse with no discernible number and with no information available – always open the jacket from the backside, where the cap should be (the backside should be very deep in the matrix, without any bone visible). The reason for this is because it has been found that sometimes if a specimen is very fragile when found in the field, or if the matrix is unstable, the collectors have been known to pour plaster directly onto the exposed specimen, which thus becomes part of the block. If this side is opened first the specimen embedded in the stabilizing plaster comes away with the jacket, resulting

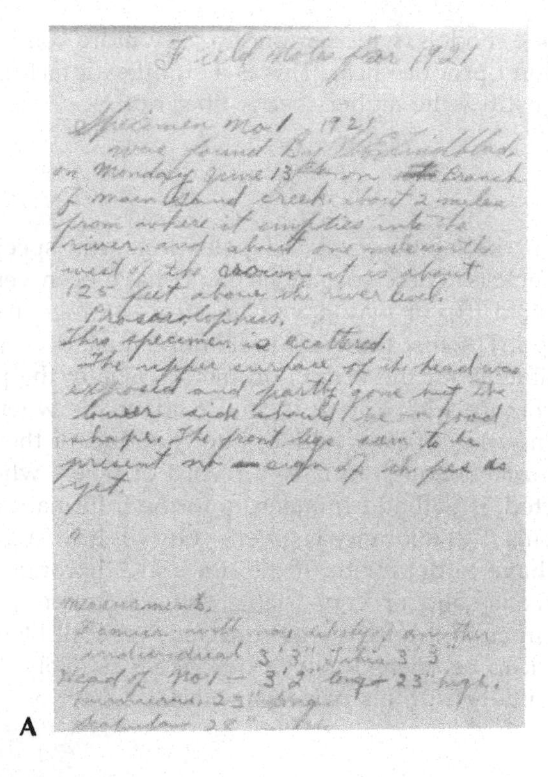

Field Note for 1921

Specimen No 1 1921
were found by ... on Monday June 13th on ... Branch of ... creek. About 2 miles from where it empties into the river. and about one mile ... west of the ... it is about 125 feet above the river bed.
Prosaurolophus.
This specimen is scattered.
The upper surface of the head was exposed and partly gone but the lower side should be in good shape. The front legs seem to be present no sign of the feet as yet.

measurements.
Femur with ... another individual 3'3" Tibia 3'3"
Head of No 1 — 3'2" long 23" high.

A

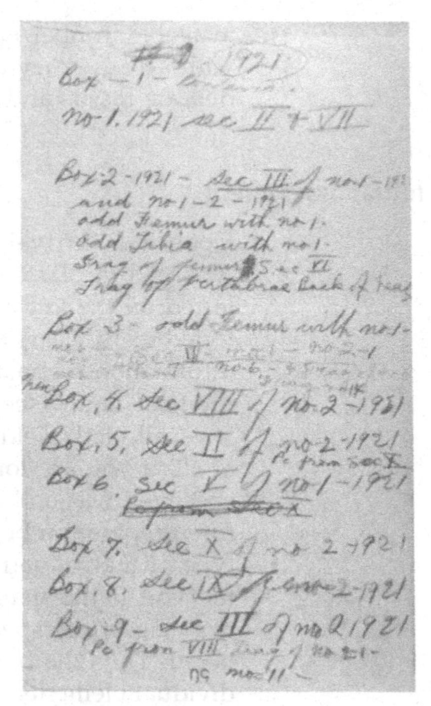

1921
Box — 1 — ...
No 1. 1921 sec II + VII

Box 2 - 1921 - sec III of no 1 - 1921 and no 1 - 2 - 1921
odd Femur with no 1.
odd Tibia with no 1.
Frag of femur Sec XI
Frag of vertebrae back of head

Box 3 - odd Femur with no 1-

Box. 4. Sec VIII of no 2 - 1921
Box. 5. Sec II of no 2 - 1921
Box 6. sec V of no 1 - 1921
Box 7. sec X of no 2 - 1921
Box. 8. sec IX of no 2 - 1921
Box 9 - sec III of no 2 1921

B

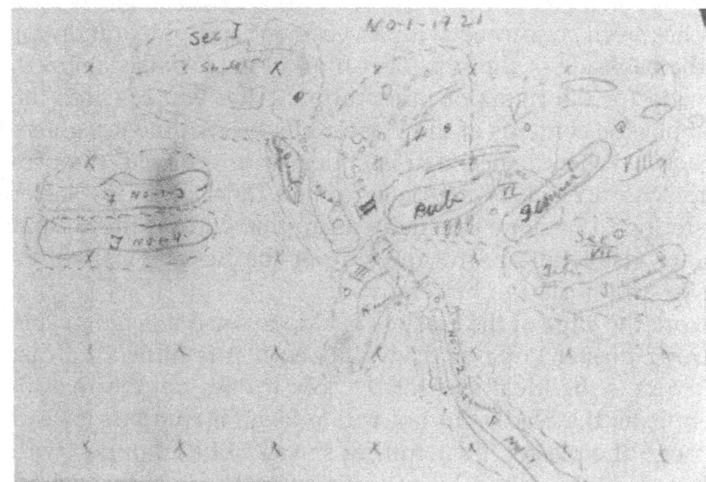

C

D

Figure 6.3. Initial assessment. (A) Sternberg's field notes from 1921. (B) Field description. (C) Drawing showing numbering system and skeleton in situ. (D) Block contents and box contents.

in breakage of the specimen. The damage to the specimen will depend on how much of the bone is in plaster and how quickly the jacket is removed. Such a situation in the hands of an inexperienced preparator could be tragic. Every section of plaster jacket removed from the block must be done slowly. It is impossible to tell what is under the plaster. A little caution can save a lot of work and sometimes the specimen, so always be careful.

Once the jacket is removed, it is time to assess the matrix and if possible the specimen (see Figures 6.4a, and 6.4b). The amount and type of matrix will determine which tools are required. If a large amount of a very hard, well-consolidated sandstone is to be removed, pneumatic chipping hammers alone or with angle grinders are very efficient. The angle grinders can be used to grind channels as thick as the grinding disk and in crosshatching the matrix the ensuing squares can then be knocked off with the chipping hammers. This is a very quick way to remove a lot of matrix, but always make sure that the dust is well controlled because the ground sandstone contains silica, which is known to cause silicosis (stone lung). (In Canada this is a controlled substance, and certain ventilation requirements and personal protection must be in place before preparation of any specimens embedded in sandstone is begun.) The rough preparation is usually done well away from any clean areas, in, if possible isolation in a soundproof room equipped with air lines, electricity, and the recommended exhaust requirements. If you are working with a very large heavy block, make sure it is on a sturdy table equipped with casters.

Another piece of equipment that can be used for quick matrix removal is an industrial sandblasting unit that comes with an enclosed cabinet. However, extreme caution and an experienced hand are required because control is difficult, and by the time exposed bone is spotted, the abrasive has also located it. Control of the nozzle can be difficult and the viewing port is not suited to fine preparation. Still, in properly trained, careful hands sandblasting can do a quick, efficient job, with the specimen being subjected to a minimal amount of vibration. This coarse preparation should bring the matrix down to within 2.5 cm of the bone. From this point on finer preparation tools come into play. Proceed with the utmost care; whenever bone is exposed, stop working and decide whether consolidants are needed.

If a specimen is embedded in a block that is very unstable, the pneumatic hammers can do a lot of damage to the specimen while the matrix is being removed. In cases of large cracks running through the block and also cutting through the bone, the vibration can actually cause the bone to fragment and the ensuing pieces become lost in the fissures. In such instances the entire block should be consolidated so that safe removal of the matrix can proceed with minimal damage to the specimen. If large open cracks are present it is best to stabilize the block by pouring a thin plaster mix down them. Melted Carbowax has also been used and permitted to harden. This method will entail a little more work during preparation but it does ensure that the specimen, which is also fractured, has no chance of disintegrating during preparation. The fit between the cracked bone will be infilled, but both the plaster and the Carbowax are easily removed, the plaster by air abrasive or mechanical means, the Carbowax by melting. In the case of Carbowax the wax has to be completely removed or the glue will not bond because the wax will act as a separator between the

Figure 6.4. Opened blocks. (A) Note fractures in matrix and loose material on top of main block. (B) This specimen has been embedded in plaster due to very friable matrix.

adhesive and the specimen. There are other methods to deal with this problem. What matters is to strengthen the areas around the cracks to ensure that the bone does not shatter at these weak spots and that the fit between the broken pieces is as tight as possible. The procedure to follow will always be dictated by the collection method and by the type of matrix that surrounds the specimen.

The chipping hammers and guns can only be used on well consolidated blocks. Extreme care must be taken to ensure that the vibrations produced by these tools do not shatter any bone. Be sure the blocks are well stabilized so that no damage is transmitted to the fossil just to expedite the job. When the first signs of bone are apparent, use the larger tools only as long as it can be done confidently with no damage to the specimen. It may take a little longer with smaller tools, but the safety of the fossil is of utmost importance.

Now the finer preparation of the specimen can begin. At this point all exposed bone should have at least one application of consolidant; a few different products can be used (VINAC, Butvar, etc,.). For adhesives to have on hand, refer to Chapter 2. The preparator should be equipped with the fine preparation tools that were described earlier above, (Airscribe, air abrasive unit, small rotary grinder, dental picks, etc.).

The preparation always proceeds from the coarser work down to finer procedures. In the case of specimens embedded in sandstone, the majority of the work can be carried out with the Airscribe. In most cases the bond between the matrix and the bone is easily separated by the vibration of the tool. On smooth long bones, such as femora and humeri, the Airscribe can be used for all of the cleaning. If the separation between bone and matrix is not very good, remove as much matrix as possible, coming as close to the bone as is comfortable. If possible, use the air abrasive unit to sandblast the remaining matrix away or use hand tools to finish the job. The air abrasive unit is very good at removing matrix, and the control is excellent between the amount of air pressure and the type of abrasive. It can be used for fairly coarse matrix removal down to some fairly fine preparation. The air abrasive unit, with a little bit of experimentation, can succeed where a lot of other tools fail. It can be very gentle on the bone because it relays no vibration at all to the specimen. Because of this type of control, the air abrasive unit is an invaluable tool in any preparation lab.

Until now, all of the work has been performed on one side of the block. The specimen will have to be turned over in order to prepare the other side. But before the specimen or block can be turned over, other procedures will require attention. The prepared portion of the specimen should be completely mapped, photographs taken, measurements made, and all notes or references (Figures 6.5a and 6.5b) should be recorded. If the block is fairly large with an associated semi-articulated specimen, it may be unwarranted to separate the elements. Every element should now be on a pedestal of matrix, and a minimal amount of matrix is left in the block. Ensure that every bit of exposed bone is well consolidated. At this time it will be evident if the specimen was collected by pouring plaster onto the exposed bone; if not, the paper towel separator that was used in the field should be visible throughout the exposed field jacket.

Now another plaster jacket is going to be applied to the side that has been prepared. A separator has to be applied over the specimen, matrix, and exposed sections of the original plaster jacket. Any potential undercuts should be filled

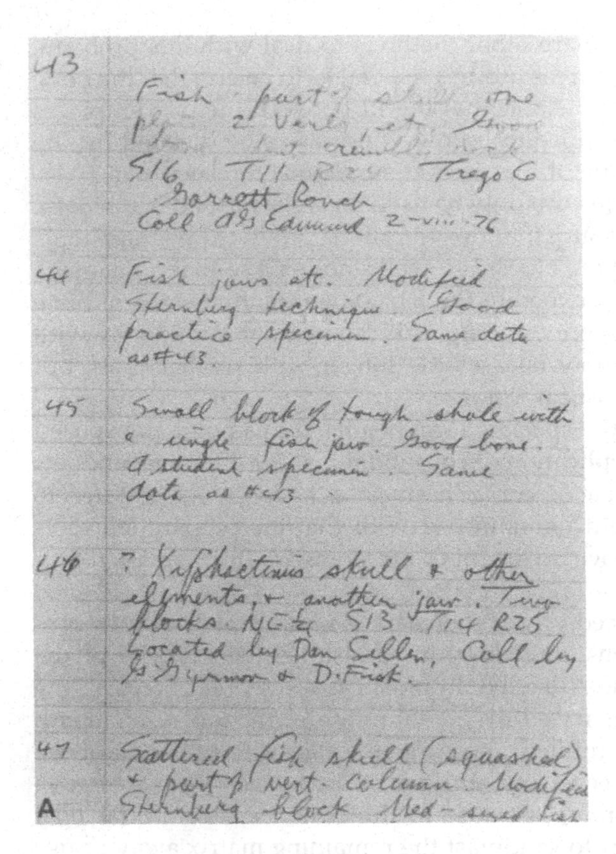

43 Fish part of skull [illegible]
[illegible] 2 Vert [illegible]
[illegible]
S16 T11 R25 Trego Co
Garrett Ranch
Coll GS Edmund 2-viii-76

44 Fish jaws etc. Modified Sternberg technique. Good practice specimen. Same data as #43.

45 Small block of tough shale with a single fish jaw. Good bone. A student specimen. Same data as #43.

46 ? Xiphactinus skull & other elements, & another jaw. Two blocks NE¼ S13 T14 R25 Located by Dan [illegible], Coll by [illegible] & D. Fisk.

47 Scattered fish skull (squashed) & part of vert. column. Modified Sternberg block Med-sized fish

Figure 6.5. (A) Edmund's notes on field number 44. (B) Fish embedded in plaster using modified Sternberg technique.

with wet paper towels so that the new plaster jacket does not wrap around the prepared elements. If the ultimate goal is to remove all elements from the jacket, this will facilitate the removal. Apply a layer of wet paper towels over the entire exposed area, then do it again. In all, three layers should suffice. Over this the plaster jacket is applied using plaster bandages. Cut strips of burlap 7 to 10 cm wide and roll them through a dilute plaster mix. Then apply them to the block over the wet paper towels. The size of the block will dictate how many layers to apply. Be sure it is strong enough to support itself; if necessary, reinforce it strategically with wood strips. When the jacket has cured, apply webbing clamps around the two blocks, making sure the two blocks are firmly secured together. If either block moves independently of the other, breakage is bound to occur. Once they are firmly clamped, flip the block over.

Now it is time to remove the original field jacket and finish the preparation. Refer to your drawings to ensure that the positioning of the elements is clear and slowly proceed to remove the jacket. First, use the cast cutter to score the jacket so that it will come off in small sections; do not attempt to pull the entire jacket off at once. When the jacket is completely removed, the exposed surface is what was exposed in the field. Now it is time to finish the preparation of the specimen in the same manner as the backside. Before any elements are removed completely from the block, be sure that the researcher has finished studying it in the block – it must be mapped, photographed, and measured. As the individual elements are removed, make sure they are numbered and cataloged and put into the collections. All information should be documented in a lab notebook and a specimen data sheet filled out.

Marine deposits

Specimens found in marine deposits are usually deposited in a completely different manner and the method of preparation varies slightly from nonmarine deposits. Specimens from chalk beds can be fairly easy to prepare by using an air abrasive unit with a gentle abrasive powder such as a fine-grain dolomite. The matrix is much softer than the bone, so it can be rapidly removed; it separates easily from the specimen, giving excellent results. Again, be sure to apply consolidant to the specimen as the fossil bone is exposed. One concern is to make sure that the block is well supported. The matrix does not supply the same strength as sandstone, and usually the collection techniques leave a very thin block because sedimentary layers are stripped off sometimes to the same level as the specimen. Be very careful when opening these blocks. In many cases plaster has been poured directly onto the specimen, especially in the cases of exposed disarticulated marine specimens (this usually ensures that all data are collected from specimens that are barely collectable).

Some marine deposits can be very difficult to work with, for the matrix can be much harder than the fossil bone. This makes for not only extremely difficult collecting but also challenging preparation. In the case of marine deposits from the late Triassic, the matrix can be a very hard oil-bearing sedimentary limestone. To be collected it requires feathers and wedges to break the blocks out of the field, and to establish the specimen's dimensions is usually a matter of taking out more than necessary to ensure that every bit of the animal is collected. In the lab the same procedure is adapted for preparation.

The blocks have to be very well documented in the field and the size of most of them make them incredibly unwieldy in the lab. The first step is to cut the blocks down to manageable size. This is accomplished by carefully removing all loose pieces and mapping every piece's location, numbering them, and checking for bone. Every crack in the blocks is used as a natural break. These are carefully separated, and all loose pieces collected, numbered, and mapped. A Polaroid camera is a very useful tool to keep an accurate record during preparation. Once the blocks are at a manageable size, they are immersed in 10% acetic acid for a few hours, to establish the stratigraphic layers in the individual sections. The blocks are then cut down along the stratigraphic layers until the matrix is within 2.5 cm of the specimen. This is accomplished by manufacturing small-scale feathers and wedges for use in the lab in the same but scaled down procedure that was used in the field.

Once the blocks are cut down to size, the preparation can begin in earnest. The technique deemed best is an acetic acid, air abrasive and mechanical combination. The matrix is too hard for purely mechanical means and an air scribe cannot remove any of the matrix efficiently, so 4-inch angle grinders are used to rapidly remove thick areas of matrix; thinner areas of matrix are reduced by a dental tool. Once within a centimeter of the bone, it is immersed in tubs of acetic acid. Polybutylmethacrylate is used to protect the fossil in the acid (see Rixon,

Lindsay) and a rubber latex/fiberglass jacket applied to the back for support and protection. Immersion in the acetic acid is for 3 to 8 hours maximum. The specimens are then washed for the same amount of time and dried and then the air abrasive unit is used to remove the weakened matrix. Any exposed bone is painted with polybutylmethacrylate and the process repeated until all matrix is removed from one side. The majority of the specimens only require preparation on the one side; to prepare both sides the techniques discussed in Chapter 7 are used. Once the specimen is cleaned of the bulk of the matrix, fine preparation techniques are incorporated.

When the specimen is completely finished, notebooks should be brought up to date and filed with any drawings and photographs, and a specimen data sheet is filled out. It is always important to keep in mind that the specimens that are prepared will be housed in the collections forever. Thus all information associated with the specimen will be very important to anyone who decides to work on it in the future.

Specimen data sheets

Making notes and keeping records of the specimens worked on in the lab are very important procedures. This information should be available to any researcher who works on the specimens. There are many instances where a specimen has been described as a fully articulated skeleton only to find many years later that it is actually a composite made up of the postcranial skeleton from one animal and the skull of another stuck on for display purposes. Over time these reconstructions are forgotten and can result in an incredible amount of scientific research going for nought, not to mention the possible embarrassment. Not many field crews come away from the field without some record of what was collected and valuable information on locality, geology and specimen data, yet when it comes to preparation of the specimen, at times information can be totally lacking.

The easiest way to provide a minimal amount of information about the specimens is by filling out a specimen data sheet that can be standardized for departmental use. A few good articles have been written on this subject and a sample is shown in Seymour, 1990. Now that computers are evident in most preparation labs it is just a matter of inputting the data at a time that is convenient. In this day and age there is no excuse for sloppy record keeping.

Bibliography

Chaney, D. S. 1989. Hand-held mechanical preparation tools, (pp. 186–203). In: R. M. Feldmann, R. E. Chapman, and J. T. Hannibal (Eds.), *Paleotechniques*. Paleontological Society Special Publication No. 4.

Converse, H. H. 1984. *Handbook of paleo-preparation techniques*. Gainesville: Florida State Museum.

Fitzgerald, G. R. 1988. Documentation guidelines for the preparation and conservation of paleontological and geological specimens. *Collection Forum* 4(2):38–45.

Hannibal, J. T. 1989. The air-abrasive technique (pp. 186–203). In: R. M. Feldmann, R. E. Chapman and J. T. Hannibal (Eds.), *Paleotechniques*. Paleontological Society Special Publication No. 4.

Howie, F. M. P. 1987. Safety considerations for the geological conservator. In P. R. Crowther and C. J. Collins (Eds.), The conservation of geological material. *Geological Curator* 4(7):455–460.

Lindsay, W. 1987. The acid technique in vertebrate palaeontology: A review. In P. R. Crowther and C. J. Collins (Eds.), The conservation of geological material. *Geological Curator* 4(7):483–488.

Rixon, A. E. 1976. *Fossil animal remains: Their preparation and conservation.* London: Athlone/University of London.

Seymour, K. 1990. Computerized specimen and preparation/conservation worksheets for fossil vertebrates. *Collection Forum* 6(2):46–50.

Stucker, G. F., M. J. Galusha, and M. C. McKenna. 1965. Removing matrix from fossils by miniature sandblasting (pp. 273–275). In B. Kummel and D. Raup (Eds.), *Handbook of paleontological techniques.* San Francisco: Freeman.

Whybrow, P. J., and W. Lindsay. 1990. Preparation of macrofossils (pp. 499–502). In D. E. G. Briggs and P.R. Crowther (Eds.), *Palaeobiology: A synthesis.* Oxford: Blackwell.

Wilson, R. L. 1965. Techniques and materials used in the preparation of vertebrate fossils. *Curator* 8(2):135–143.

Microscopic preparation

William W. Amaral

The previous section dealt with setting up a preparation lab for large vertebrate fossils. Here we discuss establishing a workstation suitable for the preparation of small vertebrate fossils and discuss the basic methods of micropreparation. The last 20 years have seen great advances in glues, lighting, and microscopes that greatly facilitate the process of mechanically developing small fossils. Today, with the proper equipment and techniques it is possible to extract exceedingly small specimens; teeth as small as 100 microns at the base and 150 microns high can be prepared (see Figures 6.6 and 6.9).

The work station

Although preparation labs are generally dusty, cluttered, and poorly illuminated, the area set aside for micropreparation should be clean, uncluttered, and well lit. Cleanliness at the workstation is vital, for it is inevitable that important portions of a fossil undergoing preparation will break away and may become permanently lost if excess clutter prevents retrieval. Sufficient electrical outlets should be readily available, and a high pressure air line capable of delivering a constant 90 to 100 psi should also be close at hand.

Once a location has been selected, a workstation needs to be appropriately outfitted. Ideally, the workbench should have a hard, smooth surface (white Formica works well) with no cracks or seams into which pieces of the fossil being prepared can disappear. Likewise a backboard fitted to the back of the bench will prevent broken pieces from escaping over the edge and becoming permanently lost. Sufficient drawer space for tools and fossils should be under the

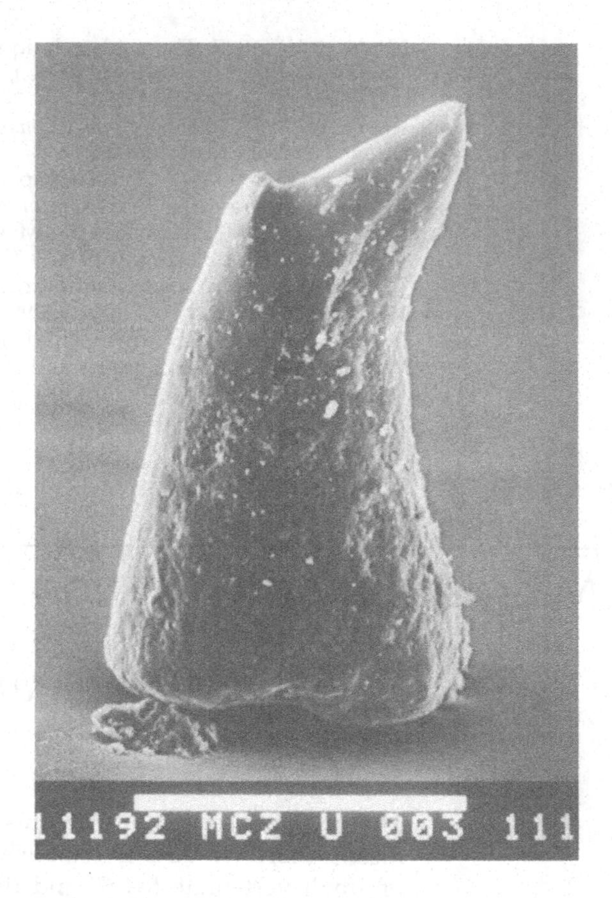

Figure 6.6. An early Jurassic caecilian tooth, prepared and extracted from a siltstone matrix. The scale bar represents 100 microns.

desktop. Consider splitting the high-pressure air line into two lines: one fitted with a regulator and two or three outlets (useful for low-pressure applications) and the other fitted with a filter, regulator, and lubricator and as many outlets as desired (for high-pressure applications). An Airscribe and various air driven grinders run off the high pressure line, and airhoses and most other tools use the low-pressure line. For a right-handed preparator, all the air outlets should be on the right side of the desk.

The microscope

Today's high-quality stereo microscopes offer excellent optics and enable one to work all day without eye fatigue. Stereo microscopes permit each eye to see an object from a slightly different perspective; the two images are then merged in the brain as a perception of depth. This stereo effect varies from microscope to microscope. Because not all models or brands provide an equal sense of depth, procure a microscope that enhances your sense of depth. An easy way to check the stereo effect is to get two, three, or more microscopes produced by different manufacturers together. Set all the microscopes to the same magnification and then go from one to another looking at the same area on a coin. The results can

Figure 6.7. An extended arm mounted to the wall. The fiber optic system is mounted above the arm with a flexible fiber optic cable running the entire length of the arm ending at its attachment point on the microscope assembly. The black line is a low-pressure air line terminating in a pair of hollow copper tubes from which a gentle airstream can be directed toward the area where the microscope's focal length and the lights converge. During preparation, specimens rest on sandbags, which in this case are filled with steel shot.

be surprising. It makes little sense for a preparator to labor with a low-quality microscope while a researcher reviews the work with a higher quality microscope. Although microscopes such as those manufactured by Wild and Zeiss have given me good service in the past, my current microscope is a Nikon SMZ 2B.

The microscope should be mounted on as long a boom arm as feasible. Ideally, the boom arm can be attached to a wall behind the desk (see Figure 6.7). Boom arms with heavy counterweights at their base should be placed as far from the working area as possible.

As the microscope is a delicate instrument, it should be kept as clean as possible; periodically blow accumulated dust off the microscope with an airhose; and when not in actual use, the microscope should be kept covered. Eventually the lenses will require a thorough cleaning. Do not use lens paper or anything else on a dry lens; rather, wash the lens under running water (with soap if necessary) and dry thoroughly with lens paper.

Lighting

As with money, the micropreparator can never have too much light. Improving the illumination system is the easiest and least costly way to upgrade marginal microscopes. Even a low-quality microscope benefits immensely from updating the lighting system. Poor lighting contributes greatly to eye fatigue, and if the lighting is not adequate the entire optical system is compromised. Modern fiber optic lights provide more than enough light but until recently had the disadvantage of requiring the illumination source to be on the actual work surface. Typically, two fiber optic cables emerge from the illumination source and are focused so that the separate lights merge at the focal length of the microscope. The disadvantage of this system is that, should the microscope be raised or lowered, the lights will no longer be focused at the focal length of the microscope and will have to be repositioned. Repeated repositioning of the fiber optic cables is an irksome and time-consuming task. However, with a little ingenuity, fiber optic lights can be grafted to the microscope so that the two cables ride in tandem with the microscope as it is raised or lowered, and the two lights always converge at the focal length of the microscope. Here is an example of how to mount fiber optics on a Nikon SMZ 2B.

Using a Dolan-Jenner System 175–1, part no. EEG 3772 (the last two digits refer to the total length of the fiber optic cable in inches), the illumination source is mounted on a wall just above the boom arm. A single flexible cable emerges from the illumination source and is then attached to the boom arm either with tape or Velcro straps. Running the entire length of the boom arm, the flexible cable bifurcates in the last 19 inches at a Y-splitter. The Y-splitter comes equipped with a mounting hole, which can be enlarged by drilling so that a quarter-inch threaded rod can be inserted into it. The other end is then attached to the microscope assembly. The Y-splitter should be about 12 inches above the microscope so that the two branches are in the recommended down-slope position (see Figure 6.7). Although the maze of twisted cables may look awkward, in practice the system works quite well and is capable of delivering 50,000 footcandles at 3 inches (Dolan-Jenner Fiber-Lite Catalog FM 103).

Tools

Airscribe

The Airscribe, available from Chicago Pneumatic, is a versatile miniature jackhammer; when fitted with a muffler and an extended carbide tipped point, it is capable of surprisingly delicate work. When using an Airscribe, it is advisable to fit a screw-in cover glass to the bottom of the microscope to prevent flying rock chips from striking the lower objective and possibly damaging it. As each fossil is different, you will have to determine whether matrix can be removed directly from bone. If this produces poor results, use of the Airscribe should be limited to bulk removal of the surrounding matrix with no attempt made to directly remove matrix from bone. The idea behind using an Airscribe for bulk removal is to thin the matrix quickly so that needles can be used to remove matrix directly from the specimen.

Miniature grinder

Air-driven pen-type grinders, fitted with various burrs and typically operated at 60,000 rpm (at 90 psi), are quite capable of removing unwanted matrix at a rapid rate. Grinding, of course, produces copious dust, and any grinding for more than a minute warrants the use of some form of dust extraction system. Miniature grinders are useful to thin matrix further when continued use of an air scribe is contraindicated due to excessive vibration. The grinder cuts rather than hammers against the matrix, with little vibration as a result. This allows the careful operator to remove sufficient matrix to get quite close to the bone, the concept being to grind until there is only a very thin layer of matrix remaining without touching the burr to the bone surface itself. Once enough matrix has been removed, so that only a thin cloaking film remains, it becomes a relatively simple matter to expose the bone with a needle. This technique works even with difficult matrices such as hematite.

Flex-shaft grinders

Electrically driven grinders such as those produced by Foredom and Dremel (see Sources) transmit power through a flexible shaft to a hand piece and ultimately to a rotating burr. These tools can be fitted with collets that accept the same burrs as the air driven grinders. As a backup for air-driven grinders, these grinders are capable of producing comparable results but are more tiring to use. When coupled with a small diamond wheel or disk, the flex-shaft grinder becomes an important addition to the array of tools used by the micropreparator, its most important use being as a touch-up wheel for broken carbide needles or to provide the final edge to a new carbide needle. Finally, the diamond wheel is useful in removing unwanted matrix or as a miniature cut-off wheel.

Carbide needles

Outside of the microscope and the illumination system, carbide needles are the most important and most useful tool that a micropreparator will use. By varying the diameter of the rods that are used to construct the needles, a wide assortment of points, from delicate to strong, can be had (see Figure 6.8). One of the great advantages of carbide needles (as compared to steel needles) is a greater sense of "feel," because carbide of a given diameter is more rigid than an equal diameter of carbon steel; it bends less when pressed against the matrix. This rigidity enhances the feel of the needle against the specimen. To illustrate this sense of feel, imagine that you are standing 2 meters from a murky stream and want to feel the bottom; to gather the greatest amount of information from this streambed, what type of 2-meter extension (a stout pole, a limber bamboo fly rod or a stiff rope) would be your choice?

Solid carbide rods are typically purchased in 12-inch lengths and can be obtained through supply houses that cater to the machinist trade. It is often economical to purchase two dozen or more of the rods at a time, as many supply houses give a price discount for large orders. The most useful diameters range from 1/64" for the most delicate of needles on up to 5/64" for more robust

Figure 6.8. A variety of needles produced from a carbide rod. The top needle is a typical tip of a round point; the middle and bottom show two sizes of half round points.

needles. The size depends on the job at hand. To make a carbide needle, the carbide rod is first scored with a diamond wheel, attached to the flex-shaft grinder; as in glass cutting, the rod will break when pressure is brought to bear at the score line. If the carbide rod is not first scored before being snapped, the rod will break into three or four pieces, thus wasting expensive carbide. Once a suitable length of rod is obtained, place it in a pin vise (pin vises are available in many sizes) and grind the carbide rod on a bench grinder until it assumes a needlelike shape. It is important that carbide only be ground on silicon carbide "green" wheels, as carbide is so hard that it will quickly eat up wheels meant for softer metal. One wheel should be of medium (60/80) grit, which is used for the initial shaping of the needle. The other wheel should be of fine (100/130) grit for the final shaping of the needle.

Two styles of points are commonly constructed: the round point and the half round. The round point is nothing more than a ground needle finished to a sharp point; this style is useful on hard matrixes, such as indurated sandstone or hematite. The half round is made like the round point, except that a flat facet is put into the point by holding the last half inch against the grinding wheel until a flat surface is produced. When correctly made, the half round will look much like a hypodermic needle; its advantage is a sharp point and the two sharp edges. Most micropreparation is undertaken with this style of versatile point that can remove matrix in the same manner as a round point and remove it like a plane with either of the two sharp edges. Further shaping is done under magnification using the small diamond wheel; this enables the micropreparator to produce exceedingly small, sharp points.

Carbide, though very hard, is extremely brittle. Even though a carbide point

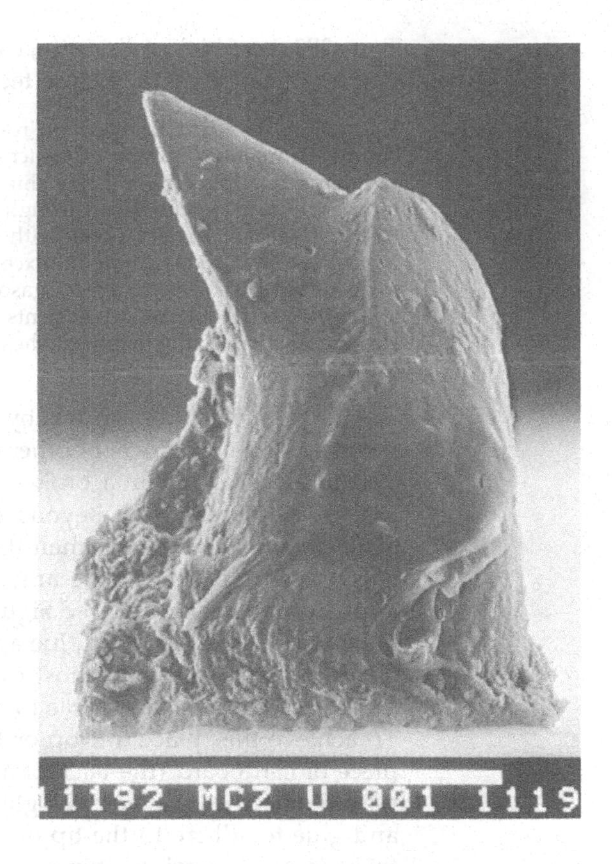

Figure 6.9. An early Jurassic caecilian tooth, prepared at ×100, illustrates the deleterious effect solvent-based glue has when viewed at high magnification. At ×500 the glue that coats the entire tooth becomes readily apparent; note the craterlike vacuities just below the secondary cusp. Scale bar equals 100 microns.

won't easily be dulled, it will break. If a pin vise with a carbide needle should inadvertently be dropped, the carbide will shatter and a new needle must be made from scratch. It is not uncommon to break the tip of a fine carbide needle a dozen or more times a day – this is to be expected! And this is why the flex-shaft grinder equipped with a small diamond wheel is so useful, for it usually takes only a few seconds to re-dress a broken tip with it.

Glues

The ability to repair a broken fossil ultimately determines how small a specimen the micropreparator can work on. Significant advances in glue technology in the past 20 years enable one to prepare and repair very small bones. Prior to the advent of epoxies and cyanoacrylates, most micropreparators relied on various solvent-based glues, with decidedly mixed results. Solvent-based glues typically have poor penetration properties and invariably leave a film of glue on the specimen; this feature is a disadvantage especially if the specimen is photographed under high magnification in a scanning electron microscope (see Figure 6.9). A good rule of thumb is to strive to get the glue *into* the specimen and not on it; thin cyanoacrylate glues, with their low viscosity, excel at this. Numerous brands of cyanoacrylates are available on the market. The MCZ lab generally uses Krazy Glue of two types; one, Ethyl Type 201, is of low viscosity, and the other, Ethyl

Type 203, is of higher viscosity, each with its own advantages and disadvantages. Quoting from the Krazy Glue technical data sheet:

> Krazy Glue is a one component, room temperature curing structural adhesive composed mainly of alpha cyanoacrylate monomers in a liquid state. When this monomer is applied in a single thin coating to the bonding surface of the material, it instantly solidifies through polymerization to produce an extra high strength, clear, colorless bond with virtually no shrinkage. Krazy Glue bonds almost any nonporous material except fluorocarbons. Resinified, Krazy Glue is not affected by solvents such as gasoline, propane, oil, alcohol, or benzene. The adhesive itself contains no solvents. The Krazy Glue bond has a high tensile strength but a relatively lower shear strength. Krazy Glue is not a gap filler. Good surface contact is essential.

Although this was written by the Krazy Glue company, it applies in equal measure to the products of other companies. It is essential to control the amount of glue used to repair a broken bone surface. When too much is applied, the excess will spread well beyond the area being repaired; this at the worst can actually damage the specimen through inadvertent breakage as the excess glue is removed. To control the amount of glue applied, use the carbide needle to apply a precise amount (see Figure 6.10) Because cyanoacrylates have a positive meniscus, a small drop of glue applied to the needle will run up the needle and away from the point. In most cases this feature is of little consequence. Sometimes, however, it is desirable to have the glue actually at the point of the needle. To achieve this, place a drop or two of low viscosity cyanoacrylate glue onto a piece of cardboard (the back of a legal pad is ideal), then scratch the cardboard with the needle tip where the glue was applied. This will cause cardboard fibers and glue to adhere to the tip of the needle (see Figure 6.11). In practice, both of these methods work well in delivering an exact amount of glue to the area to be strengthened or repaired with the latter being of greater use in applying extremely small amounts. In either case some flashing or spreading will occur; excess glue on the bone surface should be removed with a needle once it has polymerized.

Low viscosity cyanoacrylates are also useful in consolidating specimens. When carefully applied, thin cyanoacrylates will spread through cracks in the bone surface, effectively gluing incipient breaks before they can occur; as always, excess glue on the bone surface should be removed. Thicker, high viscosity cyanoacrylates are often used to repair actual breaks. One of the characteristics of the thicker cyanoacrylates is a slower polymerization rate (as opposed to the rapid polymerization rate of the thinner cyanoacrylate), and this gives the micropreparator some time to maneuver broken pieces together before polymerization occurs. If maneuvering pieces together requires more time than the polymerization process, use a two part epoxy resin. Five-minute epoxy is suitable for most applications, but 2-hour epoxy provides more working time.

Miscellaneous

Basic principles of preparation

The simple principles of preparation that follow find universal applications whether preparation is of a large fossil vertebrate or a small microvertebrate.

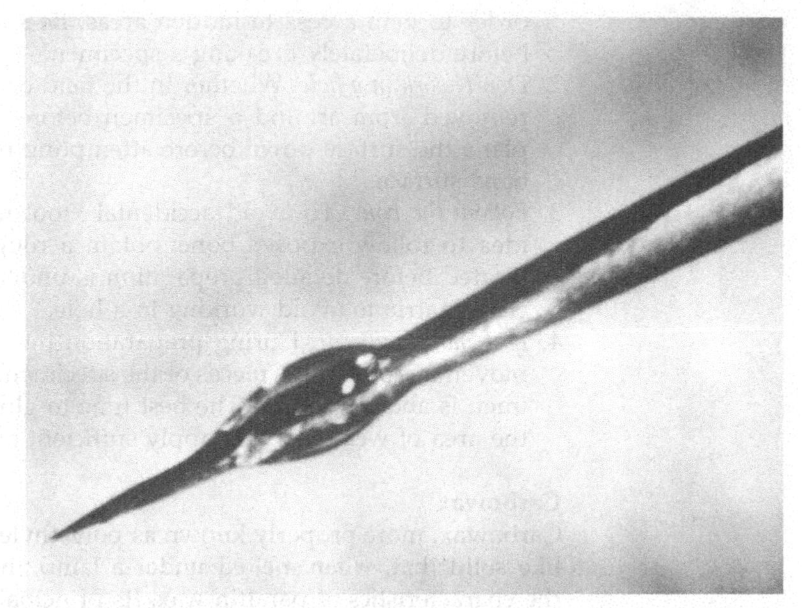

Figure 6.10. Thin viscosity cyanoacrylate glue applied to the tip of a carbide needle. The glue has run up and away from the tip of the needle, thus preventing a straightforward application.

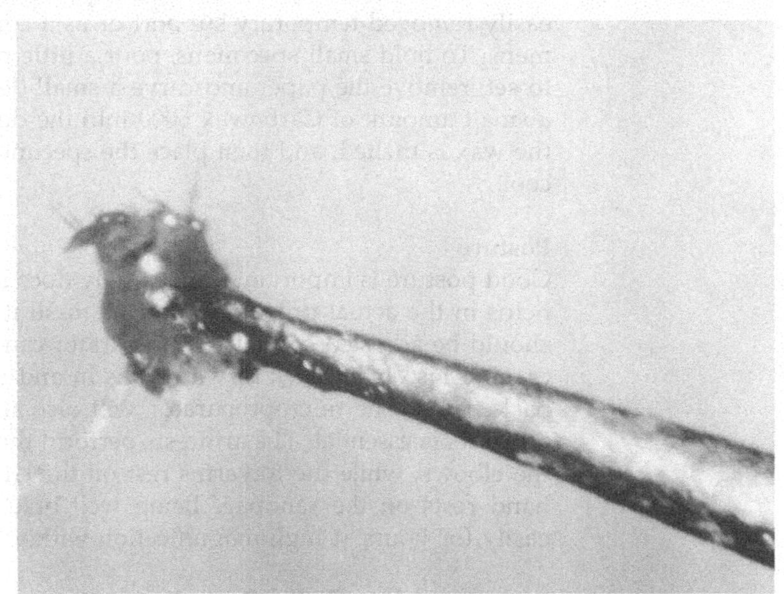

Figure 6.11. A carbide needle scratched on a cyanoacrylate saturated piece of cardboard. Precise amounts of glue can be delivered with this technique; excess glue and stray cardboard fibers are removed from the area repaired after the glue polymerizes.

1. *Take advantage of breaks.* During the course of development the micropreparator can expect that various portions of the specimen will break. The experienced preparator often finds this to be an excellent opportunity to prepare previously inaccessible areas; once prepared, the broken pieces are glued back or stored for later repair. A significant goal of any preparation is to provide the researcher with as much information as possible from the specimen; in some cases this means that portions of the specimen will be deliberately broken in

order to gain access to hidden areas. Be sure to consult with the researcher before deliberately breaking a specimen.

2. *Don't work in a hole.* Whether in the field or in the lab, overburden should be removed from around a specimen before exposing it. For the best results, plane the surface down before attempting removal of matrix from the actual bone surface.
3. *Follow the bone.* To avoid accidental "tool marks of discovery," it is a good idea to follow exposed bone; obtain a rough idea of where all the bone is located before detailed preparation is undertaken. Remember: Remove sufficient matrix to avoid working in a hole.
4. *If it moves, glue it.* During preparation the micropreparator will often notice movement of various pieces of the specimen, often an indication that the specimen is about to break. The best time to glue a break is before it occurs; find the area of weakness and apply sufficient glue to consolidate it.

Carbowax

Carbowax, more properly known as polyethylene glycol, is a water-soluble wax-like solid that, when melted under a lamp and then cooled, assumes many of the characteristics of paraffin wax. Its principal use in micropreparation is as an easily removed temporary support or as a convenient way to hold small specimens. To hold small specimens, pour a little plaster into a paper cup, allow this to set; remove the paper and carve a small depression into the plaster. Sprinkle a small amount of Carbowax 6000 into the depression, heat under a lamp until the wax is melted, and then place the specimen in the pool of wax and allow to cool.

Posture

Good posture is important, for not only does it aid in combating fatigue but also helps in the actual preparation of the fossil itself. The height of the microscope should be adjusted so that the preparator can look down the objectives without craning his or her neck forward; this in and of itself will do much to eliminate back pains. The micropreparator will also find that a comfortable chair with armrests is essential. The armrests perform the important function of supporting the elbows, while the forearms rest on the edge of the desk and the heel of the hand rests on the sandbag. Being well braced allows the preparator to work easily for hours at high magnification without tiring.

Recovering lost pieces (the value of cleanliness)

Nobody is infallible, and small pieces will inevitably be knocked off a fossil in the course of development. No matter how small these are, they can always be put back if they can be found (Rixon 1976). The odds of finding small pieces decrease proportionately with the amount of clutter and debris laying on or about the work surface. It is strongly suggested that accumulated debris be cleaned from the work area periodically. When a piece is lost, first turn off any airstream that might be directed at the specimen, carefully gather up any tools that may be on the work surface, then sweep everything that has accumulated on the desktop into a white cardboard tray. With luck, the broken piece will be in the tray with the detritus produced during preparation. The cleaner the desk, the less there will be to sort through; if the piece is very small, use the microscope at a low magnification to search it. Occasionally a broken piece will fall to the floor. If it is an extremely important piece, such as a cusp of a tooth, remain still and have a co-worker (if one is available) sweep the area under and around the desk before rolling the chair back. This minimizes the possibility of crushing the lost piece under the chair wheels or underfoot. If no co-worker is available or the lost piece is deemed only of minor importance, smoothly roll the chair back and, as before, sweep the area, placing everything collected in a white cardboard tray.

Summary

This description, though basic, assumes that the reader is familiar with the tools and techniques in general use in a preparation facility. As with many things there is no substitute for hands-on experience. Those interested in learning more about microscopic preparation are strongly advised to seek out an experienced micropreparator and work under his or her guidance until basic skills are mastered. With modern optics, materials, and illumination systems, the task of preparing microvertebrates is easier than ever – if the micropreparator has a well-grounded knowledge of the basic preparational skills required.

Sources

Airscribe
Chicago Pneumatic, 2200 Bleeker St., Utica, NY 13501 1-800-367-2442

Dotco Midget Grinder
Cooper Power Tools, PO Box 1410, 670 Industrial Dr., Lexington, SC 29071, 1-800-845-5629

Fiber Optics
Dolan-Jenner Industries, Inc., Box 1020, Blueberry Hill Industrial Park, Woburn, MA 01801, 1-800-833-4237

Krazy Glue
Mills Co., 1076 Cambridge St. Cambridge, MA 02138, 617-547-9500

Flex-shaft grinders, carbide rod, silicone carbide grinding wheels
MSC Industrial Supply Co.

Nationwide Sales/Customer Service
AT&T 1-800-645-7270
MCI 1-800-753-7900
FAX 1-800-255-5067 (nationwide sales)

Airscribe and Dotco midget grinder, sales and service
Palmak Co., 72 South Ave., Natick, MA 01760, 617-237-3550

Carbowax, polyethylene glycol powder #6000
Union Carbide Corporation, Houston Customer Center, 10235 West Little York Rd., Suite 300, Houston, TX 77040, 1-800-568-4000

See local distributors of dental equipment and supplies for carbide burrs and diamond disks/wheels

Acknowledgments

I thank Kathy Costello for inputting much of this chapter into the computer; Amy R. Davidson for compiling the source list; and Farish A. Jenkins, Jr., for reviewing the manuscript and providing the SEM pictures. And my wife Paula Chandoha for printing the pictures.

References

Dolan-Jenner, Fiber-Lite Catalog FM 103.
Krazy Glue, Industrial Grade Technical Data Sheet.
Rixon, A. E. 1976. *Fossil animal remains – Their preparation and conservation.* London: Athlone/University of London, p. 70.

Nonpetrification preparation

Kathy Anderson, Judy Davids, and Terry Hodorff

The Mammoth Site of Hot Springs, South Dakota, represents the accumulation and preservation of late Pleistocene *Mammuthus columbi* and *M. primigenius* remains concentrated in a natural death trap. This trap consisted of a spring-fed pond formed in a karst depression, which was filled with sediments, and due to reverse topography, is now a hilltop. The majority of bones are preserved and displayed in situ; some have been removed for research and storage. Due to the fragile state of these nonpetrified specimens, it is the challenge of the preparators to preserve bone, reverse previously used chemicals, re-construct the fossils, and stabilize and store the prepared specimens.

Proboscidean fossils excavated at the Mammoth Site in Hot Springs, South Dakota, are prepared, cataloged, and stored on-site. The remains of other animal species have been found here and are prepared in the same manner as the mammoth bones. Due to the warm spring water that fed the sinkhole 26,000 years ago, the specimens are neither mineralized nor collagen-intact fresh bone. Prior to 1986, during open-air excavations, bones were cleaned with acetone and treated with Glyptal. Later it was discovered that Glyptal had not penetrated

Figure 6.12. In situ preservation at the Mammoth Site.

deeply into the bone but had created a discolored, peeling, brittle shell on the surface of the specimen. This left the bones friable, dry, and untreated internally. In 1986, after the construction of a permanent building over the bone bed, the Glyptal method was discontinued and polyvinyl butyral resin (Butvar) mixtures of B-98 and B-76 were introduced. Some field specimens have been stored in plaster jackets for up to 13 years. In those years the specimens were subjected to numerous moves and were stored where they were exposed to the constant freeze and thaw conditions of South Dakota winters. Specimens excavated since 1987 are in better shape and are preserved and monitored by the trained staff hired that year.

In situ exhibit

The task of consolidating large mammoths in itself poses problems. In situ situations create additional challenges not encountered in laboratory working conditions (see Figure 6.12). It is essential to reach the complete interior of the fossil being prepared. Because the bones left in situ are only partially exposed, knowledge of the internal and external structure of each individual bone is necessary. The partly exposed specimens are still embedded in matrix (see Figure 6.13), which allows limited access to cleaning, repairing, and preserving.

Consolidants used at the mammoth site

Three types of consolidants are used at the Mammoth Site, B-98 and B-76 (Butvar) and Acrysol. Butvar, polyvinyl butyral resin powder, has been found to be the most effective on dry bone. It can be mixed at a lower viscosity, thus allowing better penetration of the specimens. For any of these chemicals, label every container or bottle as to its contents, wear disposable latex gloves, be aware of safety procedures, and always keep the work area well ventilated.

Figure 6.13. In situ preservation at the Mammoth Site.

B-98

B-98 resin should be mixed using anhydrous denatured ethyl alcohol (ethanol) as the vehicle. The slower evaporation time of B-98–ethanol is more desirable, as it allows for deeper penetration of the resin into the interior of the bone. Prepare consolidant by slowly mixing 1 part resin powder to 8 parts vehicle.

B-76

B-76 resin can only be mixed with acetone. At the Mammoth Site we only use the B-76 resin as an adhesive. To prepare the glue, slowly mix 1 part resin powder to 1 part vehicle. Note that straight acetone will reverse this glue.

Acrysol

In certain areas of the excavation the bones are damp. Because B-98 and B-76 do not interact with water and create a barrier prohibiting the penetration of the consolidants, an alternate method was tested. An acrylic copolymer dispersion resin, Acrysol, was chosen. It was diluted with distilled water and applied to damp in situ bones. This water-based resin is compatible with the damp bone and evaporates slowly. The slow curing time prevents damage to the bone (e.g., cracking, shrinking, and warping). When the specimens are judged to be dry,

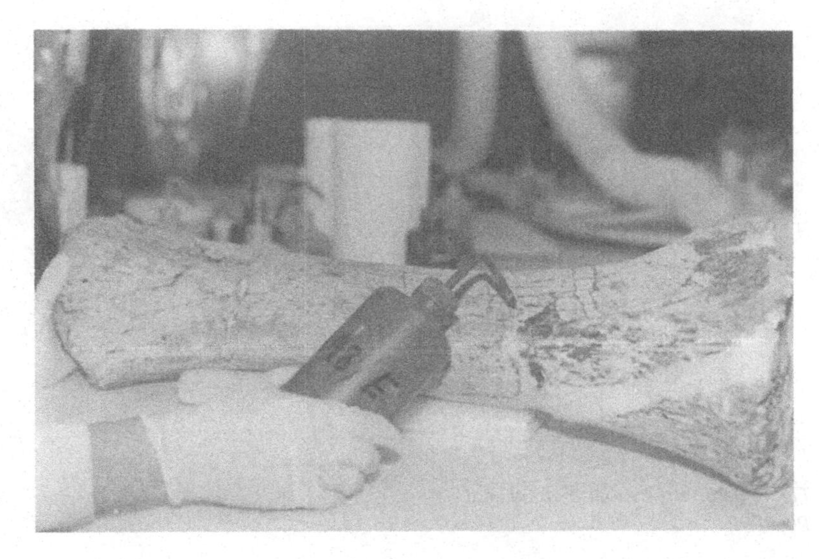

Figure 6.14. Drizzle method of application.

the Acrysol is easily dissolved with plain ethanol or acetone and B-98 used thereafter for routine consolidation.

Application procedures

B-98 ethanol may be applied to a clean, dry bone by use of one of the following methods: drizzling, immersion, or brush/soak.

Drizzling

In drizzling, wear gloves and always keep the work area well ventilated. Clean the specimen by dusting with a clean natural bristle brush to remove surface dust. Clean fissures that have soil embedded in them. The cracks are then utilized by carefully drizzling B-98–ethanol from a drip bottle (see Figure 6.14). Too much pressure applied to the drip bottle will force the preservative into the bone and cause the bone to break down. The amount applied varies with the individual specimen; bone that becomes too wet will collapse, crack, or crumble.

Prior application of preservatives (B-98 and B-76) will build up and block penetration and must be dissolved or opened periodically during the drizzling procedure. The area to be preserved is "opened" by applying straight ethanol. It is better to be prudent than too lavish with the preservative, until the preparator "learns" the bones. Preparators must always carefully observe each specimen, noting amounts of consolidants and any changes on specimen cards.

Immersion

The preparators at the Mammoth Site have had mixed results with the immersion method of consolidation. *Note:* If the bone is extremely dry and cancellous, an alternative method should be utilized.

The specimen should be solid in order to be preserved by immersion. Select a container large enough to allow complete submersion. Put the specimen carefully in the container. Gently pour B-98–ethanol into the container. Do not pour di-

Figure 6.15. Brush/soak method of application.

rectly onto the specimen but to the side of the container. Cover the container tightly to avoid fumes. Let the specimen bathe for 24 hours or until trapped air is replaced by preservative. Wearing latex gloves, carefully lift the specimen out of the bath and place on cushioned draining area. A cardboard tray filled with sawdust covered with paper towels works well. Using a soft brush, slough off old preservative while the bone is still wet. Never scrape bone with dental picks, hard brushes, or any implement that would mar the surface of the bone. Good ventilation is needed in bathing and drying of the specimen; always work under vent hoods.

Brush/soak

The brush/soak method is used when the bone is very dense, without natural cracks or openings. Clean the specimen with a soft brush and "open" the area to be worked by applying plain ethanol. Then dip a clean, soft brush into a thin solution of B-98–ethanol (1 part resin to 12 parts ethanol) and gently lay the brush against the bone (see Figure 6.15). The bone absorbs the preservative from the surface. Over the course of time, this method will cause a shiny buildup of preservative, but this can be removed by gently applying acetone to the area.

Remember: There is no absolute method; bone preservation is not an exact science, and the preparator must be flexible with when considering an appropriate method to use.

Removal of previous consolidants

The reversal of previously applied consolidants is the greatest challenge to the Mammoth Site preparators. Individual specimen records before 1987 are nonexistent, and thus amounts of substances used and application procedures are virtually unknown. We do, however, know that Glyptal and acetone were applied in working field conditions.

Acetone is the solvent used to reverse the former preservative, but this is a harsh method and must be carefully monitored. Acetone is applied to the surface of the bone, and when the Glyptal is softened it sloughs off or is gently lifted by using a soft cloth. It is up to the preparator to determine whether or not the specimen is strong enough to undergo this stressful process. The integrity of the preserved specimen is the most important consideration.

Prior to 1986, Elmer's glue and Duco cement were used as gap fillers in cracks of both in situ and study collections. These adhesives are destructive to specimens because in time they become brittle and shrink, taking the bone surface with them; thus, in 1986 their use was discontinued. In trying to reverse the glues, it was discovered that they will not separate from the bone. While attempting to soften and remove the adhesives, so much acetone or ethanol was required that the friable, spalling bone around it would become heavy and break away. We are still in the process of trying to reverse the damage done to specimens caused by these adhesives.

Repair

Whether repairing bone in the laboratory or in situ, the specimen and all broken elements must be completely consolidated. B-76–acetone mixed to a thick consistency is a good adhesive. Matching the color, grain, and cancellous sections will aid in piecing a specimen together. Remember always to clean the excess glue off the outside of the bone. Utilize sandboxes, sandbags, rubber bands, and other items to hold glued pieces together. Dried, repaired, consolidated specimens may be given extra internal support by using plaster of Paris. Reasons to plaster are to reinforce the bone for strength or to fill in the missing elements when the specimen is to be used for display.

Sometimes additional support is needed while preserving and reconstructing fragile specimens and before moving or storing them. This support can be supplied with a half-jacket constructed of plaster or plaster bandages. A separator, a layered cushion of damp toweling or toilet tissue at least 1/4 inch (6.4 mm) thick, is needed to keep the plaster from coming into contact with the bone. A lighter weight pillowlike support jacket can be made for the bone by using plaster and paper-mâché. A separator must be used with this type of support also. Pipes may be added to the jackets for easier handling and moving of the specimens.

Management of collection

Specimen record cards with all pertinent current information should be kept. These cards should list all data on the collection, including provenance, description of the specimen, museum assigned number, preparator, consolidants and adhesives used, dates of application, and storage location. Preparators must always carefully observe each specimen, making notes and changes on specimen cards. Cataloged, numbered specimens are then stored on appropriate shelving in a environmentally controlled area.

Even though the collection has been stabilized, the fragile, cancellous, interior bone never becomes completely consolidated; thus additional preservative will be routinely needed.

Monitoring the assemblage of 1100 in situ bones is an enormous task. At the Mammoth Site the majority of the specimens are in situ, and the exhibit area is not temperature or humidity controlled. Because Butvar does not penetrate at temperatures below 55°F (13°C), this currently limits the months in which the preservative may be applied.

Storage

Storage of consolidated specimens can be a problem when dealing with megafauna. The large size of the mammalian skeletal material requires very sturdy open-ended shelving. Smaller elements are best stored in metal slide-drawer cabinets. A closed cellular polyethylene foam is best for lining shelves and cabinet drawers. If closed cellular foam is not used, make sure the pad used does not allow foreign chemicals to leach into the bone. Muslin draped over the shelving prevents dust particles from collecting on the bone.

Summary

Preparation of nonpetrified bone is not an exact science, so time and patience are the best teachers. Specimens must be treated on an individual basis, with observation dictating the process to follow.

Fossilized eggshell preparation

Betty Quinn

The development of the amniotic egg during the Pennsylvanian period freed reptiles from the need to lay their eggs in water. The amniotic egg is encased in a shell, which provides the developing embryo protection from dessication, while also providing for gas exchange, support, and a mineral source for the developing skeleton.

Three major types of eggshells are known: a soft, parchmentlike shell; a pliable calcareous shell; and a rigid calcareous shell. The parchment shell found in snakes, lizards, and the monotreme, absorbs water and expands as the embryo grows. The shell is composed of a single or multiple layers of fibrous membrane with little or no crystalline calcite. The calcite typically occurs as calcium carbonate. The pliable calcareous shell occurs in the sphenodon and among most turtles (as aragonite). The shell has a thick inner membrane supporting individual calcite crystals of the outer layer. A rigid calcareous egg is found among birds, crocodiles, dinosaurs, some turtles, and geckos. The outer crystalline layer is composed of a single layer of well-defined, tightly abutted calcite crystals. Only a thin inner membrane is present.

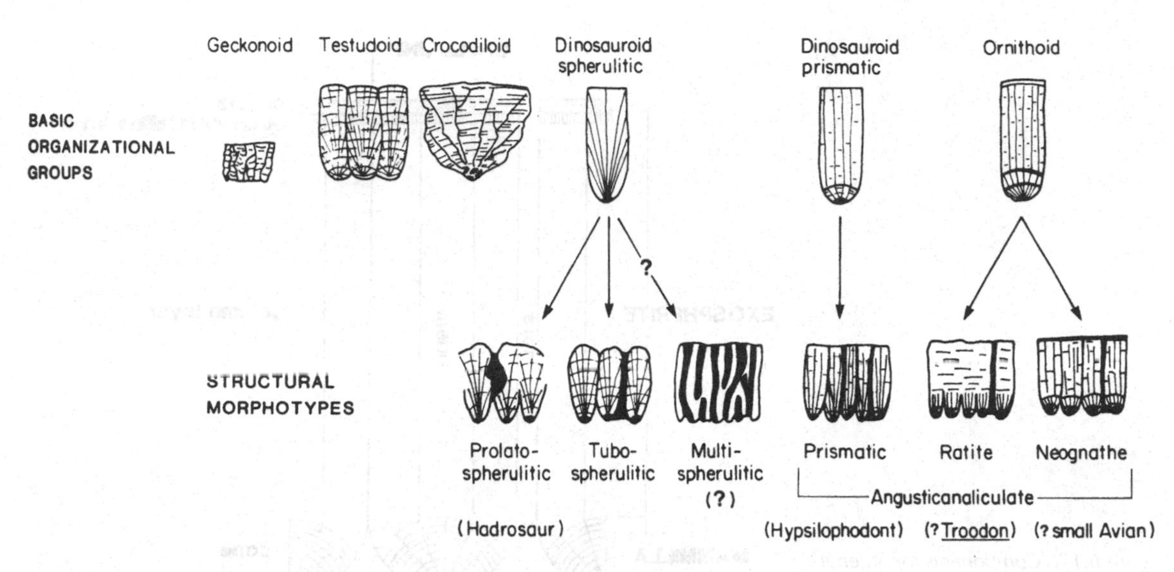

Figure 6.16. Schematic of basic organizational eggshell groups and their structural morphotypes. (Modified after Hirsch and Quinn 1990.)

The fossilization potential for the various types of eggshells varies considerably. The chance of fossilization of parchmentlike eggshell is poor because the calcite occurs in such minute amounts that it is almost impossible to trace or identify it after the organic matter has decayed. Pliable calcareous eggshells also have a poor chance of fossilization because the calcareous layer will probably dissociate with the decay of the organic matter. The chance for fossilization is best for rigid eggshells. The microstructure is often well preserved and, in some cases, the original amino acids have been isolated and analyzed.

Rigid eggshell is composed of a crystalline layer of more or less distinct, tightly abutting or interlocking shell units and a relatively thin shell membrane. A continuous net of microscopic organic matrix exists throughout this calcareous layer, and the interrelation between the mineral and organic phases during growth of the eggshell creates taxonomically important specific features. Thus, for each of the animal groups that lay rigid-shelled eggs (geckos, turtles, crocodiles, birds, and all presently known dinosaurs) a typical basic organizational shell structure is recognized (see Figure 6.16) (Hirsch and Quinn 1990; Mikhailov, 1991, 1992).

In eggshells we distinguish between: (1) the general morphology of egg and eggshell (superficial features), including egg shape and size, shell thickness, outer surface sculpture, and pore pattern, and (2) the histostructure. The histostructure consists of the general histostructure of eggshell (morphotype), including microfeatures of shell units (nucleation center, primary spherite, eisospherite, mammilla, wedges and prisms, or continuous layer) and type of pore system, and texture of eggshell, that is, sequence and composition of horizontal ultrastructural zone. The histostructure of the eggshell can best be observed in radial views. The unexposed internal structure of eggshell is not as easily affected by weathering and diagenesis as the outer surfaces. Thus freshly fractured shell

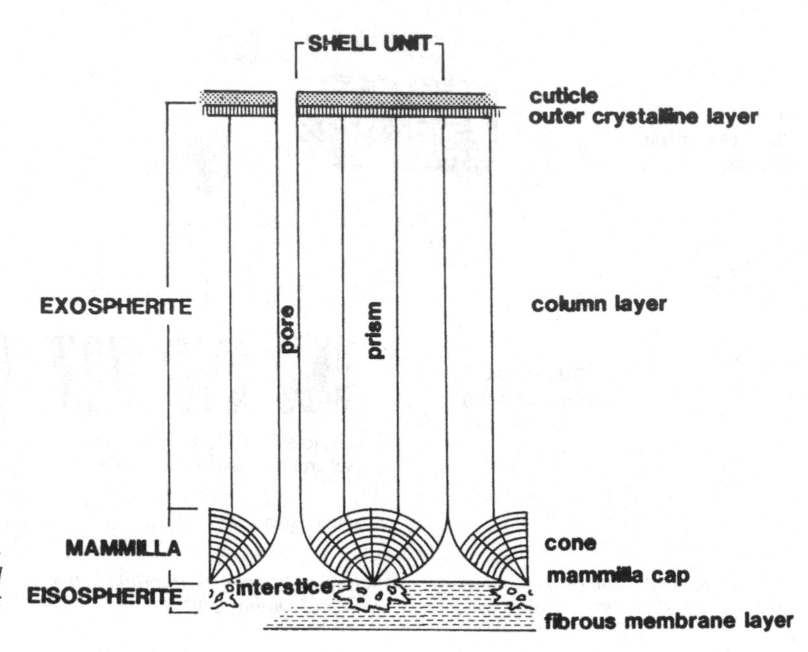

Figure 6.17. Components for a generalized avian eggshell. The cuticle and fibrous membrane layer are organic and rarely fossilize.

fragments produce the best structural views. However, to avoid being mislead by alterations due to diagenesis or weathering, a series of fragments are examined from the group being studied.

The study of fossil eggshell is based on comparison with modern eggshell. The well-studied features of the avian eggshell and the nomenclature have been used as a standard. (see Figure 6.17). Different views and sections referred to in this chapter are shown in Figure 6.18.

Preparation of specimens

Materials and equipment

Most of the supplies and equipment can be obtained from various scientific supply companies. Access to some equipment, such as the scanning electron microscope (SEM) can be obtained from some universities or oil companies. Materials needed include paper towels or Kimwipes, small labels, petrographic epoxy, frosted (dusted) microscope slides, silicon carbide polishing grits (#400, #600, and #900), and small closable plastic bags (e.g., Ziploc). Occasionally, 10% hydrochloric acid or EDTA (ethylenediaminetetracetic acid) may be used to etch the specimens. Equipment needed includes a micrometer with ball attachment, Geneva lens measure, small ultrasonic bath, 30-, 50-, and 100-ml plastic beakers, petrographic thin-sectioning machine, trim saw with extrathin 6-inch blade (the thin blade will save enough of the specimen to make more thin sections), vacuum chamber, Water Pik (or similar water spray), fine pointed forceps, fine brush, small plastic containers (pipe caps of 7/8-inch diameter) for embedding specimens, corundum or diamond-tipped identification marking pencil, a binocular and petrographic microscope, and a scanning electron microscope.

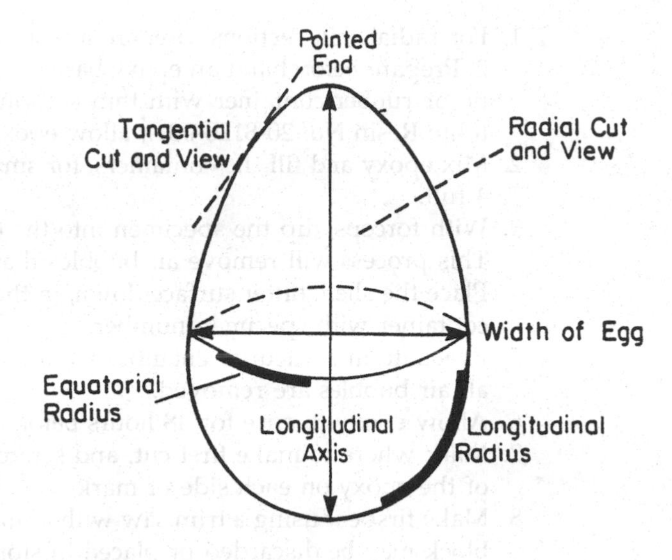

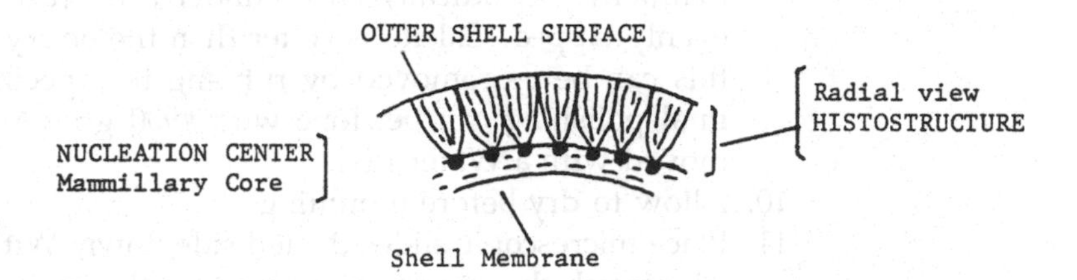

Figure 6.18. Egg geometry showing curvature components and different cuts and views.

Techniques

Initial preparation

1. Catalog individual shell fragments to be analyzed, but do not write number on shell fragment.
2. Clean each specimen in an ultrasonic bath using only water to avoid contamination. This may be done by placing the eggshell into a beaker of water and placing the beaker into an ultrasonic bath for several minutes.
3. Air-dry the specimen on a Kimwipe.
4. Using a binocular microscope, note and record surface texture (e.g., smooth, nodular, ridged, granular, etc., "oddball" is an acceptable grouping).
5. Measure and record shell thickness with a micrometer and ball attachment.
6. Estimate the curvature of the fragment using a Geneva lens measure.

Preparation of thin sections for polarizing light microscopes

Cross sectional and tangential (parallel to the shell's outer surface) (see Figure 6.18):

1. For radial thin sections, prepare a resin base. Otherwise continue with step 2. Prepare beforehand an epoxy base by covering the bottom of a small plastic or rubber container with thin sectioning epoxy resin (e.g., Buehler Castolite Resin No. 20-8120-001); allow epoxy to cure a minimum of 24 hours.
2. Mix epoxy and fill the containers: for small specimens, 7/8 full, for large, 3/4 full.
3. With forceps, dip the specimen into the epoxy and allow excess to drip off. This process will remove air bubbles that may adhere to the eggshell.
4. Place the shell, outer surface down, in the prepared epoxied container. Mark container with specimen number.
5. Evacuate in a vacuum chamber (<10 lb mercury) for several minutes until all air bubbles are removed.
6. Allow epoxy to cure for 48 hours before removing from container.
7. Mark where to make first cut, and scratch the specimen number on the top of the epoxy on each side of mark.
8. Make first cut using a trim saw with a thin diamond blade. Half of the epoxy block may be discarded or placed in storage.
9. Lap the cut face of the other half of the epoxy block, first with #400 silicon carbide grit to remove all blade marks, then with #600 grit. If lapping is done manually, periodically check under a microscope to be sure the surface is evenly lapped. Calcite is softer than the epoxy and can cause undercutting; this can be be removed by rubbing the specimen on #600 sandpaper. The final polishing can be done with #900 grit. Any embedded grit can be removed with a Water Pik.
10. Allow to dry before mounting.
11. Place microscopic slides dusted side down. With identification marking pencil, scratch the specimen number on the back (shiny side) of the slide near one end.
12. Measure and record the thickness of each slide with a micrometer.
13. Spray each slide with Ban-Dust before placing it near the corresponding specimen.
14. Warm specimen and slide under an incandescent light.
15. Mix bonding epoxy (e.g., Buehler Epo-tek) and allow to settle for 5 minutes.
16. Place two drops of epoxy in the center of the frosted side of the slide.
17. Paint a thin coat of epoxy on the lapped face of the specimen and place the specimen on the slide. Even pressure is critical to remove any air bubbles.
18. Allow to dry for 24 hours before cutting. Use the thin section machine; hold the specimen, with the aid of a slide holder, leaving about 0.025 inch thickness for grinding.
19. With a thin section grinder, reduce specimen thickness to approximately 0.004 inch (100 microns). Periodically check the specimen under a petrographic microscope until structural features and the beginning of extinction start to show. Keep in mind that glass slides vary in thickness.

Preparation of specimens for scanning electron microscopy

From the specimens that have been cleaned (see the discussion on initial preparation), choose shell fragments that are large enough to break into three pieces, if possible. Using a light microscope, select those pieces that best display the outer and inner surfaces and the cross section of the shell. Mount the specimens on clean SEM buttons with Ambroid, thick Butvar, VINAC, and the like. The specimens should be mounted with the part of the shell to be examined facing upward. If the SEM button is not clean, fine sandpaper can be used to remove any residue. Label the underside each SEM button with the specimen number.

Examination of specimens

Analysis of eggshells is performed with both polarizing light microscope (PLM) and the scanning electron microscope. The PLM relies on the transmission of light through the eggshell (usually radial or cross section). Under normal light the structure of the crystalline layer (morphotype) is seen in a single plane. Under cross nicols a typical extinction pattern is visible. However, it has to be interpreted with caution to determine the structural morphotype because the enhanced crystallographic structure may have been more or less altered by recrystallization (herringbone pattern). The SEM, on the other hand, uses a beam of electrons to reveal surface features, the histostructure in its original, three-dimensional state. Because the information provided by the two techniques differs, both should be used to complement each other (see Hirsch and Quinn 1990).

For PLM analyses the prepared thin sections have to be hand-polished with #600 grit (see step 9 in the discussion on preparation) to be ready for photography. The image of thin sections will change to a certain degree with their thickness; thus it might be necessary to take photographs, as an example, at 75 μ, 50 μ, and 25 μ. A Water Pik can be used at this point to remove any embedded grit. A polarizing light (petrographic) microscope with a camera and light meter are needed. Record all data in a notebook, including film roll and photograph number, specimen number, polarized or nonpolarized lighting, orientation of the thin section under cross nicols, light meter setting, exposure time, thickness of the specimen (subtract the thickness of the glass slide from the measurement taken as close as possible to the specimen with the micrometer), and any other pertinent information. (See Figure 6.17 for eggshell terminology.)

For SEM analyses, a drawing of the specimen highlighting typical features observed under the light microscope should be made. This will make it easier to find them after the specimen has been gold-coated for the examination. Record all pertinent data in a notebook and on the data sheet of the specimen: site name, specimen number and, most important, magnification. In regard to photographing, 35mm film is less expensive, but Polaroid 55 allows you to see immediately what has been photographed.

Interpretation

The methods used to study fossil eggshells are more limited than those for modern eggshells. For example, much of the organic matter, such as fibrous membrane layer, organic cover (cuticle), and pore coverings, is missing (Hirsch 1983). Other problems that hinder the analysis of the fossilized eggshell include the scarcity of the specimens, matrix or secondary calcite deposits obscuring the surface details, and alterations of the eggshell structure caused by diagenesis.

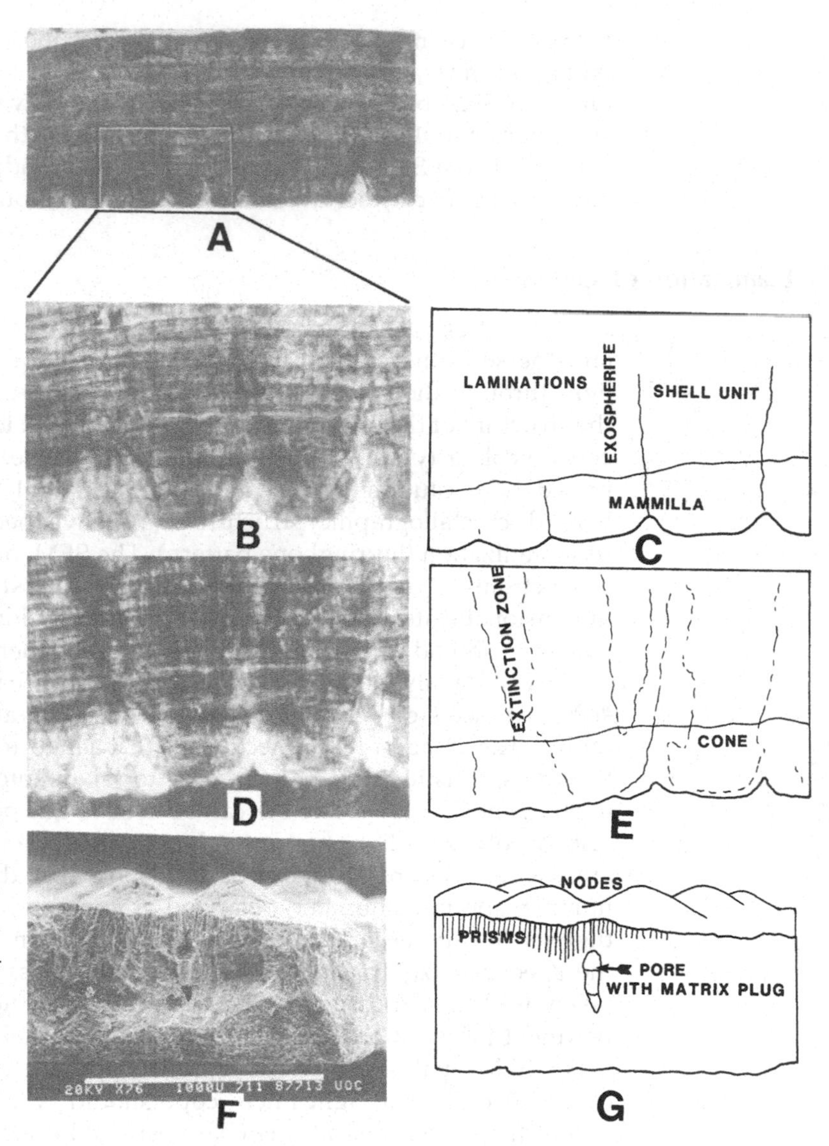

Figure 6.19. (A–E) One dinosaur eggshell that shows features of two different morphotypes. Eggshell thickness 1.1 mm. (A) Radial view, normal light. (B) As A, mammillary layer. (C) Interpretive sketch for B. Note pronounced horizontal growth lines and faint radiating lines arising from base of mammillae. (D) Same area of thin section as in B viewed under polarized light. (E) Sketch for D. Note faint extinction zones arising from mammillae. (F) Dinosaur eggshell (Troodon?), radial view, SEM, ornithoid-ratite structural morphotype. (G) Interpretive sketch for F. Note shell is covered with a secondary calcite layer which is very even in thickness. Shell thickness measured over nodes is 0.8 mm; thickness including secondary layer is 1.1 mm. The prismlike structures in outer part of shell represent a secondary layer. The continuous layer does not show any distinct structure. Very thin mammillary layer only partially visible bcause specimen is tilted to show outer surface with nodes.

Analysis and identification of the eggshell begin after cleaning in the ultrasonic bath, with the macroscopic examination of the general morphology of egg or eggshell fragment (shape and size of egg or shell fragment, shell thickness, etc.). For a whole egg, note the type of in-filling matrix and the impression the inner surface of the eggshell leaves on the matrix. If possible, determine any internal content (embryonic bone, pupae casings, etc.).

Microscopic analysis is done from the PLM and SEM photomicrographs. Emphasis should be placed primarily on structure of column layer, mammillary layer, and the pore system. Figure 6.17 provides a guide for locating the many features. Two of them, the cuticle and fibrous membrane layer, are organic and are not found in fossilized eggshells. Other features, such as relative thickness of the mammilla or column layer, varies considerably, depending on the taxon of the egg. Therefore the relative proportions of the mammillary and column layer will not hold true for all eggs.

Avian eggshell has two distinct shell layers: a column layer composed of shell units and a thin mammillary layer (1/6 to 1/7 of the shell thickness); see Figure 6.17. Pore canals are straight, narrow, and of about the same diameter. The ornithoid-neognathe morphotype (see Figure 6.16) could be assigned to this specimen. If the continuous layer has no identifiable shell units or columns, and if there is an abrupt change of structure between the two shell layers and the mammillary layer is very thin (1/10 of shell thickness), we are confronted by the ornithoid-ratite morphotype (see Figure 6.16; see also Figures 6.19F and 6.19G). At the present time this is the only modern-day morphotype found in dinosaur eggshells; all others had to be established (Hirsch and Quinn 1990; Mikhailov 1991, 1992).

Conclusion

The techniques of cleaning and preparing fossilized eggshells for study have been described. The discussion includes step-by-step procedures for making thin sections of eggshells and the procedures for obtaining comparative micrographs using the scanning electron and polarizing light microscopes. The importance of using both the PLM and SEM for the most accurate interpretation of eggshell structure is stressed.

Many other techniques not discussed here have recently been applied to the study of fossil eggs and eggshells. These include CAT scan to check for embryonic material, chemical treatments or stains to enhance certain features of the inorganic matter, elemental or mineralogical analysis, radiographs, cathode luminescence, and biochemical analyses for amino acids.

Acknowledgments

I thank Karl Hirsch for reviewing this paper and for all that he has taught me over the years. I also thank Kenneth Carpenter for his expertise as an editor.

References

Hirsch, K. 1983. Contemporary and fossil chelonian eggshells. *Copeia 1983*(2):382–397.

———, and B. Quinn. 1990. Eggs and eggshell fragments from the Upper Cretaceous Two Medicine Formation of Montana. *Journal of Vertebrate Paleontology. 10*(4):491–511.

Mikhailov, K. E. 1991. Classification of fossil eggshells of amniote vertebrates. *Acta Palaeontologica Polonica 36*(2):193–238.

———. 1992. The microstructure of avian and dinosaurian eggshell: phylogenetic implications (pp. 361–373). In R. Campbell (Ed.), *Papers in Avian Paleontology Honoring Pierce Brodkorb.* Contribution in Science. Natural History Museum of Los Angeles County.

7

Chemical preparation techniques

Ivy S. Rutzky, Walter B. Elvers, John G. Maisey, and Alexander W. A. Kellner

I. Organic acids (typically acetic and formic) are used to dissolve some kinds of rock from the bone. These acids act on limestone (carbonate material).

II. Mineral acids (such as hydrochloric) are used to dissolve the bone from the rock, thereby leaving a clean matrix mold from which a peel (which becomes the positive) can be made. This technique works well on shales, sandstones, quartzites, and oil shales.

III. As an alternative to using thioglycollic acid to remove ferric iron from the matrix, the safer Waller method is preferred. It is a nonacid technique that removes oxides and hydroxides of iron from the matrix.

Acids and certain other chemicals may be used to achieve different ends in the preparation of vertebrate fossils. Not all chemical preparation techniques are surveyed here, only those that we at the American Museum of Natural History have used extensively and we feel are of broad interest. The most common materials that require removal by chemical techniques are carbonates (especially calcium and magnesium) and ferric iron. Acid is also widely used in the preparation of invertebrate fossils (Grant 1989). All materials should be used with care as to ventilation and personal safety. As working conditions vary widely, we do not wish to predict or generalize. It is best to contact the manufacturers or suppliers directly.

Because this book is primarily for use as a laboratory manual and not as a general discussion of the topic, an outline format has been adopted for this chapter. This should make it simpler for the preparator not only to read through the procedure initially but also to skip to the essential steps when using this chapter as a hands-on guide.

Although at first glance the processes appear to be complex, basically there are just a few major stages (A, B, C, D etc.), which are broken down into several major Steps (1, 2, 3, 4 etc.). Within the steps there are "Comments": "fine-tuning" hints, observations, and refinements of methods based on experience.

I. Organic acid extraction of fossils from carbonate matrices: Direct immersion or modified transfer methods

The technique of using acids to remove limestone and other calcium carbonate–containing matrices without significantly affecting delicate fossil skeletons has been known for more than a century. Hermann (1909) mentioned the use of hydrochloric acid for this purpose, and White (1946), Toombs (1948), and Rixon (1949, 1976) reported on the use of acetic, formic, and other acids and acid mixtures. Braillon (1973) also reported in detail on the use of acetic acid. Of those acids that have been studied, acetic and formic are still widely used today, although we prefer the latter for reasons presented later on.

Frequently the carbonate matrix can be removed from large isolated bones by immersing them directly in the acid. With specimens such as fish fossils, however, dissolution of the surrounding matrix can result in a badly disarticulated skeleton that is difficult to study. This concern gave rise to the development of the special stabilizing techniques described by early workers. Some of the earlier approaches included cementing specimens to glass slides and coating them with rubber cement. Toombs and Rixon (1950) reported a transfer technique in which the fossil is embedded in a polyester resin that both fixed the skeleton in place and permitted it to be viewed through a transparent resin covering.

Described below is a transfer method modification that builds on the procedures reported by Toombs and Rixon (1950), Rixon (1976), MacFall and Wollin (1983), and others. This modification, and several variations also reviewed, produce aesthetic displays that are economical, compact, and easily studied. Staff at the American Museum of Natural History (AMNH) used this formic acid technique primarily on the limestone concretions from the Santana Formation in northeastern Brazil. We also worked on concretions from the Gogo Formation in Australia and on limestone specimens from Solnhofen, Germany. Specimens ranged in size from a few centimeters to 1 meter in length. Although the British Museum (Natural History) and others have had successful results with acetic acid, we chose formic acid. Acetic acid is very slow acting, requiring perhaps 30 immersions in acid compared with formic acid, which requires only 8 to 10 immersions. In discussions with Daniel Goujet (Institut de Paléontologie du Muséum, Paris), we felt that although formic acid was a bit stronger (putting stress on the bone), there would be more stress on the bone from the greater number of immersions required by acetic acid.

Which hardener to choose requires weighing the variables of hardness, reversibility, stability over time, safety to the preparator and other application characteristics. See the Alternative Hardening Agents section for a brief summary of properties of several hardeners currently in use. Also included are the name of each institution and the person to contact for more detailed information. As with many preparation materials, new products become available as others are discontinued, so this is an area where contacting colleagues and trading information can be very useful.

The decision to use an embedding or a free-form technique depends on the goal of the preparation:

1. The purpose of embedding fossils before an acid bath would be: for fish, which are flat, with many delicate pieces that could become disarticulated, embedding the specimen retains everything in its original configuration. How the concretion was split would determine whether we would embed one or both sides. For example, sometimes the skull might be informative on one half of the concretion, but the other side contained the rest of the fish. Other times, one side might be mostly matrix, and it would not be worth the expense of embedding and dissolving the specimen. The AMNH embedding technique uses less material than other methods, is lighter in weight, and protects the specimen during and after acid immersion. The Simpson technique allows three-dimensional viewing but affords little protection to the phenomenally delicate specimen thereafter.
2. Freestanding techniques allow for disarticulation of bones during the acid process and rearticulation after the bones are removed from the matrix.

The time sequence required when planning to use the embedding technique with the acid preparation is as follows:

1–2 days: Setting up specimen with the resin embedment procedure
1/2 hour: Pouring the resin
2–4 days: Curing time for the resin
4 weeks: Acid immersion for 8–10 complete acid/water/hardener cycles of 1 day acid/1 day water/1 day hardener)

Other, unrelated tasks could be done during this time period, as some of this time is actually spent waiting (e.g., when the resin embedment is curing, when the specimens are in acid, when the specimens are in water). A "production-line" approach could also be taken, as we did, keeping several specimens going through the procedure, each at various stages of the process. For example, there could be three specimens in acid, three more in a water bath, three drying in the morning for a first hardener application in the afternoon, and three more dry and ready for subsequent hardener applications. Keeping much more than this number going, they begin to run (swim) away from you.

For a discussion of safety, it is best to contact the manufacturers or suppliers directly. Labs could vary so greatly that we do not wish to predict or generalize working conditions.

See Aulenback and Braman (1991) for a discussion of a chemical extraction technique for removing silicified material from ironstones. This technique has been applied to fossil plants but could be applied to other silicified material as well.

The transfer procedure for matrix removal: acid preparation

Before preparation begins, a preparation plan should be worked out between the preparator and one who is knowledgeable about the anatomical structure of the fossil being prepared, so that everyone knows what to expect. The plan can be altered during the process of preparation, depending on the bones as they appear and how the preparation itself is going. It is also recommended that records be kept of the ongoing preparation procedure (e.g., immersion time, substances used, concentrations, and observations of the results obtained). This log could be useful later in identifying reasons for problems that might occur during preparation. It might also be useful for further development of the technique.

The concentration of formic acid used depends on several variables. Among the most important are the nature of the specimen and the nature of the matrix. For example, our work at AMNH was done primarily on fish: flat, resin-embedded, delicate specimens, where the carbonate content of the matrix was high. (For instructions on embedding technique, see Clear Casting Resin Embedding Technique: Preparation for Acid Treatment, following the Micropreparation section.) We used 5% formic acid as our baseline. Bob Carr and Bill Sanders at the University of Michigan Museum of Paleontology regularly use 7% to 10% solutions on three-dimensional free-form mammal specimens. And Hervé Lelièvre at the Institut de Paléontologie in Paris uses 20% formic acid on placoderms, where the mud content of the rock is high. For a discussion of three-dimensional work, see the subsequent sections.

Preliminary step (optional)

We suggest removing a sample of matrix prior to acid treatment. This may be useful in later studies of the specimen. The matrix sample can be cataloged and kept with the acid-prepared specimen (see Figure 7.1)

A. Solution

1. Prepare a 5% solution of formic acid in water.

Comment: We prepare our solutions each time, but at the University of Michigan Museum of Paleontology, carboys are premixed with 7% and 10% solutions.

Comment: Whatever concentration of acid is eventually used, the specimen can be tested in a 5% solution for a few hours. The direction in which to proceed can then be evaluated.

Comment: In the beginning of the process, when only matrix and no bone is exposed, a 5% solution is recommended. During later stages of development, it may be advisable to cut back to a 3%–4% solution on delicate specimens. Lower concentrations require more immersions in acid; one must weigh the variables. It is advisable to balance the percentage of acid versus the number of immersions based on the qualities of a particular specimen such as the robustness of the bone and how the particular matrix reacts. We have found that concentrations above 5% have reactions that were too violent for our delicate fish specimens.

Comment: Rubber or durable plastic containers are recommended for immer-

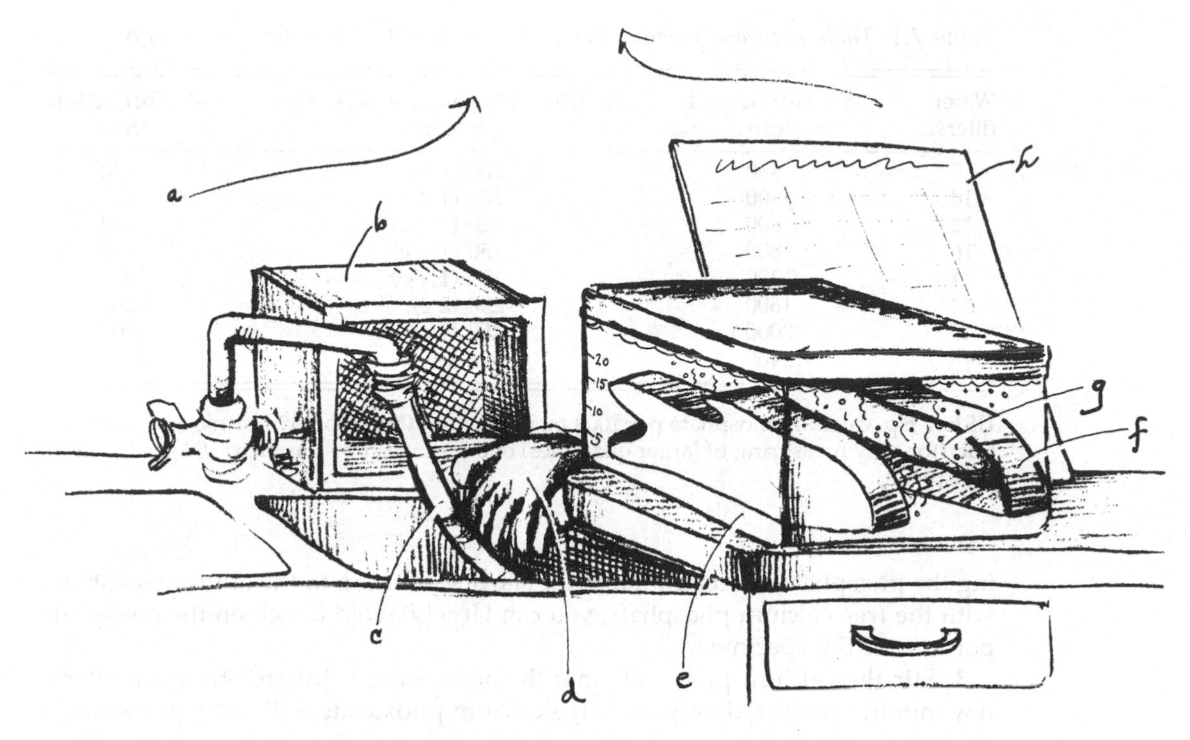

Figure 7.1. Setup, acid bath. (a) Good ventilation system. (b) Strainer to catch accidently "washed overboard" bone. (c) Plastic hose attached to faucet. (d) Rubber gloves. (e) Rubber or sturdy plastic containers. (f) Specimen (in embedded state). (g) Acid bath acting on matrix. (See text A.1.)

sion – they are cheap. On each tub mark 5-, 10-, 15-, . . . liter increments. This will subsequently make it much easier to measure amounts of water.

Likewise, if a large measuring beaker for acid is unavailable, marking increments of 100- to 200-ml units on a gallon plastic jug makes measuring the acid more convenient.

The same can be done with a measuring cup for the calcium phosphate. (See Table 7.1 for water–formic acid–calcium phosphate ratios.)

CAUTION: Do not use acetic acid and formic acid baths on the same specimen unless the specimen has been washed exceedingly well. The bone will be corroded by a bath containing both acids (Rixon 1976).

2. Add calcium phosphate at a concentration of 1 gram per 1000 ml of solution (roughly 1 teaspoon per liter) or follow Table 7.1. The important point is that there is a supersaturated solution, so that the formic acid will attack the free calcium phosphate rather than the bone.

Note: The 1 teaspoon per liter of solution of calcium phosphate was suggested by Daniel Goujet, Paris (Braillon 1973). The gram per 1000 ml measurement was empirically derived by us by adding free calcium phosphate to a 5% solution of formic acid until it no longer dissolved. (They are roughly equivalent.) One could use the same method of calculation for other commonly used percentages.

Comment: This is a crucial step, critical to a successful preparation. By increas-

Table 7.1. *Table of measurements: Water–formic Acid/Calcium Phosphate ratio*

Water (liters)	5% Formic acid (ml)	Calcium phosphate, $Ca(POH)_2$, g (cup)	4% Formic acid (ml)
8	400	100 (1 c)	320
10	500	100 (1 c)	400
12	600	120 (1⅕ c)	480
16	800	150 (1½ c)	640
24	1200	175 (1¾ c)	960
32	1600	200 (2 c)	1280
40	2000	250 (3 c)	1600
48	2400	325 (3¼ c)	1920

USE: 1 tsp. calcium phosphate per 1000 ml of solution (based on 5% formic acid solution).
OR: (for easy measuring of larger quantities) one 8 ounce cup = approx. 100 gr of calcium phosphate.

ing the phosphate concentration of the starting solution to the limits of solubility with the free calcium phosphate, you can largely avoid attack on the phosphate portions of the specimen.

3. Stir the calcium phosphate into the acid–water solution. Stir again after a few minutes. If stirred only once, the calcium phosphate will likely just settle to the bottom of the tub.

B. Acid bath procedure

1. After first wetting the embedded specimen in water, slowly immerse it in the acid solution for several hours. Watch how the matrix reacts initially; see how rapidly/violently it fizzes, how much material is removed in the first few hours. Then make a plan: what percent acid to use, how long to immerse specimen, and so on.

2. Check specimen again after 6–8 hours (if it is immersed in the acid bath at the start of a workday), or after 12 hours (if put in at the end of a workday). It may be advisable to pull the specimen out at this time, if fragile.

Comment: Immersion time variability: Depending on the hardness or muddiness of the matrix, approximately ⅛ inch of matrix will be dissolved within a 24 hour period. This time period is satisfactory in the initial stages of the immersion process, when little or no bone is exposed. If it is expected that large areas of bone may become exposed during the next bath, the specimen should be checked more frequently. Very fragile specimens or freestanding specimens may require checking even more often.

Comment: If the specimen is embedded (see the Embedment section), it should be placed on its side rim, so that the acid can attack the matrix more efficiently. As the matrix is broken down by the acid, the sediment will settle to the bottom of the tub by gravity. If the specimen is placed flat on its back, the sediment will settle down on it, slowing the action of the acid by prohibiting the acid from reaching the matrix. An uneven and pitted surface will occur. If the specimen is

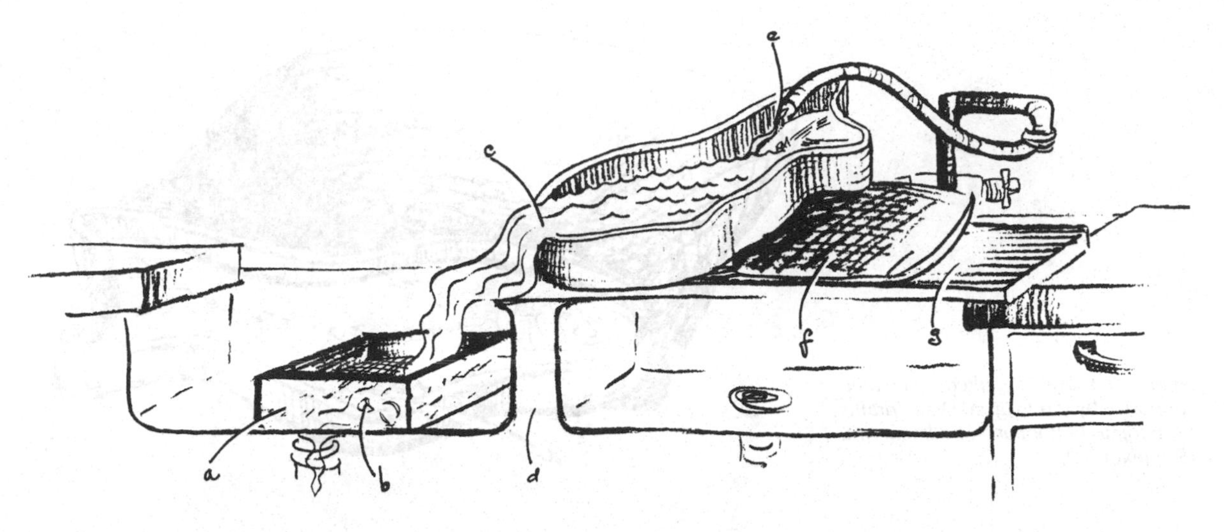

Figure 7.2. Clean specimen after acid bath. (a) Strainer. (b) Disarticulated bone fragments. (c) Fine silt skims over rim of embedment while most (heavier) bones stay inside rim. (d) Double sink support or (brick) support for board. (e) Hose positioned so water gently runs down side of embedment or on matrix, not directly on specimen. (f) Sandbag for elevation. (g) Board across sink to support specimen. (See text B.2.)

not embedded, figure out a way to prop it on its side, if sturdy enough, with sandbags or oil-based clay.

Comment: When the specimen is immersed in the acid solution, a slight fizzing reaction indicates that the acid is reacting. If the reaction is like that of Alka-Seltzer, judge whether the specimen is sturdy enough to withstand it, or reduce the percentage of acid by adding water.

Comment: In our specimens (from the Santana Formation in Brazil), the ones with muddy matrix seem to be more fragile, due to initial poor fossilization. Muddy sediments are often formed in anoxic bottom environments in which iron pyrite can form. This sometimes coats or permeates the bone and grows invasively, causing damage that only becomes apparent as the bone becomes exposed and starts to disintegrate. Filter the dissolved matrix for interesting pyritized invertebrates.

Comment: A large tub with several specimens can accommodate a relatively larger volume of acid and therefore permits longer-lasting acid action than does a small tub with a single specimen. (The volume of acid versus the volume of limestone determines when the reaction is completed.) This situation is acceptable as long as any bones that may become disarticulated can remain isolated from other specimens (the plastic embedding rim is usually sufficient, if great care is taken when the specimens are removed from the bath) (see Figure 7.2).

3. Carefully remove specimens from the bath with a scooping motion. If any of the bones have become disarticulated during the acid bath stage, the scooping motion will help contain them within the walls of the resin embedment.

4. Fill the resin embedment with gently circulating water. Tip the specimen at

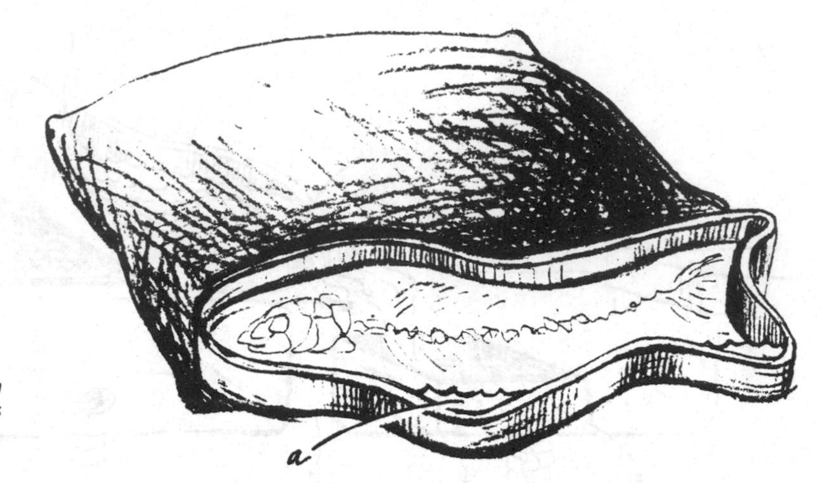

Figure 7.3. Air-dry tilted specimen supported by sandbag. (a) Water drains from top to bottom of embedment rim. (See text B.6.)

a slight angle (prop it up with a sandbag or a wedge for convenience). In this way, loose sediment will be skimmed over the lip of the embedment, but heavier loose bones will not be lost. Rocking the specimen gently while delicately brushing the matrix, you can remove much of the debris at this stage (it is much easier on the preparator and the specimen to float away the debris than to use a brush to get it off the bone later). As a precaution, this step can be done over a strainer.

Another method is to fill a tub with water. Carefully lower the embedded specimen into the tub.

Comment: Running water should not be used directly on the specimen. A setup (like that in Figure 7.2) with a plastic tubing extension from a faucet is advised. That way, the rate of water flow can be controlled, as can direction of current.

5. Immerse the cleaned specimen in a tub of water. Ideally, there should be a gentle but continuous flow of fresh water. The time in the water bath should be at least equal to the time in the acid bath.

6. Remove the specimen from the water, gently pouring off the excess.

Comment: After pouring off the water trapped in the embedding rim, set the specimen on a sandbag, tipped at an angle. This helps the water drain to the bottom of the plastic rim, thereby reducing the drying time (see Figure 7.3).

7. Let the specimen dry for several hours. It should appear damp but not slippery or wet.

Comment: Allow the specimen to air-dry. Do not use an artificial method of drying. Heat could damage the bone, and a fan aimed directly at the specimen could blow pieces away. If the specimen dries too rapidly, bones could curl (a fish operculum, e.g.) or crack.

C. Preservation

1. Initial hardener damp-coat application

While specimen is slightly damp, apply a very weak solution of hardener dissolved in acetone. Use an eyedropper (a pipette with a narrow neck ordered from a scientific supply company allows more control than a standard eyedropper). This technique helps reduce bone damage that might occur during hardener application with a brush. At this stage, the specimen may be safely left for an indefinite period of time before resuming the acid immersion process. The very thin bones (an operculum, e.g.) could otherwise curl up and fracture if left to dry without the damp-coat application.

CAUTION: Do not use Butvar as a hardener. It is said to dissolve in formic acid the next time the specimen is treated (Braillon 1973). For a discussion of various hardeners, see the introductory section and the section on hardners.

Comment: The first application is cursory. The bones need not be cleaned until the next application. They will be less fragile next time around. The acetone in the hardener solution will turn a cloudy white when applied to a damp specimen. This cloudiness will disappear with the next application of hardener.

Comment: Apply hardener as a very weak solution, so that it can better penetrate the bone. If the solution is too heavy, the hardener will sit on the surface of the bone, leaving the interior unprotected.

Comment: For smaller specimens, hardener application is best done under a magnifying lamp or a microscope with a 10× eyepiece, and a .05 reducing ring. For our purposes, 10× was a little too much magnification. The reducing ring also gives a greater depth of field and allows more working distance, about 8 inches, as compared with a 3- to 4-inch working distance without the reducing ring. The greater working distance is useful; what with the plastic rim, a brush, and an eyedropper, it gets a bit crowded under the microscope.

Comment: Cyanoacrylate glue (superglue) may also be used in this manner where the bone is extremely porous or fragile. It has a high degree of penetration. It, too, may be diluted with acetone. One advantage of superglue is that it is hygroscopic and can be used to repair or strengthen broken and flaking bones before they have a chance to dry out and deteriorate further. Use it sparingly, however, because it is not as reversible as Glyptal or the acrylic hardeners – it becomes gummy and difficult to remove. At the Institut de Palaéontologie, the application of cyanoacrylate glue is controlled by dropping a small amount just behind the point of a fine tweezer.

2. Second hardener application

Using a small brush (see Caution) in one hand and an eyedropper filled with acetone in the other, clean the remaining silt from the bones where necessary. Use a thick hardener solution as a glue to secure loose pieces (if they are to remain in situ).

CAUTION: Do not use a nylon fiber brush. The acetone will fray it. A Windsor-Newton Series 7, size 0 or 1 is recommended. Don't buy a cheap brush. If you do not have a good point on your brush, you do not have much control.

3. Additional hardener applications

Before the second immersion in acid, apply as many coats of hardener as necessary. The goal is to apply thin coats to penetrate the specimen, until a matte surface appears. Too much hardener makes the specimen too shiny (making it difficult to see under microscope lights) and obscures fine detail. Make sure the hardener has dried for at least several hours before immersing the specimen in an acid bath.

D. Repeat procedure

Repeat the acid bath/hardener application procedure (steps A–C) until the matrix is completely dissolved. (Generally 6 to 8 repetitions were required for the embedded fish specimens we used to illustrate this technique.)

Comment: The process may be stopped before all the matrix is removed if there is risk of damage to the already exposed features of the specimen.

E. Finishing

Final cleaning of loose sediment: Looking under the microscope, clean all remaining sediment from the bone.

Three-dimensional acid preparation in formic acid bath, using methacrylate resin as a fixing agent

This technique was developed in removing the matrix from concretions from the Brazilian Santana Formation, which contained pterosaur bones The technique was also successfully used on other vertebrate material, including dinosaurs and turtles (Kellner 1991a, 1991b).

Recommendations

When trying to make a three-dimensional preparation by removing all sedimentologic matrix from the bones, it is important for the preparator to treat each specimen as a special case. Unlike fishes, for example, where the entire body is normally preserved in one plane, pterosaurs and other three-dimensional preserved tetrapods are typically buried and preserved in different planar orientations.

Materials used

- Formic acid
- Methacrylate resin (Paraloid B72 [P.B72]), an acrylic plastic resin
- Plasticine, and oil-based modeling clay

Safety

Resin and acid should be used with proper ventilation, preferably under a fume hood.

Procedure

A. First cleaning

Immerse the specimen in a 10% formic acid solution with added calcium phosphate (see Table 7.1) for 5–30 minutes. Wash the specimen in running water for 2 hours. Let dry.

Comment: Normally, the pterosaur bones that are preserved in the calcareous nodules of the Santana Formation that come to us are not collected by professionals. Most of the time, the bones are broken longitudinally, and there are cases where it is very hard to tell the real limits of the bones, especially the small ones. Thus a quick acid bath helps to establish the real limits of the bones. This will help with the preparation plan.

Comment: This bath is short, with a high concentration of acid, and therefore must be watched carefully.

B. Protective covering

1. Specimen should be completely dry. The exposed bones are then covered with different Paraloid B72 solutions:

a. To 1 liter acetone, 40 g P.B72
b. To 1 liter acetone, 100 g P.B72
c. To 1 liter acetone, 300 g P.B72

Comment: It will take several hours for the P.B72 to dissolve in the acetone. It is best to leave it overnight. Because the acetone is very volatile, it may be necessary to add more acetone to thin the solution if it is being used for several weeks, but the original ratio of acetone to P.B72 should be maintained. The solutions are best kept covered, even when leaving the work area for a short time.

Comment: The purpose of the hardener is to cover the specimen and to strengthen it.

2. After the specimen is completely dry, cover the exposed bones with layers of each solution of the P.B72, starting with the most dilute. Each layer will air-dry in a few minutes. It is important to let each layer of P.B72 solution to dry well before applying the next layer.

Comment: The protective phase is the most critical of this technique. If the bone is broken and shows holes or fragile parts where the acid should not be allowed to penetrate, fill in or cover areas with Plasticine. Apply the Plasticine over a layer of P.B72 and not in direct contact with the bone. It will be much easier to remove after preparation is completed.

C. Immersion in formic acid solution

Use 3% to 10% formic acid solution in water. Add a saturated solution of calcium phosphate to the acid bath (see Table 7.1).

Comment: The immersion period is normally 3–8 hours, depending on the fragility of the specimen. Time parameters vary between a minimum of 30 minutes and a maximum of 24 hours.

Comment: It is important to examine carefully the residual solution left in the acid bath. Sometimes parts of the specimen can become disarticulated from the main material. Also, quite often, one can find parts of other animals (especially fish vertebra, scales, etc.) that were preserved with the main specimen.

Comment: If the specimen being prepared has many bones, especially articulated ones, it is useful to take pictures or make drawings at different stages of the preparation. This information could be of great value for reestablishing the original position of the bones.

Comment: A mold and cast can also be made at different stages of acid development. There will then be an exact record of the original position of the bones. This procedure establishes the original contact and the correct dimensions of the bones that might have been taken apart during collecting or preparation.

D. Neutralization

Leave the specimen under running water or in a tub where the water is changed every 3 hours. The immersion period should be at least 24 hours.

CAUTION: Water must cover the entire specimen, to avoid the acid crystallizing on the edges of the specimen. Crystallization can also cause the internal bone structure to be destroyed.

E. Drying

The specimen must dry by natural means. Oven drying should be avoided because high temperature can damage the specimen.

F. Repeat procedure

Repeat steps B–E until the desired stage of preparation is reached – that is, the matrix is completely dissolved or partially left intact to support the specimen. If necessary, P.B72 can be removed entirely or partially with acetone during any stage of the preparation process.

G. Final cleaning

The specimen is immersed in acetone to remove all or just part of the protective P.B72. Parts of the bones can then be separated.

H. Restoration

All bones that have fallen apart during the preparation process can be glued together and restored, if desired. All should be protected with a thin layer of P.B72. A thin layer of P.B72 solution should also be left on the specimen for protection. This layer should not be too thick, or it could obscure anatomical features.

I. Storage

Special care must be taken with the storage of the chemically prepared specimen. Special supports or boxes may need to be constructed.

Micropreparation with formic acid preparation technique using *rhinobatos buerleni* teeth

Rhinobatos buerleni is a ray from the Santana Formation in Brazil. The goal was to separate its teeth from the matrix for study under the scanning electron microscope. These teeth are individually smaller than the size of the period on this page. We will include the procedure as it was done, assuming the reader will be able to extrapolate to his or her own particular situation from the description.

1. Separate a small section (2 cm approximately) of the jaw from the body of the specimen using a rock saw or a chisel.
2. Fill a large jar with a 5% solution of formic acid–water. Add the usual calcium phosphate ratio of 1 teaspoon per liter of water. Be sure the jar is large enough to accommodate a fully immersed funnel (see Figure 7.4).
3. Put a coffee filter inside a funnel and then put this inside the acid bath jar.
4. Set the piece of specimen in the funnel. Check after several hours to see how much matrix is dissolved.
5. When the matrix is completely dissolved (or dissolved enough to get a sufficient number of the tiny teeth free), lift the funnel out of the jar.

- Pour used acid out of the jar.
- Place the funnel with specimen back in the empty jar (or any stand that will support the funnel).
- Dribble gently running water into the funnel. At the same time, delicately agitate the solution to bring the silt (which is even lighter than the small teeth) to the surface. The silt will skim over the top of the funnel, leaving the teeth behind.

6. Leave specimen in water for as long as it remained in acid, using the same setup.
7. Let the water drain out through the filter. Lift the coffee filter out of the funnel, carefully slit it so it will lay flat, and pick out the teeth under the microscope with a very fine brush (see Figure 7.5).

A transfer procedure for matrix removal: Clear casting resin embedding technique – preparation for acid treatment

The purpose of the procedure is to maintain the relationship of the articulated skeletal parts after the matrix has been removed. This is accomplished by fixing the exposed portion of the fossil in a transparent resin medium before undertak-

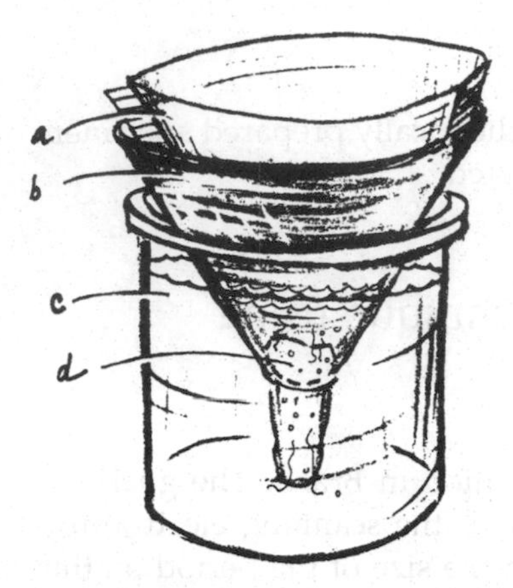

Figure 7.4. Simple micropreparation setup. (a) Coffee filter, (b) funnel; (c) acid bath to cover specimen; (d) chunk of matrix. (See text 1.)

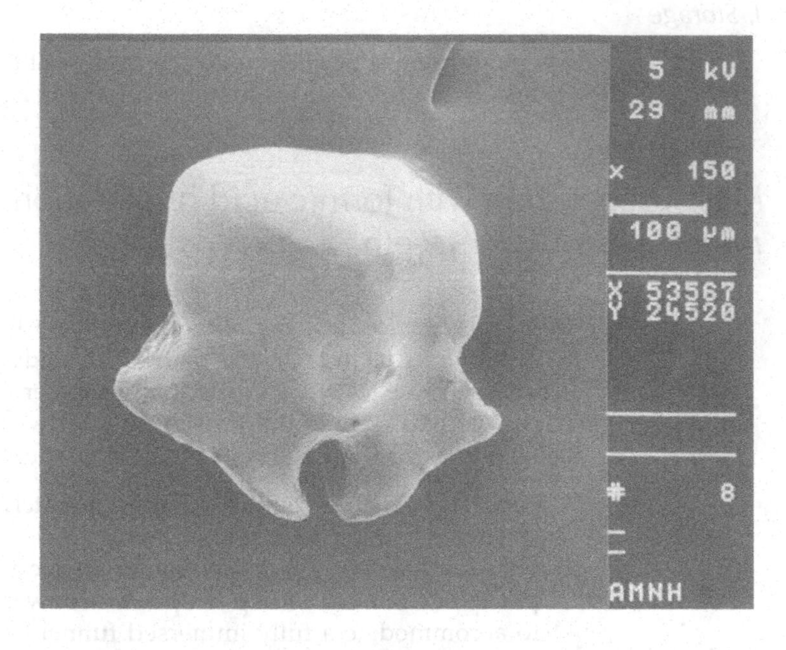

Figure 7.5. Scanning electron microscope enlargement of object under discussion, Rhinobatos buerleni tooth, after acid extraction. (See text 7.)

ing acid development. The technique developed at the American Museum of Natural History is described here in detail. Alternative methods are also outlined, along with comments as to their advantages and disadvantages relative to those of the AMNH procedure.

A. Initial preparation

1. Trim away any excess matrix with a rock-cutting diamond saw, if available. Be careful not to damage the embedded fossil.

Comments: Trimming the specimen reduces the quantity of resin needed for embedment and saves time and the amount of acid required for the acid bath.

For this embedding procedure, construction of the retaining walls is easier if the trimmed sides of the specimen slope inward from the fossil surface down to the base of the matrix and if the perimeter of the specimen is free of jagged edges (see Figure 7.6).

In instances where greater lateral access to the developed fossil is desired, it may be advisable to leave a wider rim of matrix or to construct a suitable horizontal extension out of laminate-covered cardboard (see Figure 7.7).

2. Prepare a base of fiberboard or other rigid material covered with a heavy stock acrylic-coated paper laminate, tacking it in place with rubber cement.

Comment: The choice of laminate is important, as surface properties may affect the ease of separation from the resin embedment at the end of the procedure. A suitable example is listed in the Materials section of this review, but other plastic coated laminates should be serviceable. We recommend that a brief pretest be conducted to determine the alternative candidate's release properties in this usage.

Figure 7.6 Sides of specimen trimmed to slope inward, free of jagged edges. (See text A.1.)

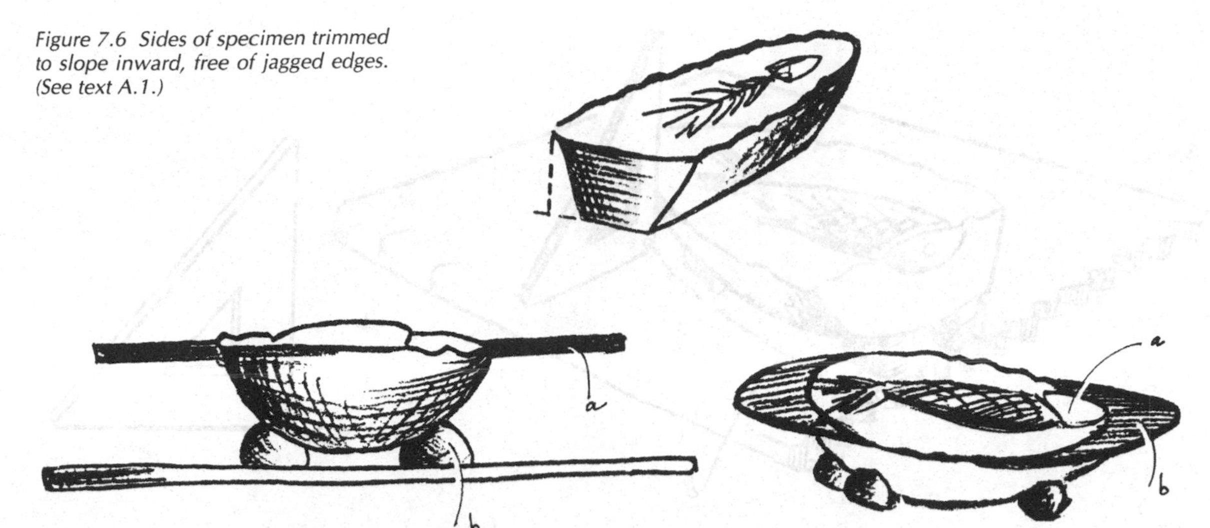

Figure 7.7 Left. Cross section of specimen showing lateral extensions. (a) Rim made of laminate cardboard. (b) Clay supports. (See text A.1.) Right. Three-quarter view of lateral extensions. (a) Bone comes to edge of matrix. (b) Laminate cardboard rim. (See text A.1.)

3. Place the fossil on the base, matrix side down, and level it with small blocks, modeling clay being among the easiest materials to use for this purpose.

4. Using a drafting triangle or other right angle guide, scribe a line on the base, duplicating the fossil outline.

Comment: This scribed line serves as the point of attachment of the retaining wall to the base (see Figure 7.8).

B. Retaining walls

1. For the inner wall, cut a strip of the paper laminate slightly wider than the distance from the highest point on the fossil perimeter to the base and (to provide for an overlap seal) 1/2 inch longer than its perimeter.

2. With a finger, press the laminate strip, coated side out, against the perimeter of the fossil to mark its outline and trim off the excess laminate above this line (see Figure 7.9).

3. Paint the outer, coated, surface of the laminate strip with the mold-release agent and let dry.

Comment: The choice of release agent is important in terms of the ease with which the laminate can be separated from the cured resin at the end of the procedure and the smoothness of the resin surface.

4. With a small brush, deposit a bead of water-based contact cement:

a. Along the line scribed on the base,
b. On the perimeter of the specimen,
c. Along all edges of the laminate strip, and
d. On the half-inch overlap areas at the end of the strip.

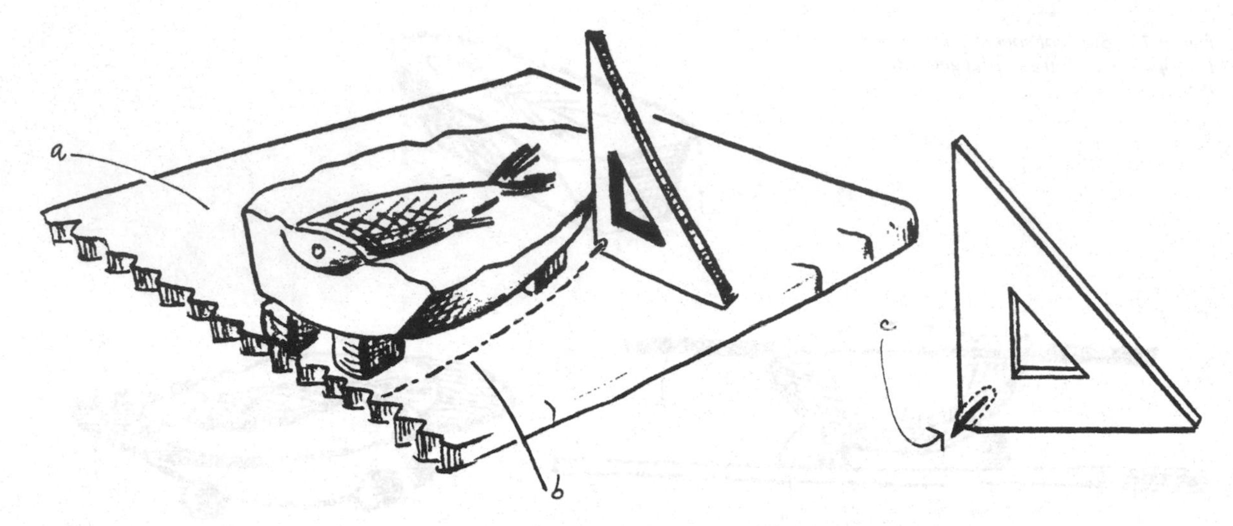

Figure 7.8. (a) Scribing a line on a base with a drafting triangle. (b) Laminate cardboard. (c) Detail of scribe; hole drilled in corner at 45-degree angle and inserted with metal point. (See text A.4.)

Figure 7.9. Inner wall, coated side facing out. Press laminate strip. (See text B.2.)

5. Allow the cement to dry to a clear, shiny film (approximately 20 minutes). *Comment:* Solvent-based cements may interact with the freshly prepared liquid resin and compromise the seal.

6. Press the laminate strip, coated side out, against the side of the fossil and the base, thus forming a vertical wall that seals off the open areas under the matrix. Press the laminate overlap closed.

7. Burnish the laminate at the edge of the fossil with a smooth instrument, to ensure a smooth seal.

8. Apply a bead of silicone rubber sealant to the seam at the base of the laminate wall to reinforce the seal at this point. Allow adequate time for the sealant to cure, preferably overnight (see Figure 7.10).

9. For the outer wall, cut another strip of the acrylic-coated laminate, this one 1 inch wider than the maximum height of the fossil above the base and 1/2 inch longer than the perimeter of the specimen.

10. Cut the top half inch of the strip along its length into narrow tabs (see Figure 7.11).

11. Paint the acrylic-coated surface of the uncut portion of the strip with a mold-release agent and let dry.

12. Apply a bead of contact cement:

a. Along the bottom edge of this strip,
b. On the half-inch overlap areas at the ends of the strip, and
c. On a line approximately one quarter inch outside the inner retaining wall.

13. Allow to dry to a clear, shiny film (approximately 20 minutes)

14. Press the laminate strip into place on the base, with its coated side toward the fossil to form a quarter-inch-wide "moat" around the fossil and press the laminate overlap closed (see Figure 7.12).

Comment: This quarter inch spacing determines the thickness of the final resin wall surrounding the specimen.

15. Reinforce the base seam with a bead of silicone rubber sealant, as in step 8.

16. Fold the precut tabs outward, to a horizontal position, and connect them with strips of tape to help support the contour of the laminate wall (see Figure 7.13A).

Comment: For large specimens, particularly with straight contours along the sides, the outer wall may require additional buttressing to resist displacement by the weight of the freshly poured resin. This can be accomplished by taping several vertically oriented strips of cardboard on the base, with their ends positioned to rest against the wall.

C: Embedment

1. If there are any small holes or cracks in the specimen that open into the walled-off space below the matrix, fill them with polyethylene glycol (Carbowax).

Comment: Such holes permit bubbles to form in the resin, partially obscuring the surface. The polyethylene glycol, being water-soluble, washes away when later exposed to acid solutions.

2. Prepare a .05% mix of a cold-setting liquid polyester resin (3 drops of hardener to 1 fluid ounce of the resin), and pour gently onto the fossil to a depth of about one quarter inch above its highest point (see Figure 7.13b)

Comment: As the proportion of hardener to resin increases from .05% to 2%, heat generation increases, resulting in a quicker set but also introducing dimensional instability and cracking of the resin.

Adding large quantities of liquid resin at one time can also generate excessive heat, causing dimensional distortions and cracking of the resin. Generally, no more than a quarter-inch layer should be poured in one setting and then allowed to achieve its initial set before continuing. Each layer will bond to the previously poured layer.

Note that if the retaining walls have not been completely sealed, a slow leak can develop. For this reason, the initial pour should be closely monitored for an hour or so before proceeding to subsequent pours.

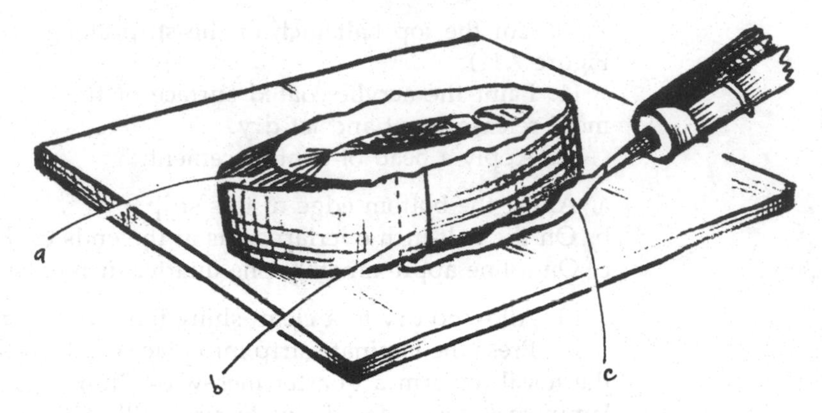

Figure 7.10. Inner wall. (a) Lateral extension rim; (b) overlap (see text B.6); (c) silicone bead (see text B.8).

Figure 7.11. Outer wall construction. Cut tabs. (See text B.10.)

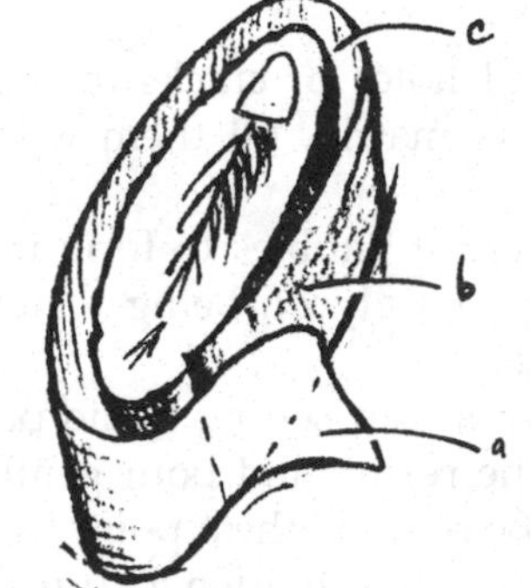

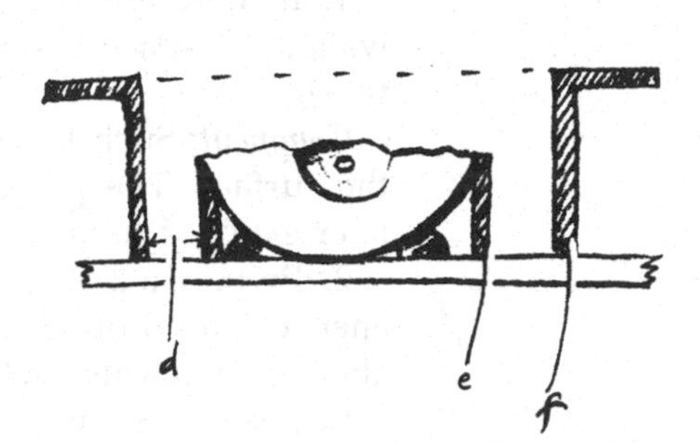

Figure 7.12A. Outer wall construction. (a) overlap; (b) outer wall, coated side facing inward; (c) 1/4-inch wide "moat." (See text B.14.)

Figure 7.12B. Cross section showing "moat" between inner and outer walls. (d) "moat"; (e) inner wall, coated side out; (f) outer wall, coated side facing in. (See text B.14–16.)

3. Once the resin has achieved an initial set, place the specimen under a lamp for at least 4 or 5 hours, preferably overnight.

Comment: Air acts to inhibit the setting action, leaving the surface tacky for days. The tackiness can be reduced by gentle application of heat (as with a lamp) to complete the curing process. Avoid too great a heat source because it may cause the surface to bubble.

4. When the resin has set completely, remove the laminate strips and trim any sharp resin edges.

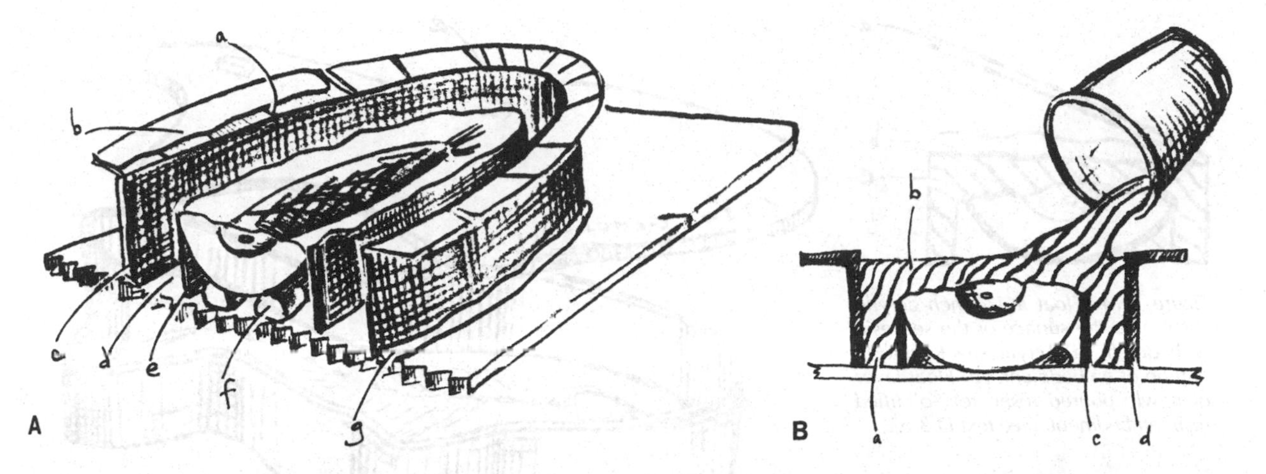

Figure 7.13A. Cross section of completed form prior to embedding. (a) Tab cuts; (b) taped edges; (c) outer wall; (d) "moat"; (e) inner wall; (f) plastiline leveling supports; (g) silicone rubber sealant. (See text B.1–16.) B. Pour resin; fill "moat," and cover specimen up to 1/8 inch above surface of highest point. (a) "moat"; (b) 1/8 inch above surface; (c) inner wall; (d) outer wall. (See text C.2.)

Comment: Small scraps of the laminate that cannot be easily removed from inaccessible areas may be left in place, to be loosened and dislodged later in the course of acid development.

D. Optional finishing step

One problem that can develop is a hazy cast on the resin surface after long exposure to the acid development solutions and water. To provide a more durable surface that retains its clarity during acid treatment, the surface may be covered with a thin sheet of acrylic.

1. Protect the sides of the finished preparation with strips of tape.
2. Scribe the outline of the preparation onto the acrylic sheet and saw it out.
3. Float a layer of resin on the solidified resin embedment surface and slowly lower the acrylic sheet onto the still-fluid resin, being careful to avoid trapping bubbles beneath it (in the same way a cover slip is applied to a microscope slide). See Figure 7.14.
4. When set, trim and sand the edges of the acrylic sheet.

E. Protective cover

1. Place the specimen on a sheet of 1/8-inch-thick acrylic, matrix side down, and pencil its outline on the acrylic's paper covering.
2. Cut out the outlined form with a jigsaw and sandpaper the edge.
3. With acrylic cement, attach four to six small blocks of acrylic along the edge of the cover to rest just inside the resin wall, serving as stops to prevent lateral movement when the cover is in place. The specimen is now ready for the acid treatment, which will remove the matrix (see Figure 7.15).

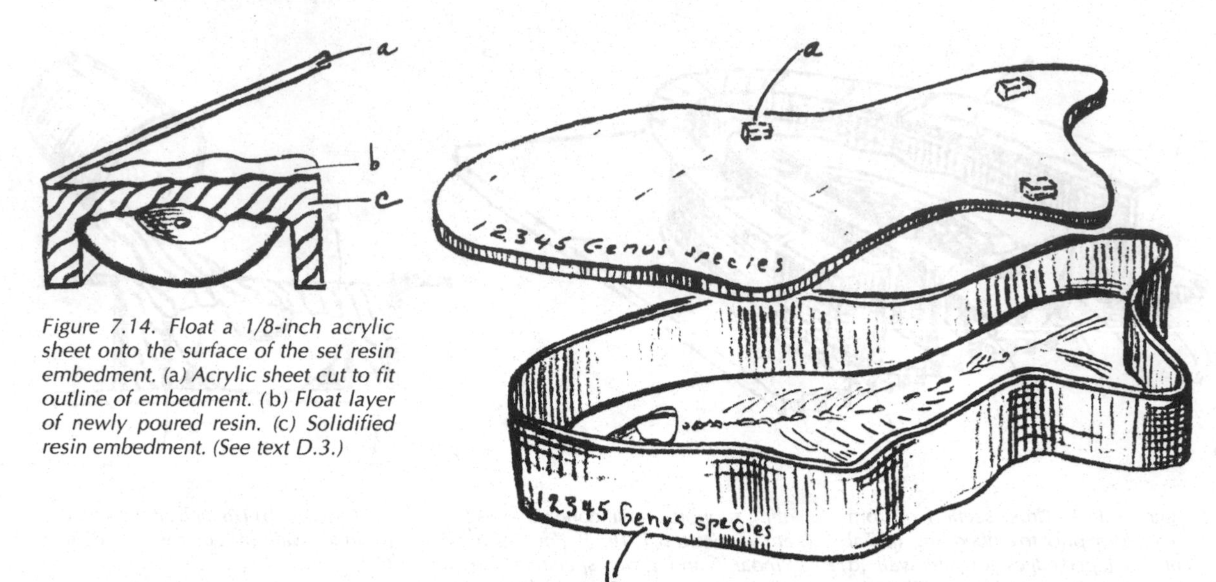

Figure 7.14. Float a 1/8-inch acrylic sheet onto the surface of the set resin embedment. (a) Acrylic sheet cut to fit outline of embedment. (b) Float layer of newly poured resin. (c) Solidified resin embedment. (See text D.3.)

Figure 7.15. Finished acid-prepared specimen with protective cover. (a) Acrylic blocks attached to inside surface of cover; (b) with motor scribing tool, scribe catalog number, and taxonomic name of specimen. Inscribe catalog number on cover as well (too often lids become separated from the specimen). (See text E.3.)

Alternative transfer methods

Consult sources, noted in the References section, for more specific details of these procedures.

1. Toombs and Rixon (1959): Clean the specimen by immersing it in dilute acetic acid followed by thorough washing and drying. Make a tray of thin aluminum foil with a quarter-inch margin all around the specimen and deep enough for the walls to project about half an inch higher than the highest point of the specimen. Place the specimen in the tray with the exposed bone facing upward and pour successive layers of prepolymer, none more than half an inch thick, around the specimen until its highest point lies one eighth of an inch below the top of the set plastic. In the case of large or irregularly shaped matrices, avoid the waste of plastic by pouring gelatin, molten paraffin wax, or carnuba wax into the tray instead of plastic to within an inch or so of the fossil and complete the preparation with subsequent layers of prepolymer. Allow each layer to harden before the next is poured. Set aside for 4 or 5 days to allow the resin to mature. Clean off the sticky surface residue with a rag dipped in acetone. Grind flat the upper surface of the plastic, through which the fossil can be seen, with a coarse abrasive on a wet band facer. Flood the flattened surface with more polymer and lower a thin glass plate onto it, taking care that no air bubbles are trapped. Cut or grind away excess plastic, wax, or gelatin to expose the rock matrix and begin the acid preparation.

Advantage

- The aluminum foil tray can be constructed easily and rapidly.

Disadvantages

- Access to the matrix for subsequent acid dissolution requires grinding away excess plastic. In the case of large or irregularly shaped matrices, an extra step is required to reduce the amount of plastic waste.
- The plastic sides of the finished specimen may not be flat enough to permit clear observation of the fossil, particularly if a wax "filler" has been used to conserve the amount of plastic.
- The sides of the plastic walls may overlay the exposed fossil to some degree, interfering with observation and/or protection of the fossil during the acid-dissolution procedure.

2. MacFall and Wollin (1983): Trim off as much of the matrix as possible without risking damage to hidden parts or weakening the slab. Clean a sheet of glass with a mold release agent (e.g., Pledge). Construct the retaining wall of a one-inch-wide strip of .005 Mylar and affix it to the glass with modeling clay, self-adhesive rubber molding strips, or masking tape, placed outside the strip. Pour polyester resin into this walled area to a depth of about 1/4 inch and allow it to set. Paint the fossil and matrix surface with a heavy coat of the resin and place the specimen face down on the hardened plastic. Add enough liquid plastic to rise up the sides from 1/2 to 1 inch. After the last layer has been added, prevent exposure to air by covering the resin with a liquid available from the supplier of the plastic. Cover and allow to set for a day. If the plastic is still tacky, heat the piece gently in an oven or electric fry pan until the plastic sets. Then remove the side walls and separate the plastic block from the glass for subsequent acid development.

Advantage

- Easier preparation of the base and retaining walls.

Disadvantages

- The sides of the plastic walls may overlay the exposed fossil to some degree, interfering with observation and/or protection of the fossil during the acid-dissolution procedure.
- Placing the specimen face down may permit some bubbles to remain, possibly interfering with later study of the fossil.
- In the case of curved specimens (convex exposed bone surface, concave matrix surface), the hollow under the matrix can fill with plastic, interfering with matrix dissolution.

3. Simpson (1990): Construct an open-topped box with walls and floor of 1/16-inch acrylic (1/8-inch acrylic used for blocks over 12" × 12"). Shape the sides of the box to conform roughly to the outline of the matrix block, thereby avoiding the need for extensive trimming. Test for water tightness to avoid resin leaks later. Apply one or more coats of resin directly to the face of the specimen to provide a smooth, bubble-free surface. When set, place the fossil face down, supported by small acrylic blocks as spacers, and pour in a 1/4-inch layer of polyester resin, being careful not to trap bubbles under it. Pour additional increments of resin, approximately 1/4 inch at a time, allowing each to set, until the sides of the fossil are covered.

Advantages

- Easier preparation of the base and retaining walls.
- A clearer window to the fossil because the face-down approach provides a smoother and harder resin surface.

Disadvantages

- The method is most appropriate for embedding flat specimens. Consequently, with curved specimens (e.g., a fossil with convex face and concave matrix surfaces), recessed areas of the matrix can fill with plastic, interfering with matrix removal in acid and with complete access to the fossil once it has been exposed.
- Placing the specimen face down requires special care to prevent trapping of some bubbles.

Suppliers

The following is a partial list of suppliers of materials referred to in this chapter. The list is intended only as a guide. All suppliers were rechecked before publication, but companies go out of business, discontinue materials, and create new materials. If you call a manufacturer and ask to speak to the chemist, that person can be very helpful in advising about safety and applicability of a particular material (as most of our uses are adapted deviations of materials and techniques). Because some materials require special precautions in handling, it is recommended that a material safety data sheet be obtained from the manufacturer for each product intended for use in the laboratory.

Laminate Walls

10pt, #2823 Bright silver foil
Hampden Papers, Inc., Holyoke, MA 01041, 413-536-1000

Release Agents

Kantstik mold release #532 (red) [low viscosity polyvinyl alcohol (PVA)]
Berton Plastics, Inc., 120 North St., Box 1906, Teterboro, NJ 07608, 201-288-7900

Investing resin

Alplex clear casting resin [styrene monomer] # 132
Adhesive Products Corp., 1660 Boone Ave., Bronx, NY 10460, 718-542-4600

Sealant

Silicone II bathroom tub and tile sealant (white or almond) or silicone II Clear Sealant
G. E. Silicones, Waterford, NY 12188, 800-255-8886

Adhesives

Weld-on #3 acrylic plastic cement
Industrial Polychemical Service, Box 379, Carson, CA 90247, 213-321-6515

Elmer's SAF-T contact cement #E810
Borden, Inc., Department CP, Columbus, OH 43215 614-225-4000, This is a little hard to find, best to contact Borden directly.

Cyanoacrylate Glue
Locally available under brand names Devcon, Super Glue, Krazy Glue, etc.

Solvent

Acetone
Locally available

Hardeners

Acryloid B-72
Rohm & Haas Co., Independence Mall West, Philadelphia, PA 19105, 215-592-3000. In New York, can be ordered from: Talas, 213 W. 35 St., New York, NY 10001, 212-736-7744

Acryloid B-67 [polybutyl methacrylate]
Rohm & Haas Co., Independence Mall West, Philadelphia, PA 19105, 215-592-3000

#1276 Glyptal clear
Brownell Electo Co., 84 Executive Ave., Edison, NJ 08817, 908- 287-3355

Paraloid B72 [methacrylate resin]
C.T.S. S.a.a., via A. da Schio, 6, 36051 Olmo Di Creazze (VI), Italy

VINAC #B-15 beads [vinyl acetate homopolymer]
Air Products and Chemicals, Inc., 7201 Hamilton Blvd, Allentown, PA 18195, 215-481-4511

Acid systems

Formic acid
City Chemical, 132 W. 22 St., New York, NY 10011

Citric acid
City Chemical, 132 W. 22 St., New York, NY 10011

Calcium phosphate (tribasic)
City Chemical, 132 W. 22 St., New York, NY 10011

Filler, magnetic stirrer

PEG 3350 [polyethyleneglycol] [carbowax], flake or powder
Fisher Scientific (Headquarters), 711 Forbes Ave., Pittsburgh, PA 15219, 800-766-7000

Alternative hardening agents

Listed here are several alternative hardening agents. We have experimented with all of them, thanks to the samples sent to us by the individuals cited. A hardener ideally should strengthen the specimen and protect against attack by acid during immersions. Each hardener dissolves to some extent in acid. Personal choice of acid depends on different variables of handling, reversibility, safety, stability over time, and other factors. For more detailed information on the characteristics and performances of the materials, write to the representatives at the institutions or companies.

Acryloid B-67
Manufacturer: See Suppliers.
This material is used at the Field Museum of Natural History, Chicago, by Bill Simpson: Department of Geology, Roosevelt Rd. at Lakeshore Dr., Chicago, IL 60605-2496. 312-922-9410.
Composition: An isobutyl methracrylate, more generally, a polybutyl-methacrylate.
Stability: Dissolves somewhat in undiluted formic acid.
Solvents: Acetone, methyl ethyl ketone.
Basic mixture: Dissolve in solvents to desired thinness.

VINAC B-15 Beads
Manufacturer: See Suppliers.
This material is used in the lab at the University of Michigan Museum of Paleontology by Bill Sanders and Bob Carr: 1109 Geddes, Ann Arbor, MI 48109.
Composition: Vinyl acetate homopolymer.
Safety: See manufacturer's safety sheet.
Stability: Essentially nonreactive.
Basic mixture: (1) Thick mixture: 1/4 beads, 3/4 acetone. Mixture takes several days to dissolve, requires occasional vigorous shaking (or use a magnetic stirrer, if available). (2) Thin mixture: 1/8 thick mixture, 7/8 acetone. Experiment and find mixtures suitable to your own needs.

Paraloid B72
Manufacturer: See Suppliers.
This material is used by Alexander Kellner: c/o American Museum of Natural

History, Department Vertebrate Paleontology, Central Park West at 79 St., New York, NY 10024-5192. It handles well. It absorbs and hardens better than acryloid.

Composition: Methacrylate resin (acrylic plastic resin).

Safety: See manufacturer's safety sheet. Use with proper ventilation, preferably under fume hood.

Stability: Dissolves somewhat in undiluted formic acid.

Basic mixture: (1) 40 g P.B72 to 1 liter acetone, (2) 100 g P.B72 to 1 liter acetone, (3) 300 g P.B72 to 1 liter acetone.

Glyptal

Many people are switching from Glyptal to resin hardeners. Glyptal deteriorates over time, leaves a darkish color, and if heavily applied, leaves a very shiny surface, which makes it difficult to see surface detail.

Composition: See Chapter 2, Adhesives and consolidants.

Safety: See manufacturer's material safety data sheet.

Stability: Discolors and deteriorates over time. Specimens treated with Glyptal 6 to 10 years ago are again in need of preservation.

Basic mixture: Thin coat–to a full jar of Glyptal, add only enough Glyptal so the (honey) color is barely visible. Thick coat–used for gluing parts together. Dilute to a workable consistency.

II. Removal of bone from acid-insoluble matrix: The use of hydrochloric acid to make a matrix mold

The following technique works well on shales, sandstones, and oil shales. Hydrochloric acid can be used to dissolve bone from matrix. This becomes the (negative) mold, and the silicone rubber peel which is subsequently made becomes the (positive) cast. This technique is recommended where bone in the matrix tends to obscure rather than clarify observation. Unlike in formic acid preparation, where the bone is preserved and the matrix removed, in the following technique the original fossil material is sacrificed. The matrix mold that remains, however, is actually far more informative. Although highly effective, the technique is irreversible. Because any information can be considered useful to some degree, it is worthwhile making a "before (immersion in acid)" silicone rubber peel. The "after" silicone peel will be the positive impression. This technique requires several immersions in the acid bath to dissolve the bone completely.

Method

A. Make a "before" silicone rubber peel

Comment: Although the "before" and "after" peels are clearly distinguishable from one another, inscribe "before" or similar notation (in reverse) along with the usual catalog number and genus and species name in the surrounding clay setup. This makes subsequent identification easier.

B. Procedure

Proportions: 10%–15% hydrochloric acid in water.

Reaction time: relatively short (approximately 20 minutes to 1 hour), rarely longer, depending on the size of the specimen and the amount of bone to be dissolved.

CAUTION: REQUIRES FUME HOOD – DO NOT BREATHE FUMES. Get safety instructions from supplier.

1. Choose a plastic tub or glass dish that will closely accommodate the size of the specimen (with formic acid, more water/acid is required in relation to size of specimen, and process requires a longer reaction time; this is not the case with hydrochloric acid).

2. Fill the tub with just enough water to cover, approximately 1/4 inch above the specimen.

3. Add 10%–15% hydrochloric acid to the water bath.

SAFETY: Wear gloves to avoid skin irritation or burns. Do not breathe concentrated fume (one possible danger being that you could inhale the fumes, have a startled, violent reaction, and drop the bottle of acid).

Comment: With the usual caution when starting a new procedure, start with the lower percentage of acid for the first bath, and if that does not seem to be sufficient, increase the percentage in the following baths.

4. Gently slide the specimen into the bath. Very small bubbles will indicate a reaction. (Do not forget that your nose precedes your eyes. Do not inhale the vapors when you look closely to see the bubbles!)

5. After approximately 20 minutes, check to see that the reaction has stopped. When the reaction is complete, remove the specimen to a rinse tub. Rinse thoroughly for at least as long as the specimen was in acid.

6. Examine the specimen under a microscope.

Comment: Probe the specimen delicately with a preparation needle or a brush to loosen up any thick fractured areas of bone. This will aid dissolution in the next acid bath. Use caution not to damage the matrix: It becomes the mold.

Comment: If matrix mold begins to get soft and fragile, dry it out slowly, then apply a hardener such as butvar or Glyptal (see the discussion on alternative hardening agents in Section I). Be careful not to apply the hardener to the bone (you may need some very tiny brushes for this). Ideally, no hardener should be used in order to get the highest fidelity peels.

7. Repeat the process, steps 2–6, several times as necessary, until the bone is completely dissolved.

C. Make an "after" silicone rubber peel

Discussion

With this technique, one is able to get information from three sources rather than just one: the original, represented by a "before" peel; the HCl-cleaned matrix mold; and an "after" peel.

We have tried this technique on specimens up to 30 cm in size. This technique has been widely applied in the study of fossil fish assemblages from shales and mudstones with a low carbonate content, for example, from the Triassic of Wapiti Lake, British Columbia, Canada (Schaeffer and Mangus 1976), from the Triassic of Angola (Schaeffer 1990), and from the Jurassic Todilto Formation of the western United States (Schaeffer and Patterson 1984). It has also been a successful technique when used on vertebrate materials from the Elgin Sandstones (Rixon 1976).

Rixon (1976) suggests using not more than 25% hydrochloric acid to water, but we found that a 10%–15% solution was adequate for these particular specimens. The reader may wish to do a test on a nonessential piece of matrix first with lower proportions and then evaluate for him- or herself.

III. Removal of ferric iron from matrix: The Waller method

This method of chemical preparation is a reducing technique, whereby oxides and hydroxides of iron are chemically reduced to a water-soluble state. This reduction technique was originally developed by mineralogists to clean crystal specimens. It facilitates the removal of ferruginous crusts and crystalline deposits. The method was first outlined by Waller (1980) for minerals and was refined slightly by King (1983), whose procedures were adopted for fossil bone (Blum, Maisey, and Lutzky 1989).

Spectacular results were obtained at the American Museum of Natural History (AMNH) by cleaning small fossil fishes embedded in laminated algal limestone (Blum 1989). Instead of a fuzzy brown outline on a dirty beige-colored matrix, it is now possible to produce crisp ginger-colored fossils against a uniform creamy white background, with far greater resolution of detail than was previously possible. Unlike an alternative reducing process developed by Howie (1974) using thioglycollic acid, an advantage of the Waller method is that it uses no acids and is essentially noncorrosive.

This technique, however, does require a well-ventilated area, preferably a fume hood, because hydrogen sulfide is released. Discuss safety precautions with your suppliers.

Method

A. Removal of hydrated iron dioxides: The Waller method

This technique takes advantage of the fact that ferrous oxide, $Fe(OH)_2$, is more soluble than ferric oxide, Fe_2O_3, over a wider pH range. Ferric oxide is dissolved by a neutral solution containing sequestering ions. The active solution contains three sodium salts. To make the stock solution (may be stored indefinitely): into 1 liter of distilled water, add 71 g sodium citrate and 8.5 g sodium bicarbonate (not very soluble in water).

Comment: The stock solution is composed of sodium citrate, $Na_3C_6H_5O_7O2H_2O$ (which sequestrates ferrous ions), and sodium bicarbonate, $NaHCO_3$ (which acts as a buffer to maintain a neutral pH) added to distilled water, because any dissolved metal salts (particularly iron) in ordinary tap water will react with the active ingredient (see next step).

B. When ready to use

1. Add 20 g sodium dithionite (1 g to every 50 ml of stock solution used).

Comment: A third salt, sodium dithionite, $Na_2S_2O_4$, is added when the solution is mixed. Sodium dithionite is highly unstable and oxidizes readily. Thus, it should be added only as needed.

2. Immerse the specimens to be cleaned.

Comment: The container should be covered but not airtight. Draping a thin plastic wrap over the solution (directly in contact with the surface) minimizes air exposure while allowing the escape of gaseous hydrogen sulfide. It is important to keep the container covered during this process, as sodium dithionite readily oxidizes on exposure to the atmosphere.

Comment: The solution should be stirred gently and continuously with a magnetic stirrer, to enhance dissolution and penetration.

Comment: A batch of 12–18 small specimens or 2–4 larger ones can be treated in a single bath of 2–3 liters.

CAUTION: This procedure should be done under a fume hood.

3. Remove the specimens after about 12 hours (the solution remains active for about 12 hours at room temperature). Discard the used solution. Wash the specimens in distilled water for another 12–24 hours before air-drying.

Comment: If a crust of iron remains, the specimens can be transferred to a fresh sequestering solution for another round, prior to washing. Washing in distilled water is desirable if the specimens are to be retreated.

Comment: There is a tendency for the sodium citrate to sequestrate calcium as well as ferrous ions, and so the limestone matrix becomes porous and chalky. King (1982) thus recommends that the Waller method should not be used on calcite or aragonite mineral specimens. However, in the case of vertebrate fossils embedded in limestone, the sequestration of calcium can be advantageous because it softens the matrix and facilitates fine mechanical preparation with a needle.

Discussion

Removal of ferric iron from the matrix around fossil bone is highly desirable in the case of terrestrial vertebrates from red beds, or wherever there is a hard ferruginous cement. Howie's (1974) thioglycollic acid technique is applicable in a variety of situations (Rixon 1976), and represents a significant advance in chemical preparation technology. There are, however, several drawbacks to the use of thioglycollic: its relatively high cost, limited shelf life, instability, and not least of all, a foul smell. We choose, therefore, to use the Waller method, which achieves similar results with less hazard to the preparator and the budget.

Our best results were obtained by cleaning specimens of a small gonorhynchiform fish (*Dastilbe crandalli*) from Lower Cretaceous Crato Member limestone from northeastern Brazil. These specimens were left embedded in their limestone matrix. Acid preparation alone was generally unsatisfactory for these particular specimens, primarily because complex iron phosphates (in permineralized muscle tissues) surrounding the bone prevented its complete exposure. The Waller method gave better resolution of skeletal morphology in this material than either freestanding or transfer acid preparations. Some problems were encountered with embedding plastics following treatment with the Waller method, primarily because our specimens were very small (5–20 cm), and the increased matrix porosity permitted excessive penetration by the plastic. Better results occurred by combining the Waller method with delicate mechanical preparation.

Hardeners

Glyptal or Butvar could only be used cautiously, as they also tended to penetrate the surrounding matrix, thereby reducing the effectiveness of subsequent treatment with the Waller solution.

Other Matrices

Experimentation with other matrix types containing a ferric component suggests that the Waller method may be broadly applicable:

1. Teeth and dermal denticles of a Pennsylvanian elasmobranch were successfully freed from a fine-grained argillaceous limestone, which was softened to a chalky constistency, then mechanically prepared with a needle. In this particular case, however, it has not been determined to what extent calcium, as well as ferrous ions, was sequestered by sodium citrate (Maisey 1989).
2. Green River fish results were less satisfactory for reasons that are not altogether clear. It may be that our particular samples were unsuitable, whereas different lithologies may yield better results.
3. Varying results were obtained from Triassic red-bed matrix. Our results are inconclusive.

Experimentation at AMNH with the Waller method was with small specimens only (mostly 5–20 cm). It is unknown whether the technique would be effective

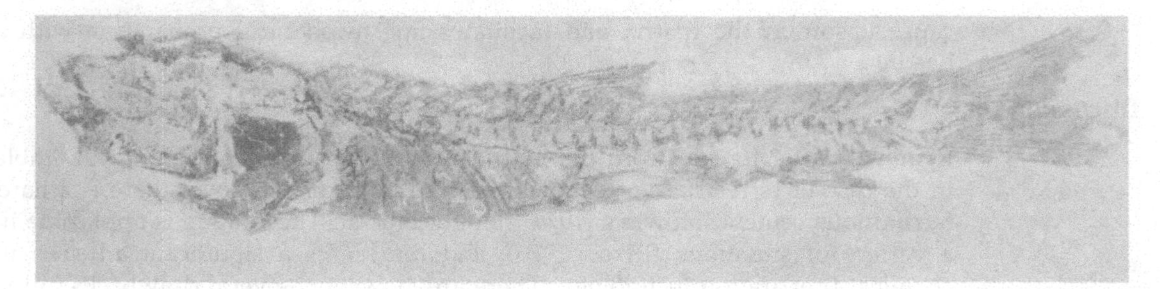

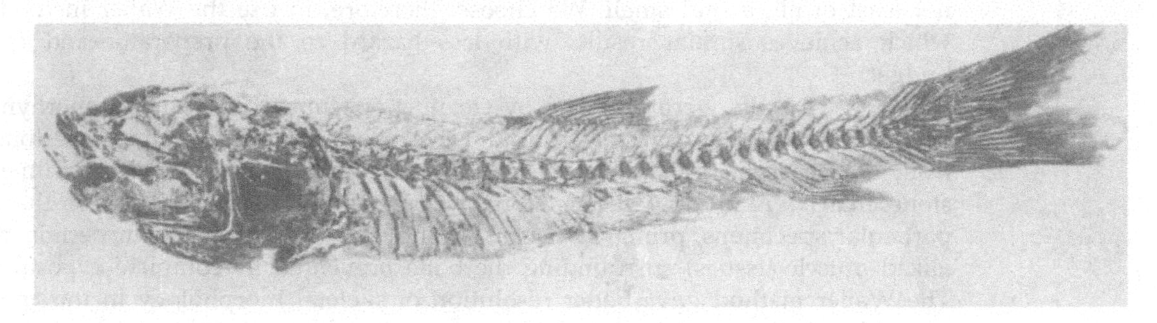

Figure 7.16. Specimen of Dastilbe crandalli. *(a) Preserved in limestone with organic and limonitic matter encrusting the bone. (b) After specimen was cleaned by the Waller method.*

on larger material. It is better suited to the removal of soft limonite rather than harder deposits of goethite or hematite.

Other Applications.
The Waller method may provide an alternative means to stabilize pyritic material (iron sulfides). The problem of "pyrite disease" and its attendant destructive consequences are well documented (Wolberg 1989 and references therein). Standard treatment involves the use of a strong base (such as ammonia or ammonium chloride) to reduce the acids produced, followed by application of an impermeable sealant such as PVA (polyvinyl acetate). At the time of writing, we intend to experiment with the Waller method to test its efficacy in pyrite stabilization.

It is worth noting that King's (1982, 1983) papers cite a wealth of methods for the chemical cleaning of minerals, many of which may be novel to paleontologists (see Figure 7.16).

Suppliers

Sodium bicarbonate: any grocery store.

Sodium citrate ($Na_3C_6H_5O_7.2H_2O$): Sigma Chemical Co., Box 4508, St. Louis, MO 63178, 314-771-5750.

Sodium dithionate ($Na_2S_2O_4$) (order as sodium hydrosulfite): Sigma Chemical Co., or Aldrich Chemical Co., Inc., Milwaukee, WI 53233.

Acknowledgments

We warmly thank Dr. Herbert R. Axelrod, who kindly donated the fossils used to develop this procedure to the American Museum of Natural History; Radford Arrindell and Michelle Murray, who helped embed and acid-prepare several hundred specimens; Daniel Goujet, whose several discussions with us were instrumental to our understanding of the nature of the process, and whose measurements helped in our development of this technique; Robert Carr, Hervé Leliévre, William Sanders, and Bill Simpson, all of whose contributions and comments are noted throughout the text; and Jo-Anne Elikann, for translation of the Braillon paper into English. We also thank Dr. George E. Harlow (Mineral Sciences Department, American Museum of Natural History) for bringing the works of King and Waller to our attention, and for his encouragement in seeking out novel approaches to chemical preparation of fossil vertebrates. Figure 7.16 was prepared by Ellen Garvens. Part of the technique of the three-dimensional acid preparation was developed at the Instituto de Geociências of the Universidade Federal do Rio de Janeiro and Seção de Paleontologia of the Departmento Nacional de Produção Mineral (DNPM), Rio de Janeiro, Brazil.

References

Aulenback, K. R., and D. R. Braman, 1991. A chemical extraction technique for the recovery of silicified plant remains from ironstones. *Review of Palaeobotany and Palynology 70*(1991):3–8.

Blum, S. D., J. G. Maisey, and I. S. Rutzky. 1989. A method for chemical reduction and removal of ferric iron applied to vertebrate fossils. *Journal of Vertebrate Paleontology.* 9(1): 119–121.

Braillon, J. 1973. Utilisation de techniques chimiques et physiques dans le dégagement et le triage des fossiles de vertébrés.". *Bulletin de Muséum National d'Histoire Naturelle.* 3 série, no. 176, (juillet-août) *Sciences de la Terre 30*:141–166. (English translation available from AMNH.)

Carr, R., 1989, 1990. Personal communication. Department of Paleontology, University of Michigan.

Goujet, D. 1988, 1990. Personal communication. Muséum National d'Histoire Naturelle, Institute de Paléontologie URA 12-CNRS, Paris.

Grant, R. 1989. Extraction of fossils from carbonates by acid. In R. M. Feldmann, R. E. Chapman, and J. T. Hannibal (Eds.), *Paleotechniques.* Paleontological Society Special Publication No. 4, pp. 237–243.

Hermann, A. 1909. Modern laboratory methods: vertebrate paleontology. *Bulletin of the American Museum of Natural History 26*:283–331.

Howie, F. M. P. 1974. Introduction of thioglycollic acid in preparation of vertebrate fossils. *Curator 17*:159–166.

Kellner, A. W. A. 1991a. Pterossauros do Brasil. Unpublished dissertataion, Instituto de Geociências, UFRJ, Rio de Janeiro.

———. 1991b. Técnica de preparação para tetræapodes fósseis preservados em rochas calcæarias. Paper presented at Congresso Brasileiro de Paleontologia, 12th Sao Paulo. Bol. Resumos, 108.

King, R. J. 1982. The care of minerals. *Journal of the Russell Society 1*(1):42–54.

———. 1983. The care of minerals. *Journal of the Russell Society* 1(2):54–77.

Lelievre, H. 1989, 1990. Personal communication. Muséum National d'Histoire Naturelle, Institute de Paléontologie URA 12-CNRS, Paris.

MacFall, R. P., and J. C. Wollin. 1983. *Fossils for amateurs – A handbook for collectors*, 2nd ed. pp. 234–237. New York: Van Nostrand Reinhold.

Maisey, J. G. 1989. *Hamiltonichthys mapesi*, g. & sp. nov. (Chondrichthyes; Elasmobranchii), from the Upper Pennsylvanian of Kansas. *American Museum Novitates 2931*:1–42.

Rixon, A. E. 1949. The use of acetic and formic acids in the preparation of fossil vertebrates, *Museum Journal 49*:116–117.

———. 1976. *Fossil animal remains: Their preparation and conservation*. London: Athlone/ University of London.

Sanders, W. 1990. Personal communication. Department of Paleontology, Museum of Paleontology, University of Michigan.

Schaeffer, B. 1990. Triassic fishes from the Cassange Depression (R. P. de Angola), III.2.2. Osteichthyes. *Ciéncias da Terra* (UNL), Número Especial, pp. 20–29.

———, and M. Mangus. 1976. An Early Triassic fish assemblage from British Columbia. *Bulletin of the American Museum of Natural History 156*(5):517–563.

Schaeffer, B., and C. Patterson, 1984. Jurassic fishes from the western United States, with comments on Jurassic fish distribution. *American Museum Novitates*, November 3.

Simpson, W. 1990. Personal communication. Department of Geology, Field Museum of Natural History, Chicago.

Toombs, H. A. 1948. The use of acetic acid in the development of vertebrate fossils. *Museum Journal, London 48*:54–55.

———, and A. E. Rixon, 1950. The use of plastics in the "transfer method" of preparing fossils. *Museum Journal, London 50*:105–107.

———. 1959. The use of acids in the preparation of vertebrate fossils. *Curator* 2(4):304–312.

Waller, R. R. 1980. A rust removal method for mineral specimens. *Mineralogical Record 11*: 109–110.

Wells, J. W. 1944. A new fish spine from the Pennsylvanian of Ohio. *Ohio Journal of Science* 44(2):65–67.

White, E. I. 1946. The genus *Phialaspis* and the "Psamosteus Limestone." *Quarterly Journal of the Geological Society, London 101*:207–242.

Wolberg, D. 1989. "Extraction of fossils from carbonates by acid." In R. M. Feldmann, R. E. Chapman, and J. T. Hannibal, (Eds.), *Paleotechniques*. Paleontological Society Special Publication No. 4, pp. 244–248.

8

Heavy liquids: Their use and methods in paleontology

Russell McCarty and John Congleton

Heavy liquids have been used as a separation medium by mineralogists, petrologists, and medical scientists since the turn of the century. However, it was not until the 1940 that paleontologists began to see the potential of heavy liquids as a tool in their discipline. Conodont researchers (Ellison 1944) were probably the first to suggest that heavy liquid separation could be used to recover microfossils. By the early 1950s, paleontologists had adapted heavy liquids to recover foraminifera (Carson 1953) and vertebrate skeletal material (Griffith 1954).

Between 1960 and 1990 was a time of rapid growth in the utilization and number of applications of heavy liquid separation in the fields of vertebrate paleontology, invertebrate paleontology, paleobotany, and such allied areas as palynology, stable isotope analysis, and zooarchaeology. It is surprising, then, considering almost half a century's successful use, that heavy liquids are not more widely used for paleontological applications. That they have not may be due to misconceptions concerning the complexity of the procedures used or to justified fears about the use of the most common heavy liquids, which are, indeed, quite toxic.

Although there are many possible applications of heavy liquids to paleontology, some of which are discussed in this chapter, as yet there has been no attempt to produce an overview of the techniques or a general procedures description that researchers can modify for their particular needs. Although this chapter deals with most heavy liquids currently in use, and the methods and procedures by which they are used to recover fossils, we strongly encourage the use of relatively safe heavy liquids such as zinc bromide and sodium polytungstate.

Note: Safety is stressed repeatedly in this chapter, and the authors and editors accept no liability for injury or damages resulting from the use of any technique described herein. Before using any equipment or technique described, readers should familiarize themselves with all of the literature available and closely follow all safety recommendations.

Principles of heavy liquid separation

Heavy liquid separation is based on the most obvious property of matter – density. Density is the mass per unit volume of a substance. One cubic centimeter of water, for example, weighs 1 gram at standard temperature and pressure. Thus its density is 1 g/cm^3. The density of water is the standard against which all other substances are compared. Lead has a density of 11.3 g/cm^3. The ratio of densities of another substance and water is defined as the specific gravity (s.g.). The specific gravity of lead is 11.3 (11.3 g/cm^3:1.0 g/cm^3); in other words, lead is 11.3 times as heavy as water.

The fact that particles of two different densities can be separated by placing them in a liquid of intermediate density can be easily verified by throwing a handful of topsoil into a beaker of water. When stirred vigorously to break up the soil into its constituent parts, and then allowed to stand for a few seconds, the sand and clay particles will sink quickly to the bottom of the beaker and the organic detritus will float to the surface of the water. Quartz sand, with a specific gravity of 2.65, sinks to the bottom of the beaker, and the organic components whose specific gravities lie somewhere between 0.2 and 0.9, float on the surface of the separating medium, which is, in this instance, water. Water, however, with a specific gravity of 1.0, is useless for separating minerals, or for separating fossils from the matrix, because these materials almost always have specific gravities greater than 1.0. Thus for most paleontological applications, liquids heavier than water are required.

Fossil bone, composed of apatite and other heavy minerals, can have a specific gravity of 2.9 or higher, depending on whether the pore spaces in the bone are filled with lighter or heavier materials. Because the particles of most matrices, such as quartz, feldspars, and calcite, have specific gravities below 2.7, a heavy liquid whose specific gravity is 2.85 would cause those particles to float on or near the surface of the liquid, and the bone, with a specific gravity greater than the heavy liquid, would sink to the bottom of the container. Successful use of heavy liquid separation in any given situation depends on the specific gravity of the matrix and the type and degree of mineralization of the fossil bone. The presence of trapped air in pores or larger spaces within a bone can also affect its buoyancy in a heavy liquid. It is probably wise to test a sample of fossil-bearing matrix with a small amount of heavy liquid before investing in large quantities of these costly substances.

Table 8.1 lists specific gravities of commonly used minerals and other materials. Without a knowledge of the specific gravities of the materials with which you are working (i.e., fossil material and matrix) it would be difficult to determine an effective gravity for the particular heavy liquid used.

Heavy liquids currently in use

Organic halogens

One of the first heavy liquids to be used, and one still frequently used, is bromoform (tribromomethane, s.g. 2.89). Its solvents are benzene, acetone, and ethanol. Tetrabromethane (s.g. 2.96) is another heavy liquid in common use. Its

Table 8.1. *Selected specific gravities*

Amphiboles	2.85–3.60	
Apatite	3.15–3.20	(remember that pores in bone can be filled with lighter material or heavier material)
Aragonite	2.95	
Bone	1.70–2.10	
Calcite	2.71	
Clay minerals	2.0–2.7	(based on water content)
Dolomite	2.89	
Dentine	2.30	
Enamel	2.90–3.10	
Feldspars	2.54–2.76	(microcline is lighest, anorthite heaviest)
Gold	19.30	
Limestone	2.70	
Hematite	5.26	
Opal	1.90–2.10	
Quartz	2.65	
Ruby	3.50–4.10	
Water	1.00	

solvents are the same as for bromoform. An even heavier liquid is diiodomethane (methylene iodide, s.g. 3.3), which is soluble in acetone and propanone. These three heavy liquids are organic compounds of the halogen elements and are true liquids that can be diluted with solvents.

It must be pointed out that these liquids are very dangerous and must be handled and used with a rigorous regard for the operator's safety. Bromoform and tetrabromethane are both easily absorbed systemic poisons, and the latter is a known carcinogen (Stone 1987). The hazards of methylene iodide are lesser known, but until more is known about this substance, we must assume its negative properties are similar to those of bromoform and tetrabromethane (Hauff and Airey 1980). The dangers associated with these halogens are compounded by the fact that the solvents themselves present a health risk to the operator. Toxic fumes mandate that all work be performed in a fume hood, and the operator should wear protective clothing, gloves, and goggles and use a chemical respirator.

In addition to their toxicity, benzene, acetone, propanone, and ethanol, the solvents used to dilute these heavy liquids and regulate their specific gravities, are extremely flammable.

Inorganic halogens

There are two inorganic compounds of the halogen family that can be made into suitable heavy liquids. Zinc chloride and zinc bromide are supplied as crystalline solids that are soluble in water and produce maximum specific gravities in the range of 2.3–2.7. Although zinc chloride and zinc bromide have lower specific gravities than the organic compounds, such as tetrabromothane, they have the advantage of being much less dangerous to use.

On the Circla ratings (scale 0–3) found on materials safety data sheets that manufacturers are required to provide users, zinc chloride has a rating of 3, the worst, due to serious problems relating to acute and chronic exposure. Gloves, protective clothing, goggles, and a dust mask should be worn, and all operations should be performed in a chemical fume hood. Zinc bromide has a Circla rating of 1, considerably safer than that of zinc chloride. However, it is recommended that the same protective measures be followed as with zinc chloride.

Sodium polytungstate (SPT)

A recent addition to the heavy liquid family is SPT (sodium polytungstate, s.g. 3.1). This compound (also known as sodium metatungstate) is manufactured in Berlin by one company and is very expensive (more than $100 per kilogram). This cost is mitigated somewhat by the fact that the solvent/dilutant for SPT is water instead of more expensive solvents, such as ethanol or propanone, which are used with heavy liquids.

The high cost of SPT is also balanced by its safety record. Stanley Krukowski of the New Mexico Institute of Mining and Technology, who has published his evaluation of SPT after working with it for one year, reports that the use, maintenance, and storage of SPT pose no known health hazard (1988). Because SPT is neither flammable nor corrosive, the safety problems associated with these properties are minimized. Some researchers have suggested that in view of the health hazards associated with bromoform and tetrabromethane, these compounds be abandoned in favor of SPT (Stone 1987).

Table 8.2 lists heavy liquids and the solvents used to vary their specific gravities.

Applications of heavy liquids

The applications and procedures of heavy liquid separation have been addressed in a number of papers found in handbooks written for different disciplines; however, the existing literature tends to be situation specific, that is, written for the recovery of a specific product such as conodonts or foraminifera. When perusing the literature on heavy liquids, keep in mind that a safer heavy liquid can, in most cases, be substituted for the organic halogens (such as tetrabromethane).

Microvertebrate recovery

To the vertebrate paleontologist, the most obvious application of heavy liquid separation is to separate microvertebrates from the matrix. More than a hundred years ago, Charles Moore washed and sieved 3 tons of clay to recover 30 teeth (McKenna 1965). Since then, screen washing has become the standard procedure for recovering microfossils, especially microvertebrates. Hibbard (1949) refined and developed screen-washing techniques to the point where a ton of matrix could be reduced to several cans of concentrated matrix. Still, screen washing is a time-consuming, labor-intensive activity that does not stop with a can of con-

Table 8.2. *Commonly used heavy liquids*

Bromoform (tribromomethane)	
Formula	$CHBr_3$
Density	2.89
Solvents	Benzene, ethanol, acetone
Tetrabromethane	
Formula	$C_2H_2Br_4$
Density	2.96
Solvents	Benzene, ethanol, acetone
Methylene iodide (di-iodomethane)	
Formula	CH_2I_2
Specific gravity	3.325
Solvents	Benzene, ethanol, acetone, propanone
Zinc bromide	
Formula	$ZnBr_2$
Specific gravity	2.6 (when made into a liquid)
Solvents	Distilled water
Zinc chloride	
Formula	$ZnCl_2$
Density	2.6 (when made into a liquid)
Solvents	Distilled water
Sodium polytungstate (sodium metatungstate)	
Formula	$3Na_2W0_4.9W0_3.H_2O$
Specific gravity	3.1
Solvents	Distilled water

centrated matrix. This end product must then be searched, or picked, under a magnifier to remove any fossil material.

In many cases, heavy liquid separation can be used to do the picking. In the 1950s, Griffith (1954) used bromoform to retrieve microfossils from Welsh bone beds. Organic halogens continued to be used by a small number of researchers for the next several decades. Murry and Lezak (1977) refined the process using tetrabromethane to recover microvertebrates. By the 1980s many researchers began to look for safer methods of heavy liquid separation, even turning to alternatives such as the interfacial method described by Freeman (1982). Congleton (1988) and Krukowski (1988) worked out techniques using sodium polytungstate as a heavy liquid for microvertebrate separation. Sodium polytungstate is a tungsten-based heavy liquid that has an excellent safety record.

Phase separation

Heavy liquid separation can also be used effectively to isolate a single phase or component of fossil-bearing material if the phase or component has a specific gravity different from the other elements that make up the material. In the application described previously, separation results in two products, the "floats," material that floats in the heavy liquid, and the "sinks," the material that sinks

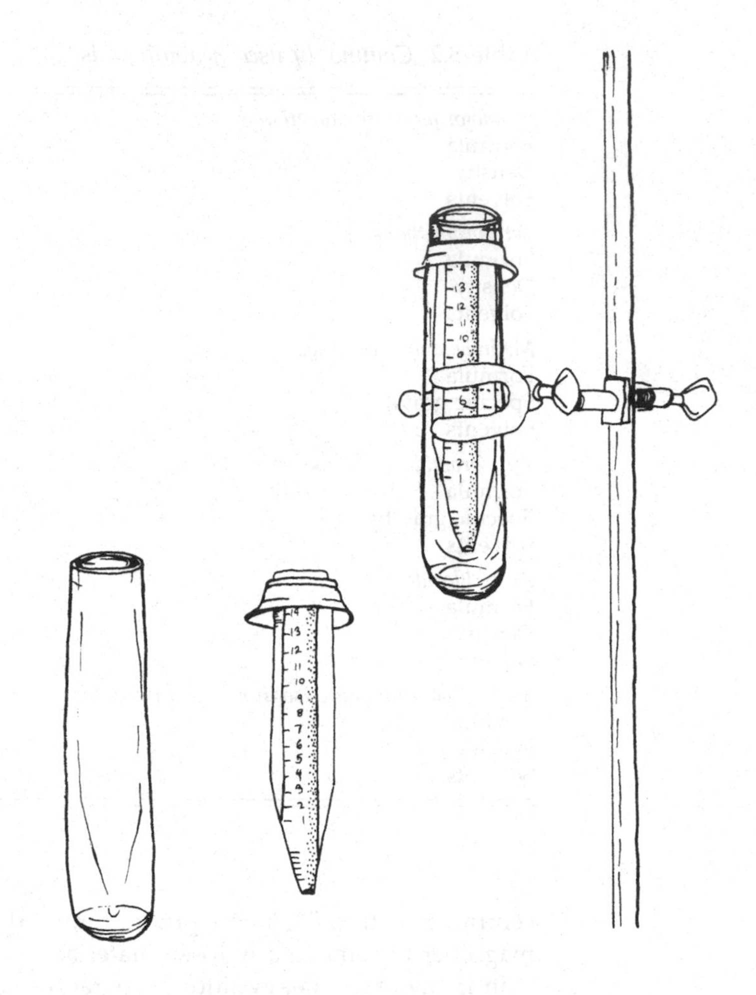

Figure 8.1. Separation apparatus for small, fine-grained samples. McCarty (1984) used this setup for separating dentine and enamel from pulverized teeth. The inner test tube has a hole in its bottom so that heavier particles will sink to the bottom and exit into the outer test tube.

in the heavy liquid medium. These end products themselves often contain certain components that the researcher wishes to analyze, and thus another round of separation is required. This is accomplished by adjusting the specific gravity of the heavy liquid to a point somewhere between that of the components to be separated and repeating the general procedure.

Such applications have been used in stable isotope analysis of skeletal material. McCarty (1985) separated dentin and enamel of fossil teeth using heavy liquid separation techniques. In this case, the entire tooth was pulverized and then passed through a #200 sieve. The resulting powder was introduced into a separating apparatus consisting of a series of test tubes, one within the other, which were filled with zinc bromide liquid (see Figure 8.1). By fixing the specific gravity of the heavy liquid at a point between that of dentine and enamel, a successful separation of these two phases of the tooth was obtained. The floats in this case consisted of all tooth components less dense than enamel, which in some species of vertebrates would be cementum and dentine. A second round of separation using a heavy liquid with specific gravity between cementum and dentine effected separation of these phases.

Other applications

Herendeen (1985) used heavy liquid separation to isolate and recover opal phytoliths from soil samples. The process was similar to that used in pollen sample recovery where a sediment sample is first macerated by acids and bases and then separated by heavy liquid. Recently researchers have found phytoliths embedded in the teeth of herbivores (Ciochon, Piperno, and Thompson 1990). Phytoliths have also been the subject of stable carbon and oxygen analyses (Piperno 1988). With these recent developments, methods of phytolith recovery utilizing heavy liquids will undoubtedly become more important in the future.

Applications of heavy liquids for conodont recovery using bromoform or tetrabromethane (as already mentioned) were among the earliest in paleontology (Griffith 1954). See Stone (1987) for a review of the process.

Laboratory setup and procedures

Although many of the materials necessary for setting up a heavy liquids facility are determined by the particular heavy liquid to be used, a number of items, such as separation funnels, are common to all methods. These materials are readily available from most commercial suppliers of scientific and medical equipment (Table 8.3). Excluding the fume hood, the setup cost of a small heavy liquids facility should be about $1000 if all glassware and equipment has to be purchased new. That cost would include enough heavy liquid to operate several separation funnels. We have placed these materials into categories such as items necessary for use with given heavy liquids and by other considerations, such as safety (Table 8.4).

Common sense is key when working with heavy liquids. Techniques vary with the type of fossil and the type of matrix. Use your best judgment to solve any problems not encountered in these instructions. The creative researcher will discover many new applications for heavy liquid separation and in the process derive more effective techniques and methods.

Before beginning the actual process of separating fossils from matrix using heavy liquids, there are several things to keep in mind:

1. The operator must be safety-conscious at all times.
2. All heavy liquids are expensive. Avoid waste and recycle heavy liquids whenever possible
3. You may have to modify the basic techniques to suit your specific needs. Let common sense be your guide. Read the literature appropriate to the heavy liquid you wish to use, and also any literature dealing with the material you wish to recover.

The maximum specific gravity of a given liquid is an inherent physical property and cannot be increased. However, by diluting the heavy liquid with the appropriate solvent, the specific gravity may be lowered substantially. Sodium polytungstate, for example, has a working range of s.g. 1.0 (= water) to about

Table 8.3. *General lab equipment*

Item	Description/comments
Funnel rack or ring stands and supports	To hold three or more funnels.
Hydrometer	For heavy liquids, s.g. 2.00–3.00.
Plastic tubing	Nalgene or Tygon brand. Should fit funnels.
Hosecocks	Large enough to close off the entire width of the tubing.
Large funnels	Powder funnels, 150 mm diameter. Stem diameter ca. 30 mm.
Small funnels	Small lab funnels to sit in recovery beakers. A tea strainer fitted into the mouth of the beaker will also work.
Wire mesh disks	Circular wire disks to rest in neck of small funnels.
Cover plates	Plastic or glass, large enough to cover the funnels.
Beakers	500 ml glass or plastic one for each funnel used.
Aspirator filter pump	Used for halogen recovery.
Filter funnel flask	Used for halogen recovery.
Vacuum flask	Used for halogen and zinc compound recovery.
Stirring rods	Glass or plastic, 2 or 3.
Graduated cylinder	For use with hydrometer to check s.g., 500 ml.

Note: When working with halogens or the zinc compounds, glassware should be used. For sodium polytungstate, Nalgene ware should be used because the salts formed are difficult to remove from glass.

Table 8.4. *Safety equipment*

Item	Description	Comments
Fume hood	Standard chemical fume hood that exhausts fumes outside the building	Must meet OSHA standards
Rubber gloves and apron	Latex protective gloves, rubberized apron	Some heavy liquids are toxic through skin
Protective goggles	Lab safety glasses	Most chemicals in the lab pose threats to eye
Chemical respirator	Mask that protects against toxic odors	With the organic halogens, do not take chances
Chemical spill protection	Absorbent material such as kitty litter or disposable baby diapers	Such protective materials minimize the hazards of heavy liquid spills

s.g. 3.0. By adding water to SPT and monitoring with a hydrometer, you can obtain the desired specific gravity.

Sodium polytungstate

The procedure described here is that adapted specifically to the recovery of microvertebrates at Southern Methodist University (Congleton 1988).

Properties of sodium polytungstate are the following:

- Soluble in water
- Maximum useful s.g. = ~ 2.90
- pH = 6.0
- High concentrations in water have low boiling point
- Decays into insoluble sodium tungstate at temperatures above 65°–75° C
- Forms insoluble sodium tungstate at temperatures above 65°–75° C
- Forms insoluble calcium or magnesium tungstate when exposed to these ions in aqueous solution
- Tends to stratify into zones of different density if allowed to evaporate even slightly
- Nontoxic

Additional laboratory equipment

You will need a supply of distilled water and an assortment of glass- and plasticware specifically for use with sodium polytungstate. SPT may be obtained from Sometu, Falkenried 4, D-1000 Berlin 33, Germany. Glass may be used instead of Nalgene; however, dried SPT salt adheres to glass and must be washed off with distilled water. This presents problems when the funnel stem becomes clogged. SPT salt flakes easily from plasticware (Krukowski 1988).

The following items will also be necessary:

- Several 1-liter beakers
- 1 or 2 large petri dishes
- Heat lamp
- Small tea strainer
- Teaspoon
- Plastic wash bottle (squeeze type)
- 5 sheets of semirigid plastic film to cover the funnels
- Fine screen cut into 2- to 3-cm diameter circles
- 20-cm diameter filter papers (coffee filters are okay)
- 3 or 4 wide-throated plastic funnels
- 4 or 5 enameled steel evaporating trays
- 1 kg sodium polytungstate (the minimum amount required to start an operation employing 3 separate funnels)

The plastic- and glassware must be free of dust and stray ions, which cause deterioration of the SPT. All plastic- and glassware must be rinsed with distilled water before and after use. Avoid using soaps or detergents on this equipment.

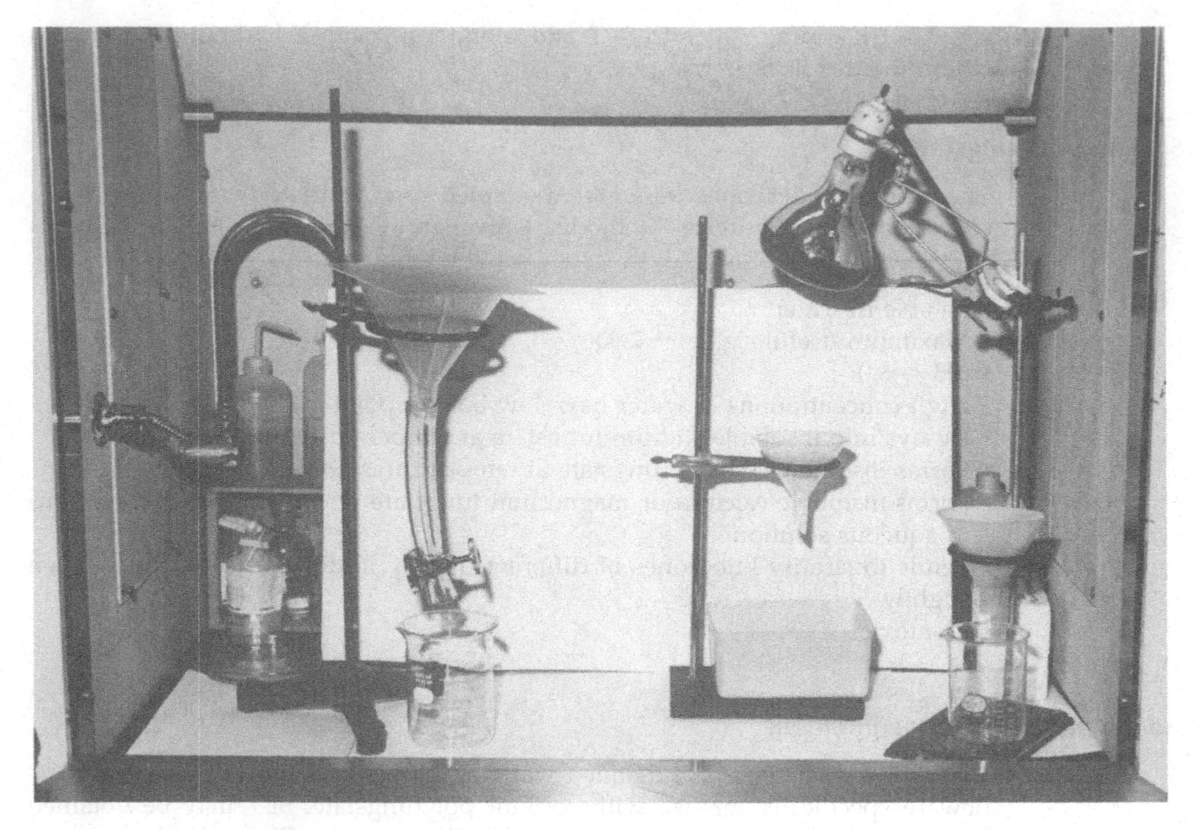

Figure 8.2. Fume hood showing a heavy liquid lab setup. Separation area is on the left and the heavy liquid recovery area is seen at the right.

Laboratory setup

The equipment will be set up in fume hood that will be divided into a space for the separating operation and a space for the recovery operation (see Figure 8.2).

Separating area

Two large ring stands (large enough to hold several kilograms of weight) with four or five rings need to be set up for the separatory and recovery funnels. All funnels should have silicon rubber tubing connected to the funnel stems and have strong tube clamps clipped near the ends of the hoses. For thc separatory funnels, the tubing must be about 20 cm long. This hose is where the fossils that sink out of the floating matrix will be collected. If the hose is fairly long, a substantial amount of fossil material can be collected before the hose fills up and must be removed from the system (a time-consuming and difficult task). Place a 500 ml beaker beneath each funnel. These beakers catch drips or recovered SPT, and are also a safety feature to catch possible spills. See Figure 8.3 for separatory funnel setup.

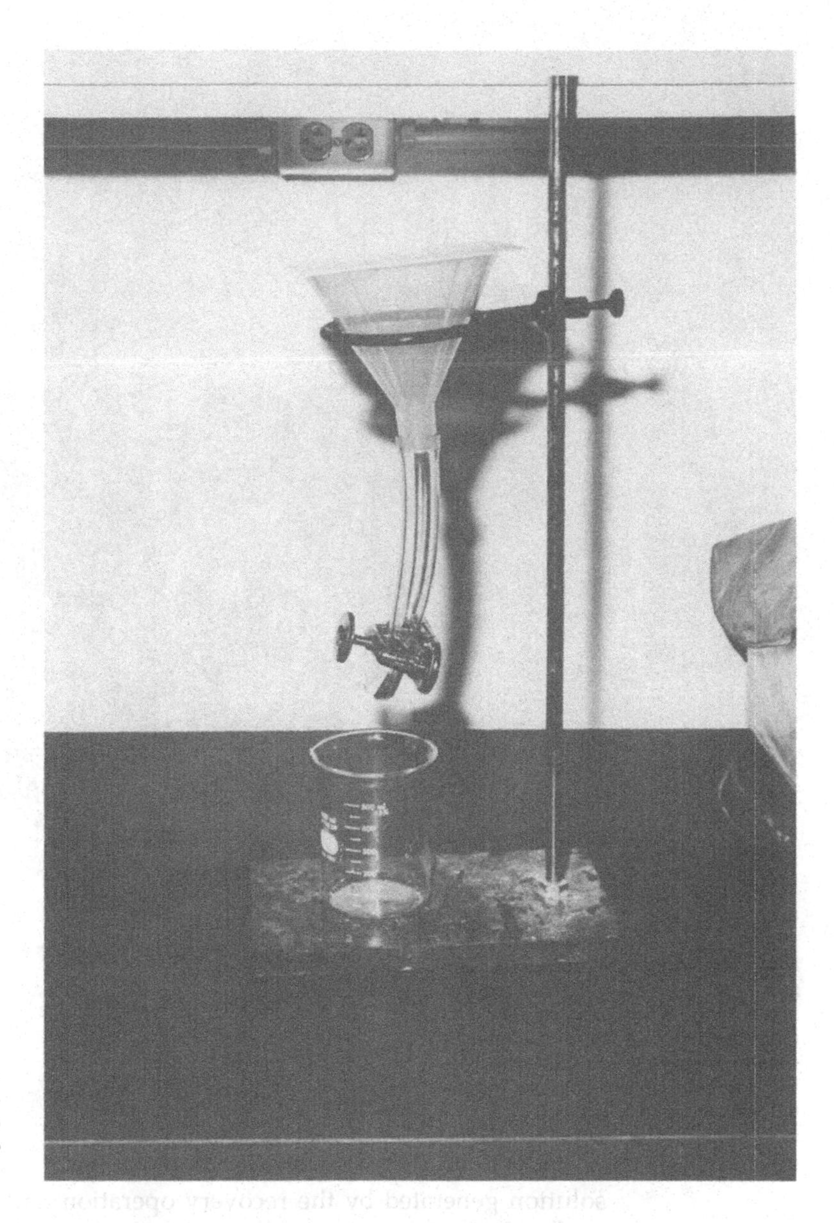

Figure 8.3. A heavy liquid funnel operation setup. The wide-stemmed funnel allows the large-diameter hose to collect fossils. The fossils in the hose are drained off into the beaker.

Recovery area

Initial recovery of the SPT is accomplished in two funnels set up on the separating ring stands (see Figure 8.4). Place a small screen in the throat of each funnel to prevent matrix from dropping through it. An additional ring stand with two small funnels and a closable container should be set up for filtering recovered tungstate.

A hot plate and a heat lamp should be available in the recovery area. The heat lamp can be effectively mounted on a ring stand, with a metal ring used to support it. Place a tray beneath the heat lamp for evaporating low density SPT

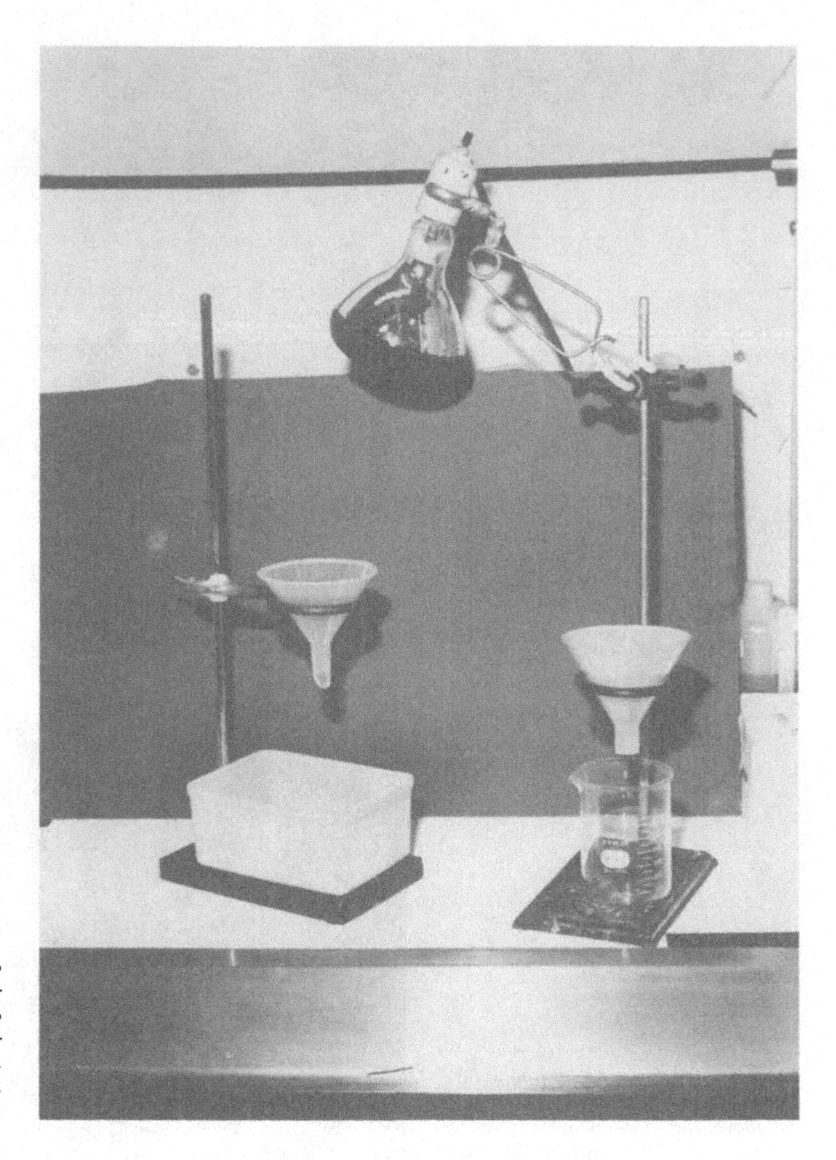

Figure 8.4. Recovery area setup. The funnel on the right is used to recover heavy liquid from spent matrix. The funnel to the left can be used to recover sodium polytungstate rinsed from fossils. The heat lamp is used for evaporating water from heavy liquid.

solution generated by the recovery operation and wash water from the rinsing of the glassware.

Matrix preparation

The matrix must be clean and relatively free of clays and calcite dust. Clay particles soak up the SPT and cause a small percentage of the expensive fluid to be lost during each separation. Thus it is wise to remove as much clay as possible before attempting separation. Kerosene treatment of the matrix may be necessary to remove clays. Calcite dust is removed by first rinsing the matrix in hot distilled water and screening in a basin of cool distilled water. Distilled water is essential because it has no calcium or magnesium ions, which will cause deterioration of the SPT into salts of calcium or magnesium that precipitate out of the solution. Screening in distilled water is necessary for all samples because this removes the finest sediment fraction, which tends to alter detrimentally the characteristics of a heavy liquid. This screening also removes sources of magnesium and calcium ions from the matrix. Considering the high cost of SPT, it is probably wise to wash all matrix in a solution of distilled water and a very small amount of Calgon, a water softening agent. This will help to ensure that all calcium ions are removed from the matrix.

Placing wet matrix in any heavy liquid can change the density of the liquid. Thus the matrix should be allowed to dry thoroughly.

SPT preparation

The sodium polytungstate should be prepared beforehand and kept ready for use in a container with the specific recorded on the outside.

If you are starting with new polytungstate, mix the powder in warm distilled water, stirring constantly and gradually adding more powder until the desired specific gravity is achieved. For example, by mixing 840 g of dry SPT powder with 160 g of water, a solution with a specific gravity of 2.94 is obtained. A quick method of estimating the specific gravity of the solution is to place a clean chip of quartz (s.g. = 2.65) in the solution. When the chip floats to the top of the solution, the specific gravity of the solution is equal to or greater than 2.65. Allow all air bubbles to escape from the liquid before measuring the final specific gravity of the liquid, as air in the liquid will influence the density of the solution.

Pour the liquid into a graduated cylinder and measure the specific gravity with a hydrometer. You will find that in an average 500-ml graduated cylinder you will need at least 300 ml of heavy liquid to float the hydrometer and get the desired reading of the specific gravity.

It is advisable first to mix the liquid to a specific gravity higher than needed and then add distilled water in small increments to lower it to the desired level.

When the specific gravity of the premixed SPT needs to be lowered, pour the liquid into a graduated cylinder and add distilled water a few drops at a time. Mix the solution thoroughly after each addition of water, because a uniform solution is necessary to measure accurately the specific gravity. Check the specific gravity after each addition of water with a hydrometer until the desired reading is attained.

Raising the specific gravity of premixed SPT will take more time than lowering it, as the only way to reduce the specific gravity is by evaporating some of the water in the solution. Evaporation is done by pouring the amount of SPT that needs to be reduced into a large beaker and allowing it to evaporate. To speed up the rate of evaporation, the beaker may be placed under a heat lamp. However, the solution must not be heated above 60° C, because higher temperatures will damage the liquid. The density should be checked periodically by pouring SPT into a graduated cylinder and measuring with a hydrometer until the proper specific gravity is reached.

Separation procedure

Fill each separatory funnel with 250 m of SPT solution. (Be sure the funnel hoses are securely clamped before filling the funnels.) Next add 2 or 3 teaspoons of dry matrix to each funnel. Stir vigorously with a plastic stirring rod to wet the grains thoroughly. Place a sheet of plastic film or plastic food wrap over the funnel to prevent evaporation. Allow the mixture to stand for 10 minutes, then stir again. After the mixture has stood for another 10 minutes, the floating matrix may be skimmed off with a tea strainer (or any similarly sized strainer). Allow a minute or two for the excess SPT to drain out of the strainer and back into the funnel. Place the drained matrix into one of the recovery funnels that has a screen disk in its throat (again, make sure the hose clamp is in position).

Repeat this procedure until all matrix from a single locality has been processed.

During the separation procedure, carefully watch the hose at the base of each separatory funnel. When it becomes filled with fossil debris to a point a little below the funnel tip, apply a clamp to the hose above the fossil material but below the funnel tip. Extra care at this step will ensure that no fossils are crushed by the hose clamp. Next, while holding the hose end in a small beaker, remove the bottom clamp. Gently tap the hose with a stirring rod to loosen the fossil material. If it does not flow out of the hose, release the upper clamp briefly to force the fossil material out by the hydraulic pressure of the heavy liquid.

At the end of each day, filter all SPT liquid that remains in the funnels. This is important, because any matrix that stays in contact with SPT for an extended period (overnight) has the potential for degrading the heavy liquid.

Recovery procedure

The spent matrix that accumulates in the recovery funnels will gradually yield SPT liquid that has adhered to the grain surfaces during separation by slowly seeping to the bottom of the funnel. Occasionally tap off the accumulated fluid and return it to the separatory funnels. Perform this task before the fluid level in the hose backs up as high as the draining matrix. The less time the matrix is in contact with the SPT, the less deterioration of the heavy liquid will occur. It is wise not to let the matrix stand in the funnel for more than 4 hours before rinsing it with distilled water for final recovery of SPT.

After the matrix has drained sufficiently (about 1 1/2 hours after the last spent matrix was added to the funnel), pour warm distilled water into the funnel. With a stirring rod, agitate the mixture to remove air bubbles. Then drain the matrix into a plastic funnel lined with filter paper, which in turn drains into an evaporating tray. Repeat this procedure at least four times. The water from the first two or three rinses should be kept separate from the final rinses, because the higher concentration of tungstate in the initial rinses will allow quicker recovery of the tungstate. Therefore keep several evaporating trays in operation, each with a different concentration level of recovery liquid in it.

The fossiliferous concentrate obtained from the separation procedure must also be washed to remove SPT adhering to the fossils. The easiest method is to rinse the concentrate in a small beaker, swirling the contents gently. Finish by decanting the liquid into an evaporating tray. Repeat this procedure several times to ensure that all SPT is recovered from the concentrate. All filters used in the separation and recovery operation should be washed with distilled water to reclaim SPT before disposing of them.

When the wash water generated by the recovery operation has been evaporated down to a density slightly higher than you require, filter it into a reclosable plastic container or directly into the graduated cylinder where its specific gravity can be measured.

Termination procedure

When the separation and recovery procedures are complete, the work area should be thoroughly cleaned of SPT spills, and any dried salts scraped away. All plastic- and glassware used in the procedures must be free of SPT and rinsed with distilled water (this rinse water should be placed in an evaporation tray to recover any SPT it may contain). After a final rinse with distilled water, the labware can be stored in a cabinet near the fume hood.

Krukowski (1988) has used SPT for the recovery of conodonts. His methodology varies somewhat from Congleton's (1988). Conodont separation takes up to 12 hours, much longer than the 20 to 30 minutes in the procedure described here. The researcher interested in using SPT should consult Kruskowski's paper.

Zinc bromide and zinc chloride

Heavy liquids made of these inorganic compounds of the halogen family are very similar in their physical properties. Their specific gravities are in the range of 2.5–2.7, lighter than the organic halogens or SPT, and thus their uses are somewhat limited. With regard to safety, both zinc bromide and zinc chloride are much safer than bromoform and related liquids, and of these two inorganic compounds, zinc bromide is clearly the better choice. Even so, zinc bromide needs to be handled with care. It can cause eye, respiratory, and skin irritation. Therefore it must be used in a fume hood, and the operator must wear protective goggles, clothing, and rubber gloves. When heated in large quantities, the fumes are quite toxic and a chemical respirator should be worn. The set-up and procedures described here are used at the Florida Museum of Natural History (McCarty 1985).

The properties of zinc bromide are the following:

- Soluble in water
- Maximum useful s.g. = ~ 2.70
- pH = 4.0

Laboratory setup

In a fume hood, set up the desired number of separation funnels, either in a wooden funnel rack or in ring stands. A short length of plastic hose (ca. 20 cm) should run from the funnel stems to 500-ml beakers placed beneath the funnels (see Figure 8.3). Hosecocks (clamps) should be securely fastened near the bottom of the tubing to prevent sinks and the heavy liquid from leaking out prematurely.

Muray and Lezak (1977) fitted suitable sized kitchen strainers in the mouths of the beakers to trap sinks, one of the two end products of the separation process.

Zinc bromide preparation

Zinc bromide can be purchased in two forms: a crystalline powder sold by medical/scientific suppliers such as Fisher Scientific, or as a prepared liquid available from Moore-Tec Industries (1185 Morris Ave., Union, NJ 07083). The crystalline form is cost-effective for limited operations that require less than a gallon of heavy liquid. However, at $500 per gallon, it becomes more economical to buy the liquid form (the minimum order is 10 gallons) where the price drops to about $70 per gallon.

To prepare a heavy liquid using crystalline zinc bromide, fill a clean beaker with 100 ml of distilled water. Add 500 g of zinc bromide and stir until dissolved. It will be quicker to place the beaker on a hot plate that has a magnetic stirrer and heat the mixture gently until all the crystals have dissolved. This ratio of 100 ml water to 500 g zinc bromide crystals should produce a heavy liquid with a specific gravity in the range of 2.5–2.7. The liquid can be poured into a graduated cylinder and the specific gravity checked with a heavy liquid hydrometer. When mixed in a ratio of 20 ml water to 100 g zinc bromide, one jar will produce 250 ml of heavy liquid (s.g. = 2.5–2.7). If a lower specific gravity is desired, add distilled water in small increments until the desired specific gravity is obtained. To raise the specific gravity to its maximum, gently heat the liquid in a fume hood to evaporate some of the water.

The liquid form of zinc bromide manufactured by Moore-Tec only needs to have its specific gravity checked before use. It can be diluted or evaporated to raise or lower its specific gravity.

Separation procedure

Fill the separation funnels to the two thirds level with the prepared heavy liquid. Add several teaspoons of clean, dry matrix (wet matrix will lower the specific gravity) to the heavy liquid. As with sodium polytungstate, heavy liquids made from zinc compounds will be absorbed by clay in the matrix. Thus it is best to remove as much clay as possible from the matrix before placing it in heavy liquids.

Stir the matrix and let sit 10 minutes. Then stir again and let stand for another 10 minutes. The floats can be strained off with a tea strainer. Sinks, which should now be trapped in the hose, can be drained off into a beaker fitted with a tea strainer (or a small funnel with a wire disk in its throat). Before the clamp at the bottom of the hose is released, the top of the hose just below the funnel stem should be pinched off with the operator's finger or with another hosecock. Care must be taken not to crush any fossils near the top of the hose. Release only as much fluid as is necessary to flush the sinks out of the hose. The sinks should be washed from the tea strainer (or from the wire mesh disk in the recovery funnel) into a beaker and further washed to remove zinc bromide. If the floats are the object of the separation, they should be washed in this manner. Both floats and sinks should be dried gently after washing.

Recovery procedure

The zinc bromide solution remaining in the funnel and that recovered in the beaker should be filtered through a filter funnel/vacuum flask setup powered by a hydroaspirator or a vacuum pump.The recovery fluid is retained for subsequent applications.

Tetrabromethane and other organic halogens

We believe that organic halogens are best left to those researchers who have a current knowledge of organic chemistry. These liquids are dangerous, as are most of their solvents, and should only be used in well-equipped labs that possess fume hoods. Those with little knowledge of chemistry or who lack fume hoods in their facilities should consider alternative heavy liquids. The geology departments of most large colleges and universities maintain heavy mineral separation facilities that utilize these heavy liquids. Researchers interested in using organic halogens should visit one of these labs before attempting to set up a system of their own.

A number of books and papers will guide researchers in their use of these materials. Hauff and Airey (1980) should be the first piece of required literature. Several geology books deal with tetrabromethane, bromoform, or methylene iodide heavy liquid separation methods. Among these, see Jones (1987) and Austin (1987). Perhaps the most informative description is Murry and Lezak (1977). This paper will guide the researcher through the separation process using tetrabromethane, including lab setup, safety measures, matrix preparation, and recovery.

Acknowledgments

We would like to thank Stan Blomely for the photographs, Laurie Walz for the drawings, and Dr. Bruce MacFadden for his comments and review of this paper. University of Florida Contribution to Paleobiology No. 403.

References

Allman, M., and D. F. Lawrence. 1972. *Geological laboratory techniques*. New York: ARCO.

Austin, R. L. (Ed.) 1987. *Conodonts: investigative technique and applications*. Chichester: Ellis Norwood, Ltd.

Brem, H., A. B. Stein, and H. S. Rosenkranz. 1974. The mutagenicity and DNA-modifying effect of haloalkanes. *Cancer Research* 34:2576–2579.

Carson, C. M. 1953. Heavy liquid concentration of foraminifera. *Journal of Paleontology* 27: 880–881.

Charlton, D. S. 1969. An improved technique for heavy liquid separation of conodonts. *Journal of Paleontology* 43:590–592.

Ciochon, Russell, D. Piperno, and R. Thompson. 1990. Opal phytoliths found on the teeth of the extinct ape, *Gigantopithecus blacki*: Implications for paleodietary studies. *Proceedings of the National Academy of Sciences* 87(20):8120–8124.

Collinson, C. 1963. Collection and preparation of conodonts through mass production techniques. *Illinois State Geological Survey Circular* 343.

Congleton, J. 1988. Heavy liquids laboratory procedures. Unpublished manuscript, Southern Methodist University.

Davis, L., and G. D. Webster. 1985. A modified funnel for heavy mineral separations. *Journal of Paleontology 59*:1505–1506.

Ellison, S. P. Jr. 1944. The composition of conodonts. *Journal of Paleontology 18*:133–140.

Freeman, E. F. 1982. Fossil bone recovery for sediment residues by the "interfacial method." *Paleontology 25*:471–484.

Griffith, J. 1954. A technique for the removal of skeletal remains from bone beds. Proceedings of the *Geological Association 65*:123–124.

Hauff, P. L., and J. Airey. 1980. The handling, hazards, and maintenance of heavy liquids in the geologic laboratory. U.S. *Geological Survey Circular* 827.

Herendeen, P. S. 1985. The alvars of the Maxton Plains, Drummond Island, Michigan: Present community composition and vegetation changes. Thesis, Michigan State University.

Hibbard, C. W. 1949. Techniques of collecting microvertebrate fossils. *Contributions of the Museum of Paleontology, University of Michigan 8*(2):7–19.

Jones, M. P. 1987. *Applied mineralogy – a quantitative approach.* London: Graham and Trotman.

Krukowski, S. T. 1988. Sodium metatungstate: A new heavy mineral separation medium for the extraction of conodonts from insoluble residues. *Journal of Paleontology 62*(2):314–316.

Krumbein, W. C., and F. J. Pettijohn. 1938. *Manual of sedimentary petrography.* New York: Appleton Century Crofts.

Lees, P. M. 1964. A flotation method of obtaining mammal teeth from Mesozoic bone beds. *Curator 8*:300–306.

McCarty, R. W. 1985. Zinc bromide: A safer alternative for heavy liquid separation. Unpublished manuscript, Florida Museum of Natural History.

MacFadden, B. J., R. McCarty, H. Converse, D. F. Williams, and R. Frith. 1984. Stable isotopic composition of bone, dentine, and enamel: Evidence from fossil and recent mammals. Unpublished manuscript, University of Florida.

McKenna, M. C. 1962. Collecting small fossils by washing and screening. *Curator 5*(3):221–235.

Merrill, G. K. 1985. Interfacial alternatives to the use of dangerous heavy liquids micropaleontology. *Journal of Paleontology 59*:479–481.

Mileson, P., and L. Jeppssom. 1983. Heavy liquid and solvent recovery – An economical method. *Journal of Sedimentary Petrology 53*:673–674.

Moore, R. C. (Ed.). 1962. Conodonts: Methods of preparation. (Pt. W, pp. 4–5). In *Treatise on invertebrate paleontology; miscellanea.* Lawrence: University of Kansas Press.

Murry, P. A., and J. Lezak. 1977. Recovery of vertebrate microfossils with tetrabromoethane. *Curator 20*(1): 15–22.

Piperno, D. R. 1988. *Phytolith analysis: An archaeological and geological perspective.* San Diego: Academic Press.

Rixon, A. E. 1976. *Fossil animal remains: Their preparation and conservation.* London: Athlone.

Stone, J. 1987. Review of investigative techniques used in the study of conodonts. (pp. 17–34). In R. L. Austin (Ed.), *Conodonts: Investigative techniques and applications.* Chichester: Ellis Horwood.

Turner, W. M. 1966. An improved method for recovery of bromoform used in mineral separation. *U.S. Geological Survey Professional Paper 550C*, 224–227.

9

Histological techniques

James W. Wilson

Histology is the biological science that studies the microscopic structure of organic tissues. Thus this chapter focuses on technical procedures that are applicable to the preparation of samples suitable for microscopic viewing and allow for or enhance the evaluation of microscopic detail of the specimen.

The specimens of interest to paleontologists present unique requirements for processing that can utilize procedures and techniques developed in both tissue histology and geology petrographic laboratories. Whereas once it was sufficient merely to obtain an acceptable photomicrograph of a paleontological specimen, it is now necessary and possible, to provide a more rigorous, detailed evaluation. To do this, one must consider the type of specimen to be studied, the location on the specimen for sample collection, the orientation of the sample, the range or diversity of microscopic changes expected, the method of viewing, possible technical enhancements to assist identification and evaluation, the methods available and suitable for quantitation of observed microscopic detail, and the equipment available to process the samples. Also, in many instances, methods and processes will need to be modified to meet the specific needs of the study.

Equipment

Saws

The specimens of interest to paleontologists are, for the most part, densely mineralized, and cannot be cut on an ordinary microtome. Consequently, the most useful method of sectioning a specimen is sawing. Until just recently the only saws suitable for sectioning were marketed through petrographic supply companies. The two most popular suppliers in the United States are Buehler and Leco, and both companies make a low-speed precision saw that is suitable for sectioning fossils. Their proprietary machines are similar. Arbor speeds are variable upward to 300–500 rpm, specimen feed is by gravity-weighted mount, lubrication is by drag application from a bath, the sample holding arm incorporates a micrometer for measuring specimen placement and section thickness, differing specimen mounting chucks are available, and 3"–5" circular blades are employed (see Figure 9.1).

Buehler markets a second saw that provides higher cutting speeds up to 5000

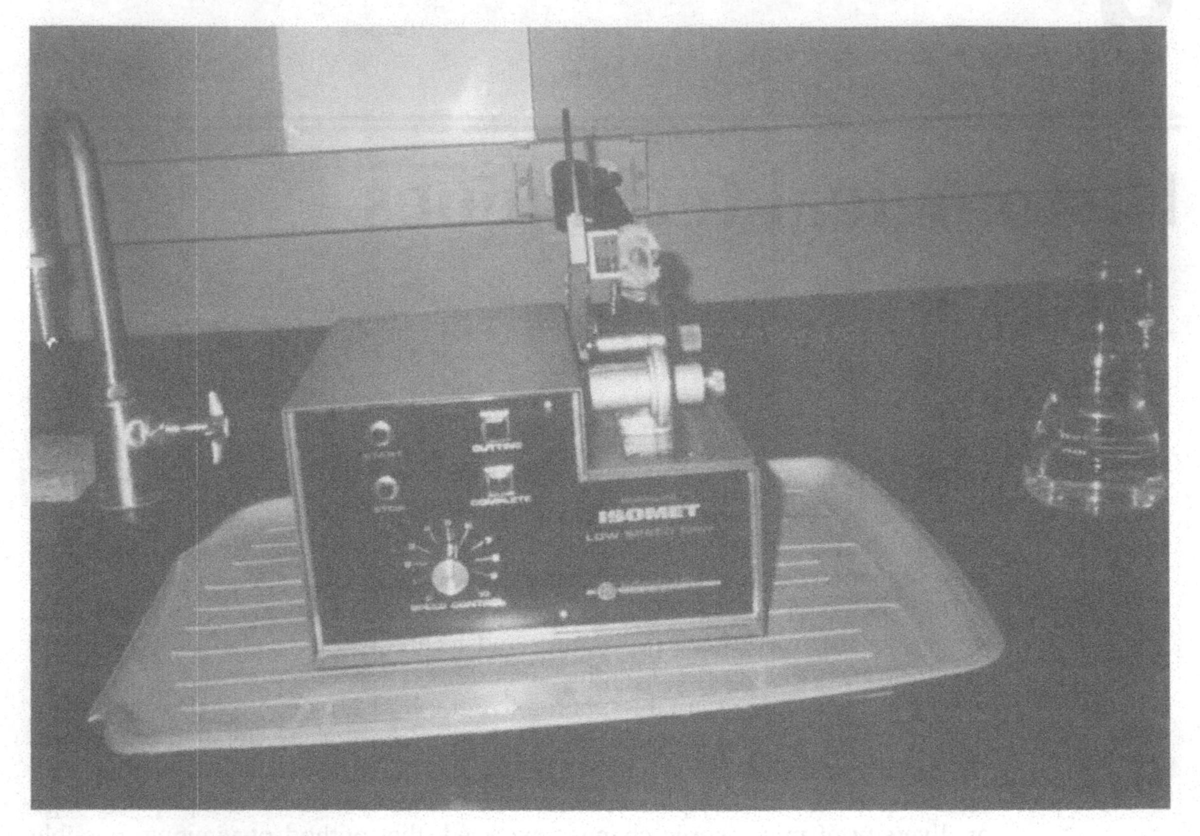

Figure 9.1. Low-speed precision saw suitable for sawing thin fossilized specimens. Circular blades 3–5 inches can be used with arbor speed variable upwards to 300 rpm. The sample holding arm incorporates a micrometer for controlling specimen placement and section thickness. Specimen feed is by weighted gravity mount. Cutting lubrication is by drag application from bath. (Isomet, Buhler Ltd., Lake Bluff, IL 60044.)

rpm and employs electronic specimen load sensing, automatic specimen arm retraction and saw shutoff at completion of sectioning, and pressure directed lubricant application and can accommodate circular blades up to 7 inches in diameter.

For larger specimens, or to block specimens down to a size suitable for cutting on one of these saws, a large industrial cutoff saw can be employed. Arbor speed is usually fixed at about 1725 rpm, cutting lubrication is by pressure-directed spray, feed of specimen can be by either gravity or hand-fed movable specimen mounting table, and blades up to 14 inches in diameter can be employed (see Figure 9.2).

The need for a precision saw capable of cutting very precise, extremely thin specimens of bone for medical research has induced the manufacture of several unique thin sawing systems. Exakt markets a modified band saw system that has a variable saw speed to 300 m/min. Individual thick specimens are mounted on glass slides, vacuum attached to the specimen mount, and then thin-sawed. Maximum cutting cross section is 190 mm × 270 mm, with a maximum usable mounting slide size of 50 mm × 100 mm. Leitz markets a saw microtome that has a rotary circular saw blade with a central cutting bore that rotates at a speed of 600 rpm. Specimens are mounted on an object holder and then spring-drawn into the central cutting surface, lubrication is by pressure-directed spray, and specimen thinness is adjusted by micrometer. Specimen size is limited to diameter of 35 mm. A third system employs a cutting wire, 0.003"–0.015" in diameter. Originally marketed by Lastec, several models are now available with cutting speeds to 2000 rpm, slurry lubrication, hand-fed specimen tables, and allowable specimen size up to 4 inches in diameter.

Saw blades

A number of blade characteristics can be specified when selecting a circular blade. Most high-quality circular blades are made of steel. The outer cutting rim can be squared, tapered, or thicker than the core blade diameter (relief cut blade). The cutting rim can be continuous or slotted. If slotted, the slots can be wide or narrow and deep or shallow. Slots carry lubricant into the cut and help with swarf removal; however, slotted blades usually leave a rough cut surface. The thickness of the blade affects both the stability of the blade during cutting and the width of the cut made. Blade thicknesses vary from 0.006 inch to 0.030 inch. Thicker blades wobble less during cutting but produce a thicker kerf. The abrasive cutting material bonded to the rim of the blade can be almost anything, with aluminum oxide, silicon carbide, boron nitride, or diamond being the most common. Bond adhesive of the abrasive to the blade affects the cutting life of the blade. Shellac, resinoid, vitrified, and metal bonds can be used. Most manufacturers will also indicate which of their bond types are more or least durable. Abrasives that wear quickly do not need extremely durable, more expensive bonds, whereas expensive, durable abrasives, such as diamond, are best purchased with very durable bond types. The actual size of the particles of abrasive used to coat the cutting rim of the blade affects both the speed at which the blade cuts and the quality of the resultant cut surface of the specimen. Particle size is either measured as grit size or range of size in microns. Sizing criteria differences between grit and microns and among individual manufacturers make correlation between sizes only approximate. Larger size abrasive particles usually cut faster and leave rougher surfaces than smaller size particles. Diamond coating is also classified as to whether the concentration of diamonds applied to the rim is low or high.

The author prefers diamond-coated circular blades over all others. If sample size is not limited, then thicker blades produce the best cuts. Continuous rimmed blades produce the smoothest cut surface and are preferred. With diamonds, the most durable type of bond is desired and will greatly increase the cutting life of the blade. Blades with diamond grits varying from 180 to 240 produce a good range of speed, quality, and finish of cut. Either diamond or boron nitride abrasive blades can be obtained for use on the Exakt band saw, with some limited options in particle size of abrasive bonded to the blade. The Leitz saw microtome utilizes a proprietary diamond blade, of which grit size is not indicated. The wire saw manufacturers provide the user with some options in regard to wire composition, wire thickness, and bonded diamond particle size. In all three cases the author knows of no generic manufacturers of blades for these specialized saws.

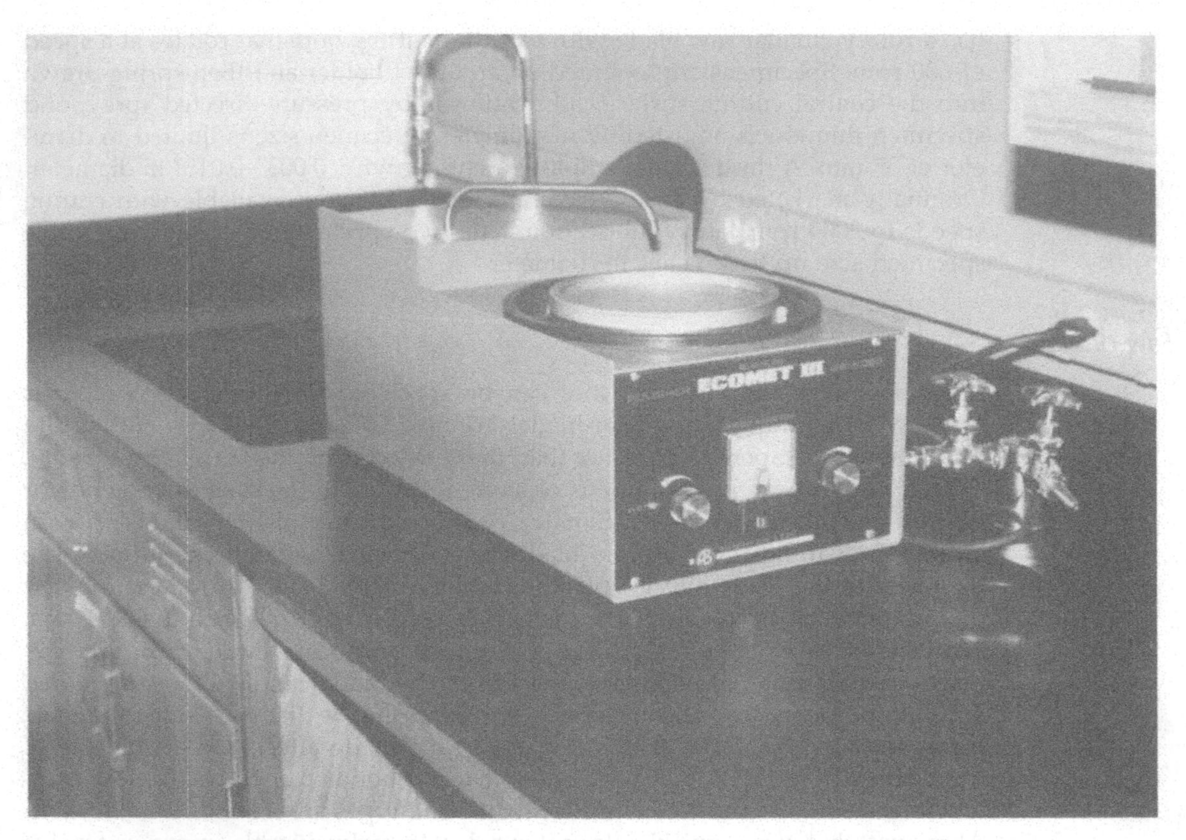

Figure 9.2. An industrial cutoff saw for large specimens. Circular blades to 14 inches in diameter can be used with a fixed arbor speed of 1725 rpm. Cutting lubrication is by pressure-directed spray. Specimen feed is by a hand-fed movable specimen mounting table. (11-BR Felker Bay State, Torrance, CA 90509.)

Grinders/polishers

In most cases it is difficult to obtain uniform, very thin sections with sawing alone. Sawn sections need to be ground further to desired thinness and the viewing surface polished. A variety of grinders/polishers are commercially available. Buehler and Leco make very comparable machines; however, unlike with saws, a wide variety of models and options are available (see Figure 9.3). The basic grinder/polisher can have a single speed of operation, two speeds, or preferably variable speed up to 500 rpm. Grinding surfaces range from 6 inches to 12 inches in diameter, can be made of aluminum or bronze, and can be single or dual. Most have some form of drip lubrication of the grinding surface on which the flow rate can be controlled and varied. Both tabletop and freestanding models can be had.

Specimens are manually held to the grinding surface and periodically checked by the operator. Automatic operation can be obtained by addition of any of a number of grinding and polishing attachments. Single sample or multiple sample grinding can be obtained with specimens held to the surface of the grinder with

Figure 9.3. Tabletop grinder/polisher. This model has a 12-foot plate with variable speed up to 500 rpm. Grinding lubrication is by a controlled drip on the grinding surface. (Ecomet III, Buhler Ltd., Lake Bluff, IL 60044.)

gauged pressure, weights, or mechanical means. Sample size is limited to the size of the specimen holder, but in almost all cases is less than 1 1/2 inches in diameter.

Recently two new grinding systems have come on the market. The makers of the modified band saw system have a grinder that utilizes an oscillating specimen holder, weight applied specimen grinding force, and 300 mm diameter diamond coated grinding plates. Grinding speeds are variable to 200 rpm. Upper limit of mounted sample size is a slide 50 mm × 100 mm. Maruto Instruments markets a speed lapping machine that utilizes 150 mm diameter upper and lower diamond lapping plates, an air counterbalanced pressure grinding force, and adjustable lapping speed to 100 rpm. Upper limit of sample size is 45 mm × 55 mm.

Basic grinders can be fitted with abrasive papers supplied with pressure sensitive adhesive or plain back, in diameters matching the diameter of the platen. The abrasive material can be aluminum oxide, silicon carbide, or zirconia alumina, and comes in a wide range of grit particle sizes. Alternatively, cloths and abrasive powders can be used. Aluminum oxide or silicon carbide are the most common powders; however, powders of iron oxide, chromium oxide, magnesium oxide, cerium oxide, and diamond are commercially available. A wide range of particle sizes is available for abrasive powder, ranging from 120 to 1000 grit and 45 microns to less than 1 micron. Grinding/polishing cloths can be obtained made of canvas, felt, wool, velvet, cotton, silk, or synthetic materials like nylon, rayon, or polyester and in a wide range of naps. Combining cloth and abrasive powder can provide the user with a wide range of grinding/polishing qualities.

The Exakt grinder offer a limited range of diamond grit platens and optional grinding discs. Maruto offers platens with at least two grit sizes.

The author has not found the automatic attachments available for basic grinders/polishers to be of much value. Limitations in sample size, variation in sample grinding/polishing characteristics, and technical/mechanical problems have not yielded the improvement in speed of operation or quality of resultant specimen hoped for. Silicon oxide paper discs provide adequate grinding characteristics for most specimens. A wide range of grinding paper grits should be kept on hand. Aluminum oxide powder is used for polishing with the 5 micron and 1000 grit powder used the most. However, specimen characteristics and personal preference may affect standard supplies greatly.

Investing versus embedding

The individual tissues that are contained in a histological sample have cutting characteristics that are determined by the hardness, density, and structure of the tissue. These characteristics affect the resistance of the tissue to cutting, the speed of cutting, and the quality of the resultant cut. These cutting characteristic differences can result in sample distortion and fragmentation and produce artifacts that complicate viewing and interpretation. To alleviate these problems, samples are impregnated with a supporting medium, which firms the character of the sample and holds the various components of the sample in normal proper relation to each other. The chosen supporting medium should be firmer than any individual component within the sample, thus imparting a uniform cutting character to the sample. In addition, the medium should be adhesive enough to hold very thin sections of the sample together. To accomplish all this, the tissue sample must be completely infiltrated with the supporting medium. The process of tissue sample preparation, infiltration with the supporting medium, and then formation of a solid block for cutting is referred to as embedding; the medium used is the embedding medium.

Paleontological samples present special problems in selecting a suitable embedding medium and for the whole embedding process as well. Most samples are densely mineralized and extremely hard. A suitable medium would need to be as hard as or harder than the specimen. Most specimens already have been impregnated with groundwater minerals filling many of the spaces within the sample. Complete infiltration of the specimen with the embedding medium would be difficult and prolonged. An extremely hard, dense embedding medium may inhibit transmission of light through an infiltrated sample, and the refractive index may alter the optical characteristics of the sample. Needless to say, a totally suitable embedding medium has not been found.

Because of this, paleontological specimens are not routinely embedded. Instead, samples are enveloped within blocks of hard material for the purpose of providing a sample size and shape conducive to sawing and to hold the sample together and prevent fragmentation during the sawing process. This process of encasing the sample rather than infiltrating the sample is preferable and is referred to as investment to distinguish it from embedding. Polyester resins are the most common investing media used for paleontological specimens: They are easy to use, cure quickly and evenly at ambient temperatures, have a suitable hardness, and exhibit suitable flexural and tensile strength characteristics. The particular resin that is used in the author's laboratory is Silmar-40, manufactured by Standard Oil Company.

To invest a specimen in Silmar it is first necessary to select a suitable size mold and prepare a base of Silmar in the bottom of the mold. Reusable rubber mounting molds, polypropylene containers, glass jars, microwave-safe kitchen storage containers, or any other suitable box can be used as an investing mold. Prepare the mold for use by spraying a light coating of Parfilm (Price-Driscoll Corporation) on the inside of the mold to facilitate removal of the hardened resin block. In a disposable paper cup or like container, add enough Silmar to provide a 6- to 12-mm base layer in the bottom of the mold. Add 0.01 g of catalyst (methyl ethyl ketone peroxide) for every 1 g of Silmar and mix rapidly for approximately 30 seconds. Pour the prepared resin into the mold, place in a vacuum jar, and gradually drop the pressure to -20 to -25 lb/sq. inch. Leave under vacuum for several minutes or until bubbles start to rise to the surface of the resin. Slowly release the vacuum and brush away risen bubbles with a disposable wooden stick. Repeat vacuum and bubble removal until a clear, smooth surface is obtained. Allow this to harden overnight. The following day, place the specimen to be invested on the base resin. Old pieces of hardened resin can be used to prop up awkward specimens into desired alignment within the mold. Mix enough Silmar to fill the mold and cover the specimen by at least 12 mm. Vacuum and remove bubbles at least twice. Again let the resin harden overnight. The hardened block can be removed from the mold after 24 hours.

Grease or oil on the surface of the specimen needs to be removed before the investment process. If present, it will act similarly to Parfilm and cause the specimen to release from the encasing hardened resin. If large amounts of resin are prepared for multiple molds or for a single large specimen, then the amount of catalyst can be decreased slightly. If too much catalyst is used, the resin hardens too rapidly and the block will crack. Also decrease the amount of catalyst used if the resin begins to harden before all the bubbles can be removed. Speed of polymerization depends on the amount of catalyst present and also on the temperature at which polymerization is allowed to occur. Thus fluctuations in ambient room temperature will alter hardening. For large specimens, or during hot days, molds can be refrigerated for 12 hours to slow the polymerization process, facilitate bubble escape, and decrease the possibility of cracking. Final hardening can then be allowed to occur at room temperature for an additional 12 to 24 hours.

If the specimen is brittle and fragile or very porous, then thorough infiltration with an embedding medium can provide the necessary internal support and uniform sawing consistency that will enhance section quality. The medium most commonly utilized is methylmethacrylate (MMA). Several procedures and modifications of procedures have been published and are currently in use; most laboratories utilize a particular procedure that works best for them. All MMA embedding protocols follow the same basic steps.

Samples for infiltration and embedding are brought to equilibrium in xylene, the solvent for the MMA. The MMA monomer, used for infiltration, is prepared by removing the dissolved inhibitor and adding a catalyst, benzol peroxide. Samples are then slowly infiltrated with monomer, under vacuum and with refrigeration, over a period of two to five days. The monomer infiltration solution may or may not be changed during this infiltration time. After the sample is thoroughly infiltrated with the monomer, it is embedded in polymer. Some protocols may include a second period of infiltration with polymer for three to seven days, under vacuum and with refrigeration, prior to embedding. Polymer is prepared by adding polymethylmethacrylate beads to prepared monomer at a ratio of 30 to 40 g/100 ml of monomer, plus catalyst. Samples to be embedded are placed in suitable molds, covered with prepared polymer, placed under vacuum, allowed to sit at room temperature for a short period, and then cured at temperatures up to 37° C for one to three days. A short period of refrigeration may be provided at the end to facilitate removal of the block from the embedding mold.

The resultant invested or embedded sample block is then ready for sawing without further processing.

Thin sectioning

To cut thin sections of paleontological specimens, one begins by following the procedures recommended for the specific type of saw employed. Modifications to the procedures can then be made to obtain the speed and quality of cut desired. It is not unlikely that several techniques will eventually be developed, each for use with individual types of specimens.

In the author's laboratory, the small, low-speed precision saw made by Buehler is utilized. We have found that most samples will fit in either the irregular shape, the single saddle, or the bone chuck. If the sample is too large to be held in one of these chucks, a small piece of acrylic plastic can be glued to the sample, and this handle is then clamped in the chuck. It is seldom that sections are cut using the fastest arbor speed. Slower speeds produce less lubricant spray, allow thinner sections to be cut, and produce smoother cut surfaces. We almost always use water as the cutting lubricant, to which a drop of household detergent is added. The largest size blade collar, or flange, should be used that the specimen size will permit. Blade collars stabilize the blade and reduce wobble. Large collars do this better than small ones. Blade diameter and collar diameter determine allowable sample cutting size, and blade collar diameter and blade thickness determine stability of the blade during cutting.

The specimen is mounted so the edge just touches the blade. The micrometer is then adjusted to place the specimen directly over the blade so that the first cut will provide a complete cut surface at which to begin thin sectioning. A minimal amount of weight is applied to the spindle on the sample arm. If too much weight is applied at the start, an uneven cut can result from slight bending of the blade and an uneven starting kerf. The sample arm is then lifted, the motor started, and the arm slowly lowered until the blade engages the sample. Once the initial kerf is made, the saw can be stopped and additional weight added to the sample arm spindle. Blade speed can be adjusted during cutting; however, it is not advisable to add weights while the blade is in motion. Always stop the machine before adding or removing weights. It is the combination of sample arm weight and blade speed that determines cutting speed and quality. Once the first cut is completed, the end of cut shutoff can be set.

Successive sections can be cut by advancing the micrometer the width of the blade, or kerf, plus the desired thickness of the specimen. The combination of physical characteristics of the saw, the blade, the cutting speed, the lubricant, the weight on the sample arm, and the specimen being cut makes micrometer settings only approximations of the final cut section thickness. The final thickness of each cut section should be measured with a hand micrometer and the actual thickness recorded. The variation between machine setting and actual thickness can then be adjusted. There will always be some variability between sections, but with practice, repeatable sections can be routinely cut.

Some specimens will tend to fill, or clog, the abrasive coating on the circular blade. Specimens will then cut slower, and the quality of cut will deteriorate. This can be remedied by frequently cleaning, or dressing, the cutting edge of the blade with a dressing stick. Dressing sticks can be purchased from the saw manufacturer. Cutting a thin slice or two off the dressing stick will clean the blade and return it to the beginning cutting characteristics. With particularly bothersome samples, blade dressing can be performed between successive or multiple successive cuts. Two methods of blade dressing are provided by Buehler. The dressing stick can be digitally held on a plastic guide plate and slowly passed by hand through a slowly rotating blade, or the stick can be mounted on a mechanically operated dressing chuck and the stick passed through the blade by manually turning a small handle. Either method works well for blade dressing between sections. Blade dressing can also be performed during cutting of a section. If the manual plastic guide is utilized, it is best to retract the specimen away from the blade during dressing so as not to disturb the kerf within the specimen. If the mechanical dressing chuck is utilized, the dressing can be performed during specimen cutting.

At the end of the sawing session, remove the blade and lubricant bath. Wash blade, lubricant bath, chuck, and blade collars thoroughly with soap and water and then rinse with alcohol. Wipe the outside of the machine clean. If the blade is not to be used for some time it is best to remount the blade in the machine and final dress the cutting edge by running the dressing stick through the blade four or five times. The blade can then be rinsed, dried, and stored.

Decalcification and demineralization

Removal of calcium phosphate salts from living bone is referred to as decalcification. Removal of the variety of minerals from fossils is preferably called demineralization. In both cases removal of mineral deposits is most commonly achieved with acid solutions in which the deposited minerals are soluble. The time required to decalcify bone or demineralize a fossil completely depends on (1) the size and thickness of the specimen, (2) the density of the mineral deposits,

(3) the strength of the solution utilized, (4) the frequency of solution change, and (5) the temperature at which the process takes place. The removal of minerals from a specimen is dependent on, and controlled by, the rate of passive diffusion of the solvent into and out of the specimen. Thus the thickness of the specimen to be processed has a greater effect on increasing total processing time than the cross-sectional size of the specimen. Thinner specimens process much faster than thicker specimens. The time to process completely progressively thicker specimens increases exponentially rather than linearly, with a point reached at which complete removal of mineral is excessively prolonged. Higher temperatures, stronger acids, and stronger concentrations of acid accelerate the process, but at a great risk of damage to the specimen or detrimental effect on further processing, such as impaired staining character.

There is no ideal method of decalcification or demineralization. Acids always produce some distortion or alteration of the specimen. The agents most commonly utilized are aqueous solutions of nitric, hydrochloric, formic, or trichloracetic acid. They may be used alone or in combination with a neutralizer to decrease detrimental effects on the specimen. Trichloracetic acid is not commonly used. Nitric acid is a rapid method of decalcification and demineralization. However, it has a reputation of being harsh on specimens and produces a marked reduction of stainability if specimens are left in the solution too long. Hydrochloric acid also has rapid action, even in dilute solutions, and like nitric acid, can be harsh on specimens and impair staining. Formic acid is generally accepted as the most suitable solution for decalcification of bone. It has good preservation of stainability and appears to have the least detrimental effect on the specimen, even when the process is prolonged or the specimen is allowed to remain in the solution after the process is complete.

Several commercial solutions are available for decalcification in routine histological processing. Most are marketed as a combination of fixative and decalcification solutions or as a rapid method of processing. Although they are probably suitable for processing paleontological specimens, the author has not found them to be as good as laboratory-prepared reagents specific for the specimen to be processed. Organic chelating agents, such as EDTA, have been recommended for the decalcification of bone specimens in which excellent preservation of the histological detail of various tissue components is needed. However, these agents work very slowly and work best only when the specimen is very small and very thin. Mineral deposits can also be removed by electrolysis. An electric current is applied by a commercial power unit to the connecting anode while the specimen is suspended in a solution of formic and hydrochloric acid. Mineral salts are caused to migrate out of the specimen and toward the nearby cathode by the electric field that is established. Although this method purports shortened processing times, a very limited number of samples can be processed at any one time, and precise control of the process is needed, as otherwise disastrous problems can arise.

The variety of minerals present in paleontological specimens is an incentive to investigate the efficacy of certain commercially available, "household" demineralization agents. These products are usually marketed as cleansers to remove accumulated scale or water spots. The author has found products containing phosphoric and/or hydroacetic acid to be effective demineralizers and may have potential use with paleontological specimens. A comparative evaluation is in progress.

Of equal importance to the selection of a decalcifying or demineralizing solution, and the one point receiving the least attention, is the removal of specimens from the solution immediately after complete decalcification or demineralization is accomplished. If this is not done, the chance of physical damage to the specimen increases sharply and the chances of detrimental effects on subsequent staining reactions are increased 10% for every 2 hours the specimen remains in the decalcifying or demineralizing solution. In fact, 90% of poor staining quality, demonstrated in decalcified tissue, is due to improper decalcification. To ensure the least damage to the specimen and the best subsequent processing result, the end point of the decalcification process should always be monitored.

Staining methods that require demineralized samples may be used on specimens that have been partially demineralized or surface etched prior to staining. This can be accomplished by immersion of the specimen (mounted or unmounted) in a demineralizing solution for a shortened time period, by surface application of a decalcifying or demineralizing solution, or by combining the demineralizing solution with the principal stain solution.

Procedure for demineralization

1. Place calcified specimen in large quantity of selected solution. Change solution daily for fastest and best result.
2. When mineral removal is complete, remove specimen from solution and rinse in running water 4–8 hours. If unbuffered acid is used, the specimen may optionally be neutralized by a bath in saturated lithium, calcium, or magnesium carbonate, followed by a rinse in running water.
3. Further process as desired.

Citrate buffered formic acid

Solution A

Sodium citrate	50.0 g
Distilled water	250.0 ml

Solution B

Formic acid, 88%	125.0 ml
Distilled water	125.0 ml

Mix solutions A and B in equal portions for use.

Nitric acid

Nitric acid, concentrated (68–70%)	5.0 ml
Distilled water	95.0 ml

Procedure for surface etching

1. Immerse specimen in etching solution for 4 minutes or longer, or alternatively, flood surface of mounted specimen with etching solution for 4 minutes or longer.

2. Rinse specimen in running tap water for 1 minute.
3. Rinse with distilled water.
4. Stain.

Osmium tetroxide buffered nitric acid

5% Nitric acid	1 part
2% Osmium tetroxide	4 parts

Osmium tetroxide is a buffer commonly used in sample preparation for electron microscope viewing, and is not specific for paleontological samples.

0.7% formic acid

Formic acid, 88%	1 ml
Distilled water	124 ml

Determining end point of decalcification or demineralization

The specimen can be tested by inserting a needle into it. If the needle enters the specimen easily, it is ready to be removed from the solution. The specimen may also be tested by checking the pliability. However, as one can imagine, this method is prone to disastrous consequences. Undoubtedly the best method to determine complete decalcification is to radiograph the specimen. Unfortunately, radiographic monitoring requires special equipment, and is expensive, time consuming, and difficult for small laboratories to perform routinely.

A fourth method is to chemically test the solution frequently for presence of mineral. Draw approximately 5 ml of decalcifying or demineralizing solution from the bottom of the specimen container. Add 5 ml each of 5% ammonium hydroxide and 5% ammonium oxalate. Mix and let stand 15 to 30 minutes. A cloudy solution indicates that there is mineral in the solution and that the specimen is not thoroughly demineralized. When a cloudy solution is no longer obtained, mineral is no longer present in the decalcifying fluid and the specimen is completely demineralized. This test can, and should, be performed frequently.

Mounting

In order to perform the necessary grinding and polishing of sawn sections easily, to facilitate microscopic evaluation, and to improve storage, it is advisable to mount thin specimens on a fairly rigid, clear, backing material. Only two materials are in common use: glass and acrylic plastics. Glass slides can be purchased from many suppliers in a variety of sizes, or they can be made to order through a local glass supplier. Manufactured glass microscopic slides come in two standard sizes: 1" × 3" and 2" X 3" and 1.0 mm or 1.2 mm thick. They can be purchased plain or frosted on one side, at one end, to facilitate labeling. Special order sizes include 3" × 4", 4" × 5", 4" × 6", and 5" × 7". The standard size petrographic slide is 27 mm × 46 mm. Glass through a local supplier is usually available in 3/32" and 1/8" thicknesses, and, of course, can be cut to any size needed.

Plastic slides are available commercially and can be made to order through local distributors. Although these plastics are commonly referred to as Plexiglass,

Plexiglass, an acrylic plastic, is the registered trademark of Rohm and Haas. Other similar plastics are General Electric's Lexan, a polycarbonate, and Du Pont's Lucite, another acrylic. Acrylic slides have been found not to block or attenuate an X-ray beam like glass and are thus especially suitable for microradiographs. An obvious advantage is that plastic slides do not crack or shatter when dropped. An obvious disadvantage is that plastic slides are especially susceptible to organic solvents, such as xylene, benzene, alcohol, and toluene. Thus special precautions must be taken if they are used for specimens that will be stained.

Mounting the thin specimen to the backing slide requires a mounting medium that is strong, durable, nonsticky, does not yellow with age, is easy to use, and dries quickly. It should also have a refractive index similar to that of the mounting slide. If used for mounting a stained section, the medium should be soluble in the agent used to clear the specimen following staining. Needless to say there is no ideal mounting medium. Fortunately, several media have been found useful, when used appropriately. Eukitt (Calibrated Instruments) has been almost uniformly accepted as an excellent medium for very thin living tissue sections. It is not too viscous, spreads evenly and quickly without bubbles, dissolves readily in xylene, hardens in less than 20 minutes, and has a refractive index of 1.510 in the liquid state. It is almost always used in conjunction with a glass coverslip. The bond strength is suitable for standard specimen mounting but has not been found to withstand the rigors of hand grinding.

For bonding specimens requiring grinding and/or polishing, mounting media with exceptional bond strengths are needed. The most popular are the cyanoacrylate glues and the epoxy resins. The cyanoacrylates, that is, super glues, can be purchased at any hardware and most retail stores. Specialty cyanoacrylates are available in several viscosities and setting times (Carl Goldberg Models). Another type of mounting medium that has been used in the past but has lost favor in recent times is thermoplastic (quartz) cement. It is often referred to by its brand name, Lakeside (Hugh Courtright & Co.). Type of adhesive or of specific brand used very much depends on personal preference gained by experience. The mounting medium used most often in the author's laboratory is 2-Ton Epoxy (Devcon Corporation).

Size of mounting slide and the size of thin section need to be appropriately matched. The edge of the section should not be too close to the edge of the slide. There should be enough room around the section for a gradual thinning of the mounting medium before it reaches the edge of the slide. Considerable force is applied to the section/mounting medium/slide interface during grinding, and a less than optimum bond can result in rapid loosening, breakage, or loss of the section during grinding. Also, after grinding and polishing is completed, less than ideal section mounting can lead to long-term loss of mounting medium bond strength. Air will begin to appear under the section and the section will loosen, peel, and crack. Because of the amount of effort that goes into making a thin section and the value of the resultant section, attention should be paid to the mounting process.

A thin, even coat of mounting medium should be applied to the slide in a sufficient quantity to coat the bottom completely and thin out around the edges of the section. Too much mounting medium is as bad as too little. The section should be gently placed onto the slide in any manner that decreases the trapping of air bubbles under the section. Once the section is in place, cover the slide with a small sheet of wax or Teflon-coated paper. A second slide or two can then be placed on top of the paper. The paper will prevent the mounting medium from adhering to the covering slide. Pressure is then applied to the mounted section to remove any trapped air bubbles and to thin the coat of mounting medium. Thick layers of mounting medium cause the mounted section to lay unevenly on the slide and result in uneven grinding. Thick mounting medium can contract with time, causing air bubbles and section peeling. It can also deter or alter the light characteristics of the slide and make for poor microscopic viewing and poor photomicroscopy. Pressure can be applied in two ways: The mounted slide can be placed on a clean, dry, flat surface and weights placed on top of the slide to compress the mounting medium interface. Alternatively, one or several small pinch clamps can be applied to squeeze the composite mounted slide. The slide is then allowed to air dry. The finished product should be visually checked and completely labeled. The mounted section is now ready for grinding or other processing.

Grinding and polishing

The specialized grinders by Maruto and Exakt have procedures tailored to the mechanical requirements of the individual grinders. In most circumstances only minor variations in recommended procedures are possible to yield still good results. Thus, when utilizing one of these pieces of equipment, it is advisable to follow the procedural requirements closely. Also, when using one of the automated attachments to the basic manual grinder, the recommended procedures and limitations should be followed and modified only after experience and with caution. The basic manual grinder is far more flexible to use, and many procedures have been developed by individual laboratories to meet general or specialized needs. A general procedure is described here and several others summarized and references given.

Variable speed grinders provide the user with the greatest flexibility. Variation in grinder speed, combined with the many available types and sizes of abrasive papers, should cover the gamut of grinding characteristics desired. Mounted specimens should be measured prior to any grinding to establish the amount of grinding needed. Coarse grit abrasives can be used initially to quickly remove excess material. Coarse grits leave a rough surface, so enough surface should be left to allow fine grinding and final polishing. Once fine grinding commences, the surface quality and specimen viewing characteristics can be checked repeatedly by microscope. Although the literature implies that one best specimen thinness exists, this is far from true. Widely disparate specimens should be expected to have different physical characteristics – and they do. Variation in quality of preservation noticeably affects microscopic detail, and variation in mineral content markedly affects light-transmitting and -polarizing characteristics. Individual studies look at or need to enhance specific and different structures. Thus the best specimen thinness is the one that illustrates best what the investigator wants to see. In most cases this will range between 100 and 300 microns. Once the specimen is at a good viewing thinness, final polishing should commence. Surface scratches can be easily removed and a lustrous surface obtained with minimal polishing if sufficient and carefully fine grinding was performed. Deep surface scratches just prior to polishing indicate excessive coarse grinding and insufficient fine grinding or use of too-coarse abrasive grit during fine grinding. As mentioned, the author's laboratory uses 5 micron and 1000 grit aluminum oxide powder for final polishing.

The actual grinding and polishing can be performed in a number of ways. The mounted specimen can be manually held to the rotating platen and pressure maintained and adjusted with the operator's fingers. Adhesive tape can be applied to the back side of the slide for the operator to grip. Slide holders are commercially available for grinding, or a similar holder can be custom-made. Commercial holders usually have boron carbide grinding surfaces provided to stop the grinding process at a preset thinness; however, the operator should realize that this set thinness assumes that a specific, exact, uniform thickness slide is used, that the mounting medium is always the same, exact, predetermined thinness, and that the slide is always mounted correctly, exactly, and never shifts position. Plastic slides have been used and advocated as an asset to hand grinding specimens (Bloebaum et al. 1989). The flexible characteristics of the slide are said to give the operator more control during the grinding process. They also are not as fragile as glass slides. A glue-glove technique has been described in which the operator dons a latex glove, applies a spray adhesive to the glove, and then presses the glued glove to the back side of the mounted or unmounted specimen, firmly affixing it to the operator's hand (Rhodes, Sanderson, and Bloebaum 1990). This method is said to firmly secure and maintain control of the specimen during grinding, allow fine control of grinding pressure, and protect the operator's fingers during the grinding process. Because hand grinding is very operator dependent, procedure used is prone to personal preference and experience far more than technical aspects.

Staining techniques

Staining quality of paleontological specimens is extremely dependent on the type of the specimen, the age of the specimen, the quality of preservation of the specimen, and the local environment in which the specimen was found. There is no single staining technique that is applicable to all specimens, nor is there any one, preferable staining procedure. There is also no assurance that staining techniques developed and used on fixed living tissues produce identical results when used on paleontological specimens. As tissue collagen progressively denatures with increasing time after death of the organism, the binding sites for stains in living tissue may disappear or alter in paleontological specimens. Also, as the tissue collagen denatures, the quality of the soft tissue sample that remains after demineralization deteriorates; thus the quality of sample available for staining procedures that require demineralization or work best on demineralized specimens may be poor or obviate use of the staining procedure.

As indicated in the discussion on decalcification and demineralization, staining methods that require demineralization may be used on specimens that have been partially demineralized or surface etched. Also, several of the staining procedures for undemineralized specimens discussed here combine an acid in the staining solution or include in the procedure a short rinse in an acid solution prior to staining. One of these methods or stains should be tried on a specimen in which complete demineralization damages the sample.

The following staining procedures either have been used with some success on paleontological specimens or have the potential to be useful. Staining times and procedures will need to be varied to obtain the best results depending on the characteristics of the specimen to be stained.

For undecalcified or undemineralized specimens

Basic fuchsin

Results

- Mineralized bone is red.

Procedure

1. Stain specimen in basic fuchsin solution for 48 hours.
2. Polish surface of specimen.
3. Wash gently in distilled water with 0.1% household detergent.
4. Wash in 3 changes of distilled water.
5. Air-dry or place in an oven overnight.

Solutions

Basic fuchsin	1 g
80% ethanol	100 ml

Filter and store in a brown bottle.

Kossa's method

Results

- Calcium deposits are black.

Procedure #1

1. Hydrate specimen to distilled water.
2. Place specimen in a petri dish and flood with 5.0% silver nitrate. Place the dish under a 60-watt bulb for 30–60 minutes, until calcium deposits are black.
3. Rinse well in distilled water.
4. Treat section with 5.0% Sodium thiosulfate for 2 minutes.
5. Rinse briefly in 70% alcohol.
6. Counterstain with eosin for 30 seconds.
7. Dehydrate, clear, and air dry.

Procedure #2

1. Stain specimen in 5% silver nitrate for 30 minutes in the dark.
2. Rinse in 3 changes of distilled water.
3. Reduce with sodium carbonate/formaldehyde solution for 2 minutes.
4. Wash gently in running tap water for 10 minutes.
5. Counterstain in methyl green pyronin for 20 minutes.
6. Wash twice with cooled freshly boiled distilled water for 1 minute.
7. Dehydrate once with 95% alcohol for 1 minute.
8. Dehydrate in 2 changes of absolute alcohol.
9. Clear in 2 changes of xylene and air dry.

Solutions

5% Silver Nitrate

Silver nitrate	2.5 g
Distilled water	50 ml

5% Sodium thiosulfate

Sodium thiosulfate	2.5 gm
Distilled water	50 ml

Eosin

Eosin Y, water-soluble	1.0 g
Distilled water	100.0 ml

Dissolve and add

95% alcohol	300.0 ml
Glacial acetic acid	2.0 ml

Sodium carbonate and formaldehyde

Sodium carbonate	5 gm
Formaldehyde	25 ml
Distilled water	75 ml

Methyl Green Pyronin (PolyScientific)

Toluidine Blue

Results

- Metachromatic shades of blue

Procedure #1

1. Stain specimen in 2% toluidine blue solution for 10 minutes.
2. Rinse in 2 changes of buffer, 1 minute each.
3. Blot dry.
4. Dehydrate in alcohol, clear in toluene, and air dry.

Solutions

Toluidine blue stain

Toluidine blue (Fisher)	2 g
Buffer	100 ml

Filter and adjust pH to 3.7 with 1 M NaOH.

Buffer

Citric acid	0.63 g
Disodium phosphate	0.30 g
Distilled water	400 ml

pH is approximately 3.7.

Procedure #2

1. Rinse specimen thoroughly in distilled water.
2. Stain in dilute toluidine blue solution for 24 hours.
3. Dehydrate in alcohol, clear in xylene, and air dry.

Solutions

Working solution

1% Toluidine blue stock solution	75 ml
Distilled water	100 ml
1 Buffer tablet	

Toluidine blue stock solution

Toluidine blue (Fisher)	1 g
Distilled water	100 ml

Filter and keep in a dark glass bottle.

Buffer tablet: Lot no. 13, formula no. 30, catalog no. 001–0060, Perkin-Elmer Corp., Coleman Instruments Division, Oak Brook, IL 60521.

Toluidine blue (surface stain)

Results

- Metachromatic shades of blue

Procedure #1

1. Place in 0.1% formic acid for 2 to 4 minutes.
2. Rinse in tap water.
3. Place in 20% alcohol for 120 minutes.
4. Rinse in tap water.
5. Stain with 1% toluidine blue solution for 2 to 3 minutes.
6. Rinse in tap water.
7. Rinse well in distilled water.
8. Blot and let dry.

Solutions

Toluidine blue stock solution

Toluidine blue (Fisher)	1 g
Distilled water	100 ml
Borax (sodium tetraborate)	1 g

Mix well and filter twice before use.

Procedure #2

1. Stain specimen in toluidine blue/formic acid solution for 20 seconds or longer.
2. Rapidly dehydrate in alcohol, clear in xylene, and air dry.

Solutions

Toluidine blue stock solution

Toluidine blue (C.I. 52040)	1 g
0.1% formic acid	100 ml

Ph of solution is approximately 2.6.

For decalcified or demineralized specimens

Van Gieson's

Results

- Collagen is red.

Procedure

1. Hydrate specimen to distilled water.
2. Stain with Wiegert's hematoxylin for 10 minutes.
3. Wash in distilled water.
4. Stain in Van Gieson's solution for 1 to 3 minutes.
5. Dehydrate and mount an unmounted specimen or air dry a mounted specimen.

Solutions

Wiegert's hematoxylin

Solution A

Hematoxylin crystals	1.0 g
Alcohol, 95%	100.0 ml

Solution B

Ferric chloride (anhydrous), 29% aqueous	4.0 ml
Distilled water	95.0 ml
Hydrochloric acid, concentrated	1.0 ml

Use equal parts of solution A and solution B. Mix immediately before use. Shelf life, 6–12 months.

Van Gieson's solution	
Acid fuchsin, 1% aqueous	2.5 ml
Picric acid, saturated aqueous	97.5 ml

Picro-sirus red hematoxylin

Indication

- Enhances birefringence of collagen.

Procedure

1. Place specimen in running tap water for 15 minutes.
2. Stain in 0.1% picro-sirius solution for 30 minutes.
3. Wash in water.
4. Stain in Mayer's hematoxylin for 15 minutes.
5. Place in running tap water for 15 minutes.
6. Dehydrate and mount an unmounted specimen or air dry a mounted specimen. The staining process reaches saturation with time periods in excess of 1 hour.

Solutions

0.1% Picro-sirius solution	
Sirius red F3BA	1.0 g
Saturated picric acid	1000.0 ml

Allow solution to ripen for a few days.

Mayer's hematoxylin	
Ammonium or potassium alum	50.0 g
Hematoxylin crystals	1.0 g
Sodium iodate	0.2 g
Citric acid	1.0 g
Chloral hydrate	50.0 g
Distilled water	1000.0 ml

Dissolve the alum in water, without heat; add and dissolve the hematoxylin in this solution. Then add the sodium iodate, citric acid, and chloral hydrate; shake until all components are in complete solution. The final color of the stain is reddish violet. Shelf life, 12 months.

Schmorl's method

Results

- Borders of lacunae and canaliculi stain bluish black.

Procedure

1. Stain in Nicolle's carbol-thionin 3 minutes or (thionin solution with 1 or 2 drops of ammonia water).

2. Transfer to phosphotungstic acid solution for a few until seconds until sections become blue, green, or gray.
3. Wash in water 5 to 10 minutes until sections become sky blue.
4. Place in dilute ammonia water for 3 to 5 minutes. If the ground substance is stained too deeply, treat with acid–alcohol for 5 minutes and wash in water before dehydrating.
5. Wash in 90% alcohol, several changes.
6. Dehydrate, clear, and mount an unmounted specimen or air dry a mounted specimen.

Solutions

Ebner's fluid	
Sodium chloride, saturated solution	100 ml
Hydrochloric acid	4 ml
Distilled water	100 ml

Add 1 or 2 ml of HCl each day until bone is soft.

Nicolle's solution	
Hionin, saturated solution in 50% alcohol	10 ml
Carbolic acid water, 1%	100 ml

or

Thionin solution	
Thionin, saturated solution in 50% alcohol	2 ml
Distilled water	10 ml

Phosphotungstic acid solution	
Phosphotungstic acid	to saturate
Distilled water	100 ml

Dilute ammonia water	
Ammonia, 38%	10 ml
Distilled water	100 ml

Sudan black B

Results

- Liquids stain black.

Procedure

1. Hydrate specimen to distilled water.
2. Dehydrate to alcohol.
3. Stain in 2% Sudan black B solution at 37° C for 1 to 12 hours.
4. Rinse in xylol and mount an unmounted specimen or air dry a mounted specimen.

Solutions

Sudan black B

Sudan Black B	2 g
Acetone	100 ml

Toluidine blue (all purpose)

Indication

- Metachromatic stain.
- Tissue is all shades of blue.

Procedure

1. Hydrate specimen to distilled water.
2. Stain in 1% toluidine blue for 5–10 minutes.
3. Wash well in running tap water.
4. Blot to almost complete dryness.
5. Dip once in the last absolute alcohol, then dip immediately in the xylene.
6. Air-dry.

Solutions

Toluidine blue stain (1%)

Toluidine blue (Fisher)	1 gm
Distilled water	100 ml

Shelf life, 6 months

Methods of evaluation

Microscopes

The heart of any histology laboratory is its microscope(s). Two varieties of microscopes find use in the paleontology laboratory: stereo and system research. The stereo microscope provides the user with a wide field of view, long working distance, and a zoom range of magnification usually from less than 1 × upward. The system research microscope (see Figure 9.4) provides high-quality optics for a variety of modes of observation usually at magnifications higher than 4×. System research microscopes have smaller fields of view and very short working distances.

The author strongly advises against purchase of single-use specialty microscopes, such as a petrographic, polarizing, or metallurgical microscope. Although these microscopes provide excellent features for their specific intended use, they are seldom adaptable to other uses. Thus, if the user's needs change, another complete microscope will need to be purchased. Instead, the author prefers a basic system microscope such as the Olympus BHS or BHT series microscope. With these types of microscopes the light source can be easily upgraded, individual objectives can be picked and combined with accompanying condensers, stage and slide holder can be selected, and specialty intermediate tubes and attachments interchanged. With the addition of interchangeable parts, the micro-

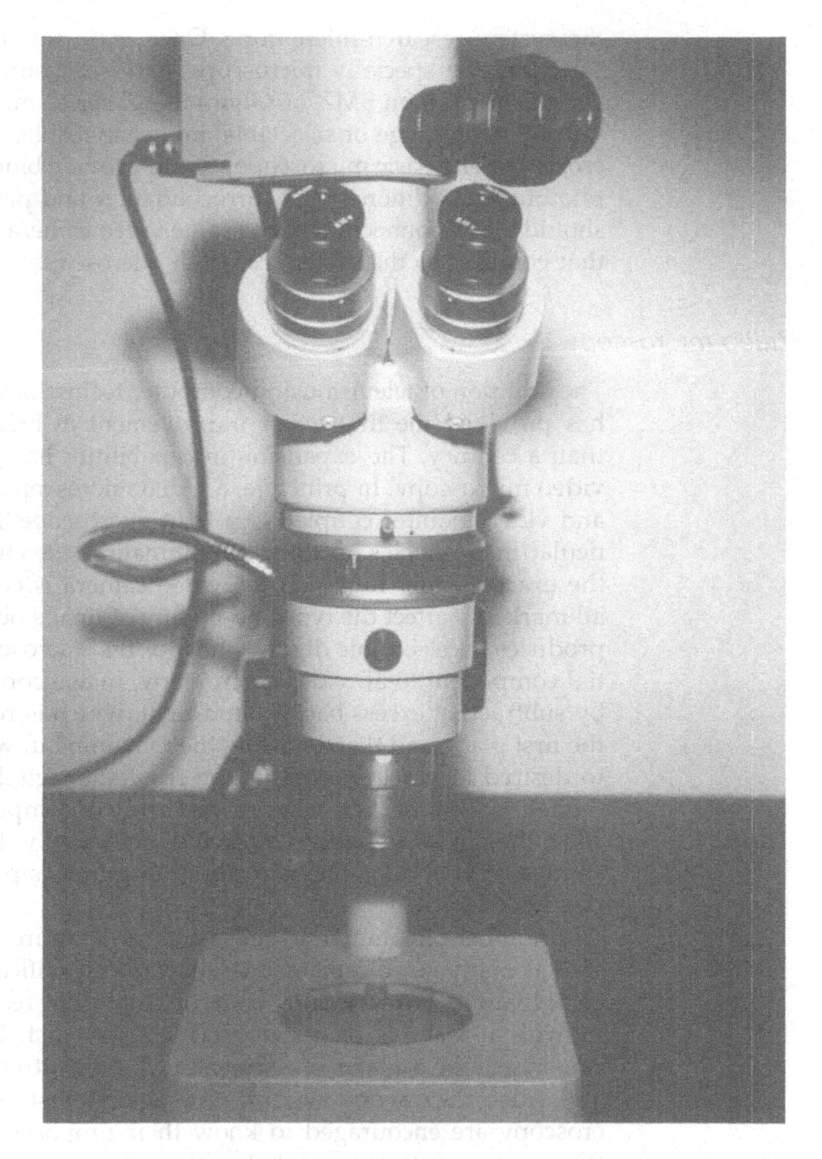

Figure 9.4. System research microscope. (BH-2, Olympus Corporation, Lake Success, NY 11042.)

scope can be converted from normal bright field to phase contrast, differential interference contrast, polarizing, dark field, or reflected light fluorescence microscopy. Optics can be adjusted to the type of microscopy employed and the degree of chromatic and spherical aberration correction selected for each objective. Addition of an adapter and projection lens to the phototube allows connection of a video camera or photomicrographic system. Thus many options and features can be conveniently available at a single operator station at a low overall cost.

No combination of interchangeable parts on a single microscope will provide all the magnifications and features needed in a paleontology histology laboratory. The stereo, or dissecting, microscope can provide the features lacking in the system research microscope. Once again the investigator is advised against selection of a specialty microscope, and is encouraged to select a modular unit such as the Nikon SMZ or Olympus SZ series microscope. Either of these provides a wide range of selectable, interchangeable, and upgradable modes of microscopy. The basic microscope should provide binocular viewing and a separate phototube. Addition of the correct adapter and projection lens to the phototube should allow connection of the same video camera or photomicrographic system that connects to the system research microscope.

Video microscopy

The addition of television and computer technology to the basic light microscope has provided the first major improvement in bright field microscopy in more than a century. The expansion in capabilities has given rise to a field called as video microscopy. In principle, a video microscope is simply a television camera and video monitor coupled to a light microscope. In practice, however, the particular microscope's optics and illumination, the choice of video equipment, and the exact manner in which the video camera is combined with the microscope all markedly affect the type and quality of image obtained. The television camera produces an electronic digital signal of the microscopic image. This is the essential component to all video microscopy. Image contrast can be improved simply by subtracting excess background light, which is responsible for low contrast in the first place, and then boosting the remaining unwanted background free signal to desired levels. Importantly, the resultant digital signal can then be sent to a digital image processor where a vast array of computer enhancements is possible. Also, the digital signal can be sent directly to a videocassette recorder for image storage or to a video image recorder or graphics printer to produce a hard copy picture.

A number of commercial television cameras are adaptable to a microscope, as well as many video cameras manufactured specifically for video microscopy. The combinations of black and white or color, line resolution, light sensitivity, and individual features of a video camera are vast. Because purchase of a video camera can be a major investment and is an extremely important component of any video microscopy system, investigators just entering the field of video microscopy are encouraged to know their proposed uses well and become more than just casually acquainted with techniques and hardware options. Several articles and texts are available on the subject of video microscopy, and current material can usually be obtained through microscope or image analysis system distributors. The possible enhancements and expanded uses of video microscopy make it, in the author's opinion, the single most important addition to any histology laboratory available today.

Image processors and analyzers

A digital image processor is a device that can enhance image attributes, such as contrast, signal-to-noise ratio, or edge features, and can shadow or highlight detail. It can also display image differences, detect motion, measure area brightness, perform image ratios, and generate three-dimensional displays. Image analyzers generate data based on morphometric parameters, such as number of objects, area, shape, length, perimeter, and boundary, etc. There are also a number of shared features, such as ability to store, sort, retrieve, edit, cut, paste, and annotate images. All of this is possible because the image exists as an array of close to a quarter-million individual pixels (picture elements) generated by the video camera. Each pixel is further stored as one of 256 gray values or as a representative shade of one of the three primary television colors. Once this array is generated, then image processing or analysis is possible, depending on the hardware and software installed in the base computer.

Full use of the enhancements and expanded uses of video microscopy are only possible with the assistance of an image processor and/or analyzer. Several programs are currently on the market and new ones are being added regularly. Most programs combine features of both image processing and analysis; however, each program tends to specialize in one or the other. Thus, when purchasing an imaging system, the investigator needs to decide whether analysis or image enhancement is of principal importance. This will then naturally narrow the field of suitable systems considerably.

An image analysis system can be composed of any one or a combination of three basic programs: planar morphometry, densitometry, or grain counting. Planar morphometry generates the myriad geometric data: area, length, distance, shape, and so on. Densitometry generates numerical counts and percentage data based on the analysis of a user-set range of pixel density or gray scale. Grain counting provides rapid counting and analysis of user-defined similar-sized objects. Basic features, such as area, percent area, and others, are shared by each program.

In simple terms, image processors and analyzers allow the user to enhance a microscopic image, highlight detail of interest, and then obtain quantitative data on the image. All of this can be performed quickly and with little user bias. As mentioned at the beginning of this chapter, paleontology has evolved to the point where merely obtaining a histological picture is not enough. Data about the image are important and necessary. Computer-assisted image processing and analysis can provide the user with the means to accomplish this rapidly, efficiently, and accurately.

Photomicroscopy

There are two methods for photographing microscopic images. The most common is use of any of a variety of camera bodies attached to the phototube of a trinocular microscope. The other is to attach a video image recorder or graphic printer to the output of an attached video camera.

The simplest of the camera body methods is to purchase an compatible adapter and attach a standard single-lens reflex camera body to the microscope. Most microscope manufacturers can provide an adapter that will fit Nikon, Olympus, Pentax, or other commercially available cameras. The normal camera lens is removed, and the whole assembly of camera, adapter, and extension tube is attached to the eyepiece or preferably the phototube of the microscope. The biggest

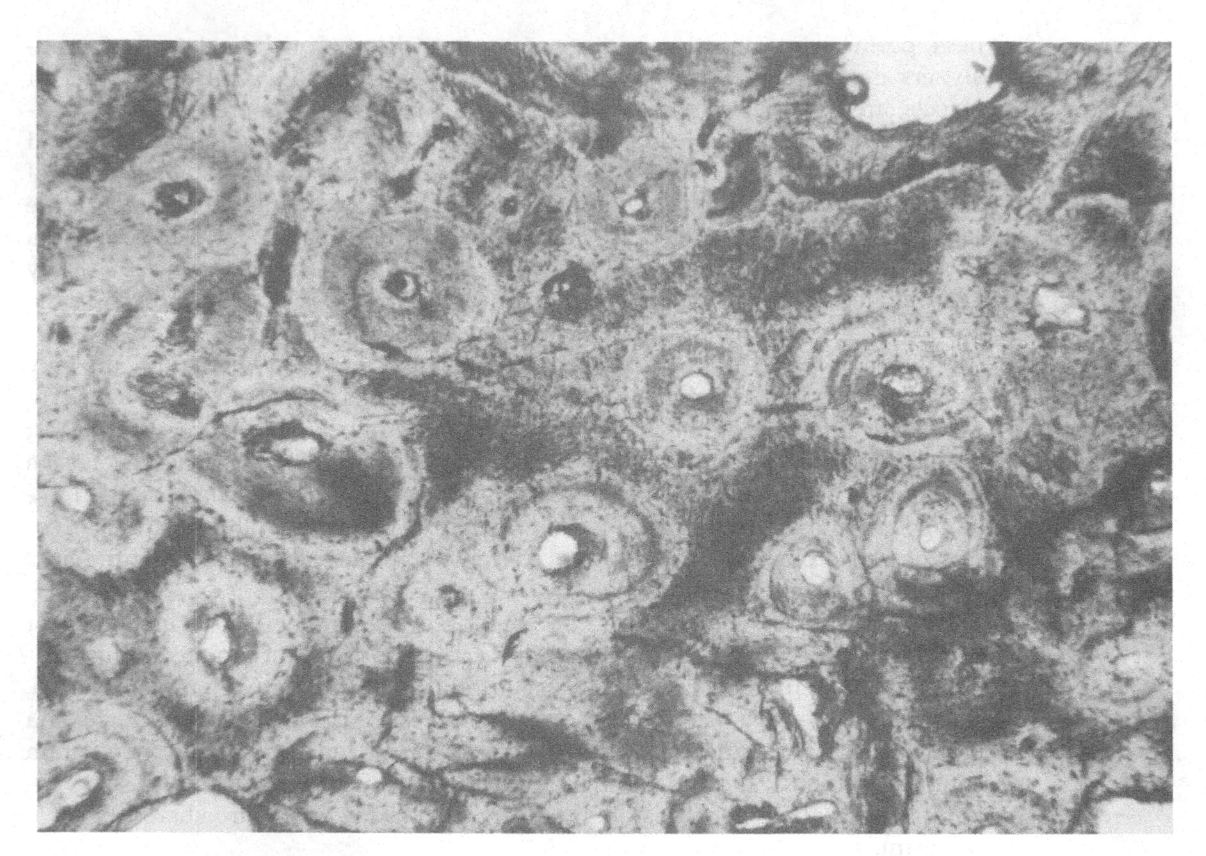

Figure 9.5. Photomicrograph of hand ground, approximately 250 microns thick, section of fossilized dinosaur bone taken with a photomicrographic system. Note that legends, arrows, and scales need to be added to the print manually.

disadvantages to this method are the lack of sensitive focusing, the lack of accurate light metered exposures, and the lack of accurate camera-timed exposures longer than 1 second.

The other standard camera body method is to purchase a photomicrographic system through one of the microscope manufacturers. Unlike many of the basic parts of the microscope, camera systems and microscopes are fairly easily interchangeable among manufacturers. If the camera system or microscope manufacturer does not have a standard catalog item, most distributors have or can provide suitable mounting adapters. Photomicroscopic systems offer many extremely valuable features, such as spot or full field light metering, exposure memory, multiple exposures, and beam splitter focusing; as well as convenient adjustment of exposure and ASA, automatic wind and rewind, and light meter resets. The biggest advantages to these systems are accurate light metering with timed exposures from 0.01 to 999.0 seconds and precise focusing (see Figure 9.5).

Camera vibration caused by microscope manipulation, mechanical shutter activation, and the microscope stand can be a problem and has to be eliminated

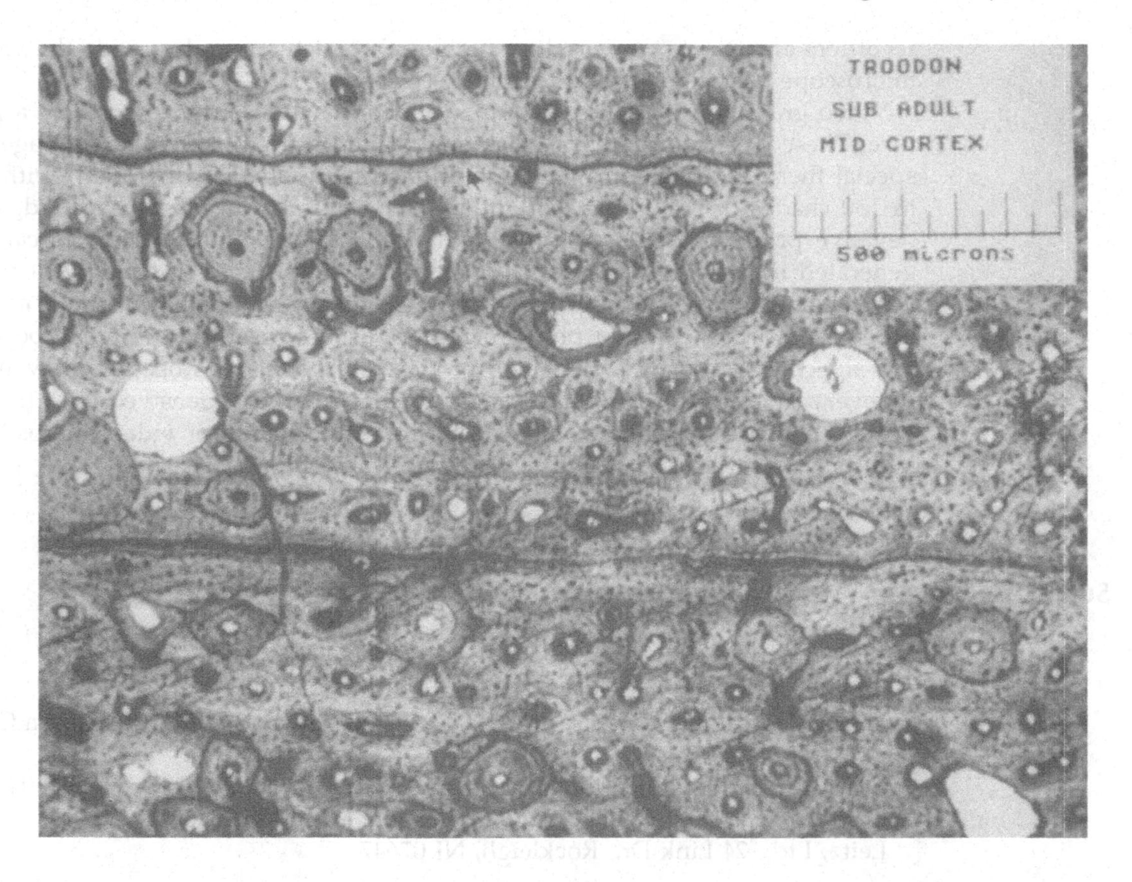

Figure 9.6. Photomicrograph of hand ground, approximately 200 microns thick, section of fossilized dinosaur bone. The image was obtained with a video camera, imported into an image processor/analyzer, and photographed on a video recorder. The label and scale were added to the video image prior to photographing and are a permanent part of the negative.

in all standard camera body photomicroscopic systems. Exposures are frequently more than a second and not uncommonly several seconds in duration; thus vibration can be the single most troublesome problem in camera body photomicroscopy.

A video image recorder requires a video image in order to produce a picture; thus video microscopy is a necessity. If a video system is available, then generation of high-quality photographs and slides can be as simple as pushing a button. Most video recorders provide the user with a selection of film types, including standard slides and prints, and instant, or Polaroid, slides and prints. Film speed, or ASA, is usually selectable, although the range of selection may be somewhat limited. Because of the method of picture generation, very slow speed, high-definition film can be utilized without the limitations inherent in standard camera body photography, and is usually preferred. Thus limited ASA selections are not a limitation. Depending on the mode of signal provided to the recorder, image brightness, contrast, color, tint, and sharpness can be altered. If image alteration is provided, an output is provided to preview the altered image on a separate channel on a video monitor prior to taking the picture. Video image recorders are easy to use, produce high quality pictures, and provide the video microscope user many useful features (see Figure 9.6).

Video graphic printers also require a video image in order to produce a picture. These devices produce a black and white printout of the video image on special thermal-sensitive paper. Resolution is high (about 8 dots/mm) with up to 130 shades of gray. Image brightness and contrast can often be varied, and negative prints as well as positive prints can be made. Finished prints can be generated in a matter of seconds. Operation is simple, and the quality of print is excellent; however, no negative is produced and only one size of print is made.

Selection of a method of photomicroscopy is very user- and microscope system–specific. No single system will provide all the features needed or wanted by every investigator. If no microscope system exists or extensive upgrading is possible, then the author would encourage consideration of video microscopy and its accessories.

Suppliers

Saws

Buehler, Ltd., 41 Waukegan Rd., Lake Bluff, IL 60044

Exakt Medical Instruments, Inc., 1000 S. McKinley, Box 83343, Oklahoma City, OK 73148-3343

Laser Technology, Inc., 10624 Ventura Blvd, North Hollywood, CA 91604

Leco Corp., 3000 Lakeview Ave., St. Joseph, MI 48085-2396

Leitz, Ltd., 24 Link Dr., Rockleigh, NJ 07647

Saw Blades

Buehler, Ltd., 41 Waukegan Rd., Lake Bluff, IL 60044

Felker Operations, Dresser Industries, Inc., 1900 S. Crenshaw Blvd., Torrance, CA 90509

Norton Abrasives, One New Bond St., Box 15008, Worcester, MA 01615-0008

Raytech Industries, Inc., Industrial Way, Stafford Springs, CT 06076

Struers, Inc., 26100-T First St., Westlake, OH 44145

Grinders/Polishers

Buehler, Ltd., 41 Waukegan Rd., Lake Bluff, IL 60044

Exakt Medical Instruments, Inc., 1000 S. McKinley, Box 83343, Oklahoma City, OK 73148-3343

Leco Corp., 3000 Lakeview Ave., St. Joseph, MI 48085-2396

Maruto Instrument Co., Ltd., Kakuta Bldg., 1-1-10, Yushima, Bunkyo-ku, Tokyo 113, Japan

Investing

Price-Driscoll Corp., 75 Milbar Blvd., Farmingdale, NY 11735

Silmar Division, Standard Oil Co., 12335 S. Van Ness Ave., Box 5006, Hawthorne, CA 90250-0590

Embedding

Aldrich Chemical Co., 940 W. St. Paul Ave., Milwaukee, WI 53233
Fisher Scientific, 711 Forbes Ave., Pittsburgh, PA 15219
Polysciences, Inc., 400 Valley Rd., Warrington, PA 18976
Rohm and Haas, Independence Mall W., Philadelphia, PA 19105

Mounting media

Calibrated Instruments, Inc., 731 Saw Mill River Rd., Ardsley, NY 10502
Hugh Courtright & Co., Ltd., 6135 W. 65 St., Chicago, IL 60638
Carl Goldberg Models, 4734 W. Chicago Ave., Chicago, IL 60651
Devcon Corp., Danvers, MA 01923; Scarborough, ON Canada; Shannon, Ireland

Microscopic slides

Brain Research Laboratories, Box 88, Waban Station, MA 02168
Buehler, Ltd., 41 Waukegan Rd., Lake Bluff, IL 60044
Scientific Products, Baxter Healthcare Corp., 1430 Waukegan Rd., McGaw Park, IL 60085-6787
Wasatch Scientific, Salt Lake City, UT

Video Cameras

Dage-MTI, Inc., 701 N. Roeske Ave., Michigan City, IN 46360
Sony Corp. of America, 10833 Valley View St., Cypress, CA 90630

Image processors and analyzers

Buehler, Ltd., 41 Waukegan Rd., Lake Bluff, IL 60044
Leco Corp., 3000 Lakeview Ave., St. Joseph, MI 48085-2396
Microscience, 31101 18th Ave. S., Suite A, Federal Way, WA 98003
Olympus Corp., 4 Nevada Dr., Lake Success, NY 11042-1179
R & M Biometrics, Inc., 5611 Ohio Ave., Nashville, TN 37209
Southern Micro Instruments, 120 Interstate North Parkway E., Suite 308, Atlanta, GA 30339
Universal Imaging Corp., 502 Brandywine Parkway, West Chester, PA 19380

Photomicroscopy systems

Nikon, Inc., 623 Stewart Ave., Garden City, NY 11530
Olympus Corp., Precision Instrument Division, 4 Nevada Dr., Lake Success, NY 11042-1179
Polaroid Corp., Box 607, W-95 Century Rd., Paramus, NJ 07652
Sony Corp. of America, Sony Dr., Park Ridge, NJ 07656

References

Anderson, C. 1982. *Manual for the examination of bone.* Boca Raton, FL: CRC Press.
Baron, R., A. Vignery, L. Neff, A. Silverglate, and A. Santa Maria. 1983. Processing of undecalcified bone specimens for bone histomorphometry (pp. 13–35). In R. R. Recker (Ed.), *Bone histomorphometry: techniques and interpretation.* Boca Raton, FL: CRC Press.
Berenbaum, M. C. 1954. Staining of bound lipid. *Nature 174*:190.

Bloehaum, R. D. 1989. Sanderson C, McCarvill S, and Campbell P. 1989. Plastic slides in the preparation of implant and tissue for interface analysis. *Journal of Histotechnology* 12:307–310.

Humason, G. L. 1979. *Animal tissue techniques*, 4th ed. San Francisco: Freeman.

Luna, L. G. (Ed.). 1968. *Manual of histologic staining methods of the Armed Forces Institute of Pathology*. 3rd ed. New York: McGraw-Hill.

Pawlicki, R. 1977. Topochemical localization of lipids in dinosaur bone by means of Sudan B black. *Acta Histochemica* 59:40–46.

Rhodes, D. M., C. Sanderson, and R. D. Bloebaum. 1990. Glue-glove technique for use in grinding and polishing thin specimen. *Journal of Histotechnology* 13:63–64.

Roush, J. K., G. J. Breur, and J. W. Wilson. 1988. Picrosirius red staining of dental structures. *Stain Technology* 63:363–367.

Schenk, R. K., A. J. Olah, and W. Herrmann. 1984. Preparation of calcified tissues for light microscopy (pp. 1–56). In G. R. Dickson (Ed.), *Methods of calcified tissue preparation*. New York: Elsevier.

10

Molding, casting, and painting

Molding and casting: techniques and materials

Mark B. Goodwin and
Dan S. Chaney

The accurate replication of fossil vertebrates is a safe and effective means of providing other scientists access to rare and important specimens. The replication process consists of two distinct but interdependent procedures: mold making and casting. Molds are made from a variety of rubbery materials. Latex, polysulfide, polyurethane, and silicone room temperature vulcanizing rubber are the most commonly used in paleontology. Casts are made from poured gypsum plaster, epoxy, polyester, or polyurethane resins, fiberglass-reinforced resin, or plaster. The choice of molding and casting materials is determined by consideration of a number of variables. These are the most important: condition, size, and shape of the specimen, use of the cast (research, teaching, or exhibition), specified weight requirements of the cast, and relative cost of materials. The preferred casting material may dictate which molding compound to use (e.g., some molding rubbers are not compatible with some casting compounds). The type of mold to be fabricated also is a factor in choosing appropriate materials. Molding and casting procedures are most effectively learned by working with an experienced preparator.

The first part of this chapter is on molding; casting is discussed in the second parts. The reader is given a list of suggested molding materials with instructions on their use in paleontology. Accessories (parting agents, clays, tools) used in mold fabrication are also listed. Techniques covered include peels, molds of microvertebrates and small fossils, two-piece molds, pour molds, slit molds, and multipiece flexible molds. A brief description of molding rubbers and their use with a variety of casting compounds is provided. Instructions for making specific types of casts (i.e., solid, hollow, fiberglass reinforced, or polyurethane foam filled) are discussed separately for each casting compound. A glossary of key words and physical properties is provided. A list of molding and casting materials and vendors is at the end of the chapter.

A variety of molding and casting methods and materials have been employed by paleontologists. Changes in techniques often correspond with the arrival of new materials. Donald Baird (1951) pioneered the use of molding paleontological specimens in liquid latex from a process described by O. F. von Fuehrer (1939), an artist at the Carnegie Museum in Pittsburgh. Waters and Savage (1971) first reported on the use of silicone rubber molds and epoxy casts in paleontology. For a review of recent molding techniques and materials, see Rixon (1976), Chase (1979), Converse (1984), Chaney (1989), Chaney and Goodwin (1989), Smith and Latimer (1989), and Chaney, Kroehler, and Lewis (1991).

Numerous molding compounds are available commercially: latex, silicone, polyurethane and polysulfide rubbers and alginate. For most specimens, the authors prefer latex and silicone. Many of the reactive materials used in molding and casting compounds have been reported to cause skin irritation, dermatitis, and/or respiratory sensitization in susceptible users. Pay strict attention to material safety data sheets and manufacturers' recommendations for proper handling and storage of all materials. Store respirators and filters in sealed plastic bags when not in use.

The most important step in molding and casting is the preparation of the specimen. In paleontology, the mold maker must have a thorough understanding of fossil preparation. A student of fossil preparation can learn mold making by hands-on experience. A thorough understanding of molding materials and methods is required. You cannot learn molding from a book, just like you can't learn cooking from a cookbook. This chapter will assist the reader, but it is strongly recommended that the student of paleontology undertake work under the supervision of an experienced preparator.

Key words

Like most specialties, molding and casting have their own jargon. These terms and their definitions are provided to assist the reader in understanding the methods and materials in molding and casting.

accelerator (or **promoter**): chemical used to speed up and control curing of resin.

cast: the positive replication of a specimen made from a mold. Normally made of plaster or a thermoplastic.

catalyst: compounds added to molding and casting materials that cause them to cure. Resins are usually cured by the addition of an organic peroxide, silicone rubbers by an ethyl silicate solution.

demold: the removal of the mold from the specimen or a cast.

detail surface: the surface of a mold that was next to the specimen, on which the features of the specimen are replicated in the negative.

dike: the portion of the mold that confines the fluid casting material prior to its setting.

gel coat: a thin layer of unreinforced catalyzed resin. Inside a mold, this is the surface of the finished cast.

gel time: time required for a liquid resin to set to a gel.

lip or flange: extensions of the mold lateral to the specimen, used on multipiece molds. They are sandwiched between pieces of the mother mold to hold the pieces of the mold in their proper positions.

mold: the negative replication of a specimen used to produce a cast.

mother mold: exterior covering or jacket applied to the outside of a mold for support and to prevent deformation. Suggested materials are plaster, fiberglass-reinforced plaster (FGR), and fiberglass and resin.

outer surface: the surface of the mold away from the specimen. If a mother mold is required, it will contact the outer surface of the rubber mold.

release agent: material applied to the surface of the specimen or mold to facilitate demolding.

room temperature vulcanizing: often abbreviated as RTV, the term refers to the vulcanization or curing of the molding rubber by the addition of an activator to the base compound. This term RTV is used by manufacturers of silicone rubbers and sometimes for urethane elastomers.

seam line: a raised line, composed of casting material, on a cast, produced where two adjacent mold pieces meet.

separator: material applied to the specimen or mold to prevent damage to the specimen or mold by chemical or physical interaction between the specimen and molding material or the mold and casting material. Separators often serve as release agents, but that is not their primary purpose.

sprue or vent: the opening in an assembled multipiece mold through which casting material is poured; the pour hole

Molding

Molds can be very simple one-piece peels or complex, multipieced affairs. Remember, the most important consideration is the fossil specimen. Irreparable damage with a corresponding loss of scientific information can easily occur if the preparator's responsibility is handled lightly. Take the time to study the physical shape and condition of the fossil under consideration.

The specimen chosen for molding must be prepared and consolidated properly, rehardened when necessary with all glue joints and fillers inspected. Inspect and identify compounds on the fossil to ensure proper compatibility with the molding medium. Undercuts and fragile areas on a fossil need special attention to prevent unnecessary breakage. Consider these points before molding a fossil specimen:

1. Can the specimen be molded without serious and irreparable damage?
2. Purpose of the cast: research, teaching, or exhibit? A cast produced for research requires particular attention to ensure detailed replication in the mold and later reproduced in the cast; casts for teaching and exhibits may warrant less detail but more durability.
3. Consider cost and time factors. Budget constraints and time considerations are part of any job. However, molding should never be rushed into; avoid shortcuts.
4. Casting medium: Determine mold compatibility with the casting compound.

After removing a mold from the fossil, proper care must be given to both. Check the fossil for loose cracks, old glue joints in need of repair, or broken fragments that need to be glued back in place. Seam line and mold residue should be removed. Reapply museum/catalog numbers to the fossil. A record should be made of when the mold was made, by whom, and materials and consolidants used. Wash the mold in a mild solution of dishwashing liquid to remove separator residues, clay, and matrix. Label the mold for future reference and keep an updated mold catalog. To ensure a long shelf life, store the mold in a cool, dry, dark place. Fill molds with a cast for short-term storage, exercising caution, as some casts are difficult to remove after many years in a mold. Molding compounds may soften, shrink, or become brittle with age. A recommended alternative is to keep a master cast and discard the mold after pulling a suitable number of casts. Molding the master cast when needed reduces the wear and tear of remolding the original fossil specimen.

Physical properties of molding compounds

Data sheets describing the physical properties of molding compounds are available from the manufacturer or vendor. Most major companies have toll-free telephone numbers for customer service and technical support. This information is useful as a guide to give you an idea of what the material is like and is helpful when trying new products or making comparisons with old ones. Some distributors have samples available for first-time users to test.

The following terms are defined in order to understand better the physical properties of molding compounds. Many of the terms are used in the text.

hardness: hardness or "durometer" is listed as "Shore A" from 0 to 100: 1–10 is very soft; 20 is medium soft; 40 is firm (= rubber eraser); 60 is hard (= rubber tire); 90 is harder, less flexible. Shore D from 0 to 100 measures hardness beyond the values of Shore A.

tensile strength: ability of a material to resist a pulling force.

tear strength: force required to pull apart a piece of material that has a tear: 50 or less is poor; 75 is fair; 100 or more is good. Softer rubbers have the poorest tear strength.

elongation: percentage a material stretches before breaking.

shrinkage: dimensional stability of rubber measured in inches of shrinkage per inch or as a percentage.

viscosity: viscosity refers to the ease with which a material flows. The unit of viscosity used is the centipoise (cps). One centipoise is defined as the viscosity of water at room temperature. For comparison: kerosene 10 cps, motor oil 10–100 cps, glycerin 1000 cps, corn syrup 10,000 cps, molasses 100,000 cps.

working time or **pot life:** minimum time for the molding material to set up in the mixing container.

demold time: the shortest time from application of the rubber on the specimen to when the mold may be removed from the specimen; prior to curing, some rubbers can be demolded from the specimen without permanent distortion.

cure time: time required for a molding compound to fully set up to its specifications. The time before casting material should be applied in the mold.

Molding compounds

Latex rubber

Latex rubber is a solution of prevulcanized rubber particles suspended in a mixture of water or water and ammonia. A variety of latex molding rubber is available commercially. The material most suitable for molding fossils is a creamy white rubber containing between 70% and 75% solids. Latex rubber that dries amber in color should be avoided, as it is less elastic and flexible with a shorter shelf life. Latex with a high ammonia content drys out, cracks, and darkens in four to five months.

Latex is an easy to use, economical molding material. It is applied by painting separate layers over a specimen. Care must be taken that the latex is not applied too thick, or an uneven cure may occur. As the cured latex shrinks, it pulls away from the surface of the original specimen if it is not painted on in a series of thin, even coats. Areas of thick latex will form a skin on the surface that inhibits a cure beneath. In order to minimize shrinkage (1%–5%), apply the latex in alternating directions (checkerboard fashion) over a specimen and allow to cure between coats. As the layers of latex are built up to a suitable thickness, 2–3 mm, depending on the specimen, a layer of cheese cloth, gauze, or nylon is applied. To make this step easier, cut the material into patches or squares, dampen with water, and apply to the nearly cured latex layer below. A thin solution of latex is painted over the material. The mold is built up with another series of latex painted on. Presoaking the brush in detergent or applying dishwashing liquid directly to the brush will make cleaning it much easier after applying each layer. Dried latex can be removed easily by soaking the brush in kerosene overnight. This swells the latex residue for easy removal with a wire brush or tweezers.

Because latex is applied in layers, the first or detail coat is critical in order to avoid permanent bubbles in the mold, which later show up on the surface of the cast. Rixon (1976) suggests first coating a specimen with a 5% solution of ammonium hydroxide before molding. This helps the latex penetrate into small details and reduces bubbles. After the detail coat is applied, the surface can be blown with compressed air at low pressure to pop any bubbles.

An alternative to painting the latex on the specimen with a brush, is to use an airbrush (Heaton 1980). The use of an airbrush is highly recommended for making one-piece peels of a specimen. A small, external mix unit works very well (Paasche model H). The latex is thinned by adding a small amount of water one drop at a time to prevent curdling. This technique has several advantages: (1) the layers dry almost instantly and can be built up in short time; (2) air bubbles do not occur when the latex is applied under pressure; and (3) no brushstrokes remain in the mold. Liquid latex is white or yellowish in color when cured. It can be tinted with acrylic paints before application to the specimen. Tinting improves the contrast of surface details in the latex peel. Heaton (1980) recommends red, orange, or green acrylic paint for good color contrast of surface detail. Gray and black peels absorb too much light, reducing surface contrast. The cured latex will dry darker than it looks premixed in the jar.

Latex has the major disadvantage of shrinkage. It is still a good, economical molding compound and recommended for casts used in exhibits and teaching rather than research. A separator must be used when casting with fiberglass and resin. Apply an even coat of carnauba wax and polish with a brush. A coat of polyvinyl alcohol (PVAL) is brushed or sprayed on over the wax. Avoid puddling the PVAL in undercuts. A separator is not necessary when casting with plaster. The shelf life of a latex mold can be improved if it is stored in a cool, dark place, away from sunlight and moisture. In some cases a plaster cast can be stored in the mold to prevent further warping and shrinkage. Note that over the long term, deterioration of the mold will cause it to stick to the cast and make its removal from the mold very difficult. Latex will not set up against polyethylene glycol (PEG) on a specimen. Some plastilene clays will inhibit latex from curing. This can be avoided by painting a separator coat of PVA over the PEG, clay, and the specimen. Do not use PVAL where uncured latex will contact it, as PVAL is water soluble and will have adverse effects on the latex.

Latex can be distorted purposely with some interesting results. By soaking latex in kerosene, a latex mold or cast can be enlarged up to 50% with little distortion. The time suggested for soaking in kerosene is varies depending on the thickness of the latex mold. Be prepared for uneven enlargement if the mold is not of uniform thickness. The latex must be completely cured before soaking in kerosene or it will deteriorate and become jellylike (Chase 1979). This method was first reported by Fuehrer (1939) and most recently by Wilson (1989). It is important to submerge the latex completely in the kerosene so that the enlarged specimen will remain covered. If possible, place the detailed side facing upward. When the desired size is reached, remove the latex and rinse thoroughly with water. The size of the specimen will determine the time needed to double in size. Wilson (1989:283) reports that small, thin latex molds or casts may double in size in less than 2 hours; thicker specimens may need 24 hours or more.

Silicone rubber

RTV silicone rubber is a room temperature curing, two component synthetic rubber that is highly recommended as a strong, flexible molding compound. Silicones are semi-inorganic polymers (organopolysiloxanes). Vulcanizing agents and fillers are mixed with these silicone polymers to produce silicone rubbers for molding. Thermosetting silicones, silicones capable of being cured into an insoluble product by heat or chemical means, were discovered in the late nineteenth century. It wasn't until the 1940s that silicone was considered more than a laboratory curiosity and the commercial applications of this polymer were developed (Cherry 1967).

The use of RTV silicone rubber for making accurate duplicates of fossil vertebrate specimens was pioneered by Waters and Savage (1971). Although materials and techniques have changed slightly, it should be noted that Waters and Savages's accurate and useful description of molding with RTV and casting with epoxy resin still serves as the standard guide in most preparation labs. Later references with useful information on RTV silicones include Chase (1979), Reser (1981) and Converse (1984); all offer some updated information on RTV molding. The reader is referred to Chaney (1989) for a recent description of materials and procedures. Chaney and Goodwin (1989) conducted a comparison of popular RTV silicone rubbers used to mold fossil specimens and report on the use of various activators.

The decreasing price of some RTV silicone and the availability of a slow cure activator that lengthens mold life to five or more years, has eliminated two of the major disadvantages attributed to silicone rubber molds: high expense and short shelf life. One advantage is a flexible, strong rubber that reproduces the microscopic detail of a specimen. Another is that separators are generally not necessary to remove plastic casts from a silicone mold.

Tests have shown that silicones capture microscopic features between 0.1 and 0.25 micron at 1500–2000× magnification under the scanning electron microscope (SEM) (Walker 1980; Rose 1983). Because of this level of resolution, they are employed by paleontologists, anthropologists and others for SEM studies of tooth microstructure and bone analysis.

The recommended RTV silicones for molding usually consist of a two part compound: a white, opaque "base" material and a clear or colored "activator." Most are mixed 10 parts base to 1 part activator. The activator is tin- or platinum-based and usually colored (blue, red, green, pink) as a guide for proper mixing. The tin-based activator is the recommended catalyst because of its stability and wide range of compatibility with various substrates and materials (i.e. clays, glues, hardeners, separators). Platinum catalyzed RTV silicones are susceptible to cure inhibition by contact with tin-catalyzed rubbers, sulfur, or amines. Inadequate stirring can result in an uneven cure of the mold and eventual mold loss. A mechanical mixer (KitchenAid KSM series, Jiffy mix attachment for variable speed drills) is recommended. Silicone rubber pigments are useful for silicone positives or color coding molds. Manufacturer's instructions recommend evacuation of the silicone mixture in a vacuum chamber to 27–29 in. Hg to remove air bubbles from the mixture. When the vacuum is applied, the silicone will increase four to five times its original volume, crest, and collapse to its original level. Deaerating is complete approximately 2 minutes after the frothing action ceases. Be sure to use an oversized container in the vacuum chamber.

Slow-cure activators with reduced amounts of tin are now available for increased shelf life of the molds. These activators have an extra long cure time of up to three days because of the lower tin levels. If time is of the essence, one can mix accelerators with the activators. These ultra-fast catalysts have the ability to reduce working times of 1–2 hours to 5–60 minutes and cure times of 16–18 hours down to 20 minutes to 5 hours. Exercise caution when using these products. The high percentage of tin in the activator will severely reduce the shelf life of the mold. Silicone molds cured entirely with the ultrafast catalyst may not separate from epoxy resin and polyester and fiberglass casts without a separator in the mold. There are high temperature–resistant catalysts available for some silicone rubbers. These are recommended when making solid casts of microvertebrate fossils from quick-setting, exothermic epoxy plastics.

Major differences of base compounds are found in their physical properties of viscosity, tensile strength, hardness, and working time. Some are pourable; others are thixotropic for reduced flow on vertical and overhead surfaces. Thixotropic silicone can be painted on. Fumed silica fibers can be added to make low-viscosity silicones thixotropic. This may weaken certain silicones, however, by decreasing their tensile strength.

Certain materials and products will cause inhibition of the silicone to cure fully. Choose clays, adhesives, and consolidants carefully. Neoprene and polysulfide rubbers and other products containing sulfur will inhibit silicone. Many plastilene clays contain sulfur. A low oil-based plastalina clay (Life Plastalina) is available that is 100% compatible with silicone, will not stain the bone, stays soft, and is reusable by melting and screening out any impurities. Beware of materials containing moisture, like potter's clay and some waxes. Some silicones will adhere to the surface of low fire potter's clay (often used in model making and sculpture). Apply an even layer of paste wax to seal the porous surface of the clay. When in doubt, make a test patch of the silicone against the surface of the suspected inhibitor. This is also recommended when molding a fossil that has been treated with an unknown consolidant. Manufacturers often change product chemistry or ingredients. Though such changes may not affect the exact specifications, it is best to test any product thoroughly.

Some silicones still have a tendency to tear if they are not reinforced with nylon, fiberglass cloth, or gauze. After a series of coats is applied to a specimen, a layer of nylon, fiberglass, or gauze cloth is "sandwiched" into the layers of silicone. Strips of cloth can be applied to silicone when the surface is still tacky, just before it fully cures. This enables the cloth to "lie down" easier. A layer of thin silicone is painted over the cloth layer. Repeated layers of silicone are added until the mold is built up to suitable thickness (3–5 mm). Recycled nylon stockings and lightweight fiberglass cloth are recommended. Silicone reinforced with nylon or fiberglass is quite tear resistant, even around deep undercuts. For a more flexible mold, nylon stocking material is recommended over fiberglass cloth. Cotton gauze tends to tear inside the mold along areas of stress and deep undercuts.

To build up the mold faster, thixotropic silicone can be brushed on. A liquid additive is available that thickens low-viscosity silicones. This may be a good alternative to the use of fumed colloidal silica and microballoon fillers, which are health hazards when inhaled.

Rixon (1976) reports that silicone rubber will expand one and a half times when soaked in kerosene for 4 hours. He reports that silicone is more suitable than latex for enlargement.

Silicone caulking sealant (be sure that the brand is 100% silicone) is useful for filling undercuts and repairing torn silicone molds. These caulks should never be used as the primary molding compound due to the pressure of acetic acid. It is available in 10.3 fluid oz. containers that fit into a standard caulking gun. An emulsion of silicone caulk with water (3–4 drops per ounce) may be used to fill undercuts on molds. To repair a tear, clean the surface with acetone and apply the caulk and let cure 24 hours. A piece of cloth applied with extra silicone to the exterior surface of the mold will aid in keeping the tear closed. This emulsion cures in about 10 minutes. Large undercuts and voids can be filled more efficiently by first applying the silicone to a precut piece of flexible foam and sticking this silicone covered "plug" into the undercut. For a removable plug, apply a parting agent before inserting the plug.

Vinyl polysiloxane

Vinyl polysiloxane is a silicone impression material formulated for use in dentistry with applications in paleontology. It is available as a two part putty that is mixed by hand or as a paste in a syringeable dispenser kit that combines the two components as they extrude through a disposable mixing tip. The combined product is applied directly to the fossil after mixing. In keeping with its original use in dentistry, the product has very low mixing, working, and set times, in the range of 30 seconds to 3 minutes. Certain gloves, glove residues, and hand lotions will inhibit the setting of the putty. Vinyl polysiloxane is useful for molding on brief visits to museum collections (Leiggi 1989) or on overseas trips when molding materials may be in short supply or are difficult to transport. Their high price makes their everyday use somewhat prohibitive. They are available commercially as 3M Express and Colteen President putty.

Polyurethane elastomer

Polyurethane elastomer is a two part "elastic" thermoplastic resin. A variety of polyurethane resins plastics are also used in casting (polyurethane coatings, flexible and rigid polyurethane foams). Polyurethane elastomers produce a flexible, tear resistant molding compound, with good fidelity. Cure time is relatively short. Most polyurethane elastomers require the use of a parting agent when casting porous materials or fossils. They are excellent adhesives and extremely tough, making specimen removal risky if separation is difficult (Chaney 1985, 1989). A parting agent should be used on the polyurethane mold when molding porous specimens or casts. Apply a barrier film to models or the fossil to prevent sticking. The specimen will break before the molding rubber gives it up. Polyurethane elastomers are recommended for clay models for exhibits and teaching. They should be avoided when working with fossils.

Combinations used by different manufacturers produce polyurethanes with a variety of properties, working times, and specifications. Follow manufacturer's instructions in regard to mixing the components in the recommended proportions. Avoid trying to obtain a faster cure by adding more hardener. It does not work like a catalyst. The polyurethane may be thickened by adding fumed silica fibers or microballoons. Some polyurethanes will not cure against latex rubber and some plastilene clays.

Moisture has a damaging effect on polyurethanes, causing bubbles and foaming in some cases, and it affects the curing ability in the hardener. The resin crystallizes into unusable solids. By keeping the compounds in tightly sealed containers and storing unused amounts full to the top in smaller cans, some of these problems can be avoided. For larger quantities, fill partly empty containers with nitrogen gas to remove the unwanted air. Silica gel packets taped to the underside of the can lid may also help decrease moisture levels. Adequate tests for prolonged shelf life are not available. Until recently, their relatively low cost made them an attractive alternative to the once higher priced silicone rubber.

Isocyanate is a major ingredient of polyurethanes. Effects of overexposure include eye and skin irritation, allergic sensitivity, and respiratory problems. Prolonged exposure to its vapors is also hazardous and should be avoided. Read carefully and observe all cautions stated in materials safety data sheets.

Polysulfide

Polysulfide rubber is a synthetic compound that cures at room temperature. It is catalyzed by mixing with a polymer and curative compound. It is imperative that they are mixed according to the manufacturer's instructions or they will not set up properly. Disadvantages are some shrinking and a long term storage problem where the rubber deteriorates and "weeps" or precipitates volatiles, causing the mold to become gummy and soft. A separator is recommended when casting in plastic. A thin film of dishwashing liquid will aid in the separation of plaster casts. Some polysulfides cannot be used with thermoplastics. Polysulfide rubbers react with polyvinyl acetate and acrylic resins commonly employed as consolidants in paleontology. Proper testing for compatibility is recommended. For the most part, the use of polysulfide rubber in paleontology has been replaced by RTV silicones and polyurethane rubbers.

Alginates

Alginate powders are made from algae or seaweed. They were developed in the 1940s for use in dentistry. Mixed with water, they gel quickly to form a rubbery material. They are safe and nontoxic and reproduce a highly detailed but fragile mold that needs to be kept moist to prevent shrinking and cracking. Though flexible when cured, this material will not easily pull out of deep undercuts before tearing. The material will not stick to anything, including itself, after it cures. A bonding agent, sodium carbonate solution, is used to "etch" the cured surface of the alginate to facilitate building up multiple layers in a mold. In special circumstances, alginates are useful in paleontology. Their impact on a specimen is minimal. Because an alginate mold is so fragile, a master cast should be pulled immediately. Plaster performs well with alginates for casting. A durable silicone or latex rubber mold can be replicated from the master cast for multiple casting.

General instructions for the use of alginates are provided by the manufacturer and are easy to follow. They are usually mixed with warm water. Colder water retards setting and warmer water accelerates it. Alginates are available commercially as Jeltrate and Dermagel. Moulage, another type of alginate, is made from agar, a gelatinlike product of certain seaweeds. It comes in a gellike form and is reusable. The desired amount is melted in a double boiler. The first time it may not be necessary to add water; however, this is suggested for repeated use. Moulage is stronger after being used once or twice. Do not let the moulage cool when applying to a specimen. When moulage cools, it will not adhere to any surface, including itself. No parting agent is required.

Parting agents

Parting agents include separators and mold releases and are available in liquid, paste, or film form. There are numerous products available. Some are recommended for use with specific molding and casting compounds. Many are formulated to be either sprayed, wiped, or brushed onto the surface of a mold. When a release agent is required, it is advisable to test the product for compatibility and effectiveness with the molding and casting materials. Parting agents can be separated into these general categories: hard waxes, paste waxes, silicones, parting films, and polymer consolidants.

Hard waxes (mold release wax)

The best hard waxes for use as a release agent are those containing carnauba wax. Waxes come as hard or presoftened pastes and are also available in solution as aerosol sprays. They are applied with a brush or cloth rag and can be buffed to a gloss for improved results. Hard waxes are used by themselves or as an undercoat in combination with other separators such as PVAL. They are recommended when separating plastic from latex rubber molds, plastic from plaster, and plaster on plaster. Hard wax is a good general purpose parting agent and surface sealer. Meguiars Mirror Glaze (Mirror Bright Polish Co.) is recommended.

Paste waxes

Paste waxes are available as presoftened liquids or creams. The wax is soft to the touch and appears wet. Unlike the hard waxes, most paste waxes cannot be buffed to eliminate surface buildup. They are recommended for use with some low density plastics, such as urethane elastomers. Their use with higher density plastics, like epoxy resin, should be avoided, as these compounds tend to displace the wax. Most paste waxes are available commercially as automotive waxes.

Silicones

Silicone release agents are generally available as aerosol sprays. They are used as a parting agent for resin casts from silicone rubbers if needed and will work with most plastics and molding rubbers. Silicone has the disadvantage of not drying or evaporating off the mold surface, which can make painting casts from silicone-sprayed molds difficult. Silicone remains wet to the touch. Silicone is also used as an ingredient in other types of parting agents.

Parting films

Parting films provide a barrier coat between the mold and cast. Polyvinyl alcohol is a water-soluble solution produced by the hydrolysis of polyvinyl acetate. It is used mainly as a parting film when laying up polyester resin against latex rubber molds. It is easily applied by brush, spray bottle, or airbrush. PVAL is a good separator for latex on latex and latex against polyester resin; it cannot be used for plaster on plaster. The film of PVAL can be removed from the mold with warm, soapy water. For improved results when casting resin in latex molds, an even film of hard wax is first applied to the latex and then sprayed with two to three coats of PVAL. PVAL should be reapplied between casts. Avoid the use of PVAL over releases that contain silicone.

A second group of parting films contains petroleum jelly (Vaseline) or wax dissolved in a solvent like methylene chloride. It is applied to the mold as an aerosol or by brushing. The solvent evaporates quickly, leaving a dry barrier film. Some may appear greasy or oil-like. It is an excellent parting agent for rubber on rubber multipiece molds and is recommended as a separating medium for silicone, polyurethane, and polysulfide rubber molds. It should never be used on latex, as petroleum products cause latex to swell and distort. It is useful as a separator between plaster pieces to prevent bonding. Petroleum jelly can be used as a separator between epoxies and polysulfide simply by wiping a smooth thin layer over the mold.

Polymer consolidants
Acrylic (Acryloid), polyvinyl acetate (VINAC), and polyvinyl butyral (Butvar) resins are consolidants and adhesives used on fossil specimens (see Chapter 2). They are also useful as parting agents and sealants for rubber on rubber surfaces and plastaline clays. They work well as a barrier coat for silicone on silicone and latex on latex. They do leave a thin film after parting of the mold, which easily peels off or can be removed with gentle scrubbing in warm, soapy water. Building up excessive amounts on the mold will increase the thickness of the flashing left on the cast.

Miscellaneous separators

Vegetable oil aerosol sprays (e.g., Pam) are used in the same manner as silicone sprays, but tests should be made on individual combinations of materials to test for compatibility. Aluminum foil and cellophane may be applied over the exterior surface of a rubber mold to act as a separator from the mother mold. This works well on latex molds with fiberglass jackets. Cellophane film is useful as a barrier for plaster on plaster in multipiece mother molds. Dishwashing liquid provides an all purpose barrier film for plaster on plaster and plaster mother molds over rubber.

Additives

Fumed colloidal silica (available commercially as Cab-O-Sil) is useful as a thickening agent to make resins thixotropic. Mix into the resin in small amounts until the desired consistency is reached. Fumed colloidal silica must be used with care. Always wear a respirator with proper filters.

Fumed silica is inert and will not chemically react with any of the plastics typically used in the preparation laboratory. It is used to thicken runny plastics to a thixotrophic gel, which allows the plastic to be applied with brush or spatula to vertical surfaces and overhangs. Fumed silica will add strength to the plastic. However if too much is used, it will slow the setting of and weaken the plastic, as fewer polymeric chains will form due to interruption by the silica particles. It also helps dissipate heat caused by the increased exothermic reaction in thick resin casts. Chase (1979) recommends using up to 30% to absorb some of the heat generated during curing.

Powdered limestone and dolomite are also used as fillers in some plastics to increase the number of casts possible from a fixed amount of material. Although the limestone does not react with the plastic, it is heavy and adds considerably to the weight of large casts. It also absorbs a lot of the excess heat generated during curing.

Molding methods

Proper preparation and evaluation of the fossil must be completed before molding begins. The following mold-making methods are covered in this section: peels and pour molds, single and multipiece molds using latex and silicone rubbers.

The work surface for the mold is important. In mold making, a portable surface that the specimen rests on facilitates moving the mold around the laboratory. Potter's stands and lazy susans facilitate rotating larger molds. Remember to place a film of cellophane, aluminum foil, or clay over the work surface and under the fossil before molding begins. This makes removing the fossil or turning over a multipiece mold much easier. Few molding compounds stick to Formica. Remember, don't rush; shortcuts inevitably make more work for the preparator and increase the risk of damaging the fossil.

Peels

Peels can reveal details that are difficult to observe on the original specimen. They are commonly made of natural molds or casts of fossils, footprints, and skin impressions. Peels may be made in the laboratory or in the field.

Examine the specimen carefully and determine exactly what is to be reproduced. If the specimen is soft or friable, harden it with a thin (10%–20% is good) solution of polyvinyl acetate (PVA) or similar consolidant. After this dries and the surface of the specimen is stable, the specimen is ready to be coated with rubber. Latex or silicone is recommended.

Latex rubber can be tinted with acrylic artist's paint. This enhances detail in the peel by providing an opaque surface and shadowing. A large selection of acrylic colors is available; they do not fade or discolor over time, and they do not affect the latex. One teaspoon added to 4 ounces of latex is adequate. The latex will cure a few shades darker than the color of the pigmented solution.

Latex is brushed on or sprayed on with an external mix airbrush. When using an airbrush, the latex must be thinned by adding water one drop at time to prevent curdling. Application of latex with an airbrush eliminates brushstrokes, lost bristles, and bubbles. The latex penetrates into deep orifices and undercuts. It dries almost instantly and the coats build up faster. Once a suitable detail coat is applied, thicker coats of latex can be painted on. Be sure to let the latex dry adequately between coats.

When brushing on latex, the first, or "detail," coat is the most important. Use a high-quality nylon or pig bristle artist's brush. Sable or camel hair will catalyze liquid latex; cheap brushes lose their bristles and don't last. Coat the brush with dishwashing liquid to prevent the latex from building up on it. Apply the latex evenly in thin coats. Avoid over brushing, as drying latex will be lifted off the specimen by the brush. Repeat, allowing the individual coats of latex to dry, until the specimen is no longer visible through the latex. This will occur after three to four layers.

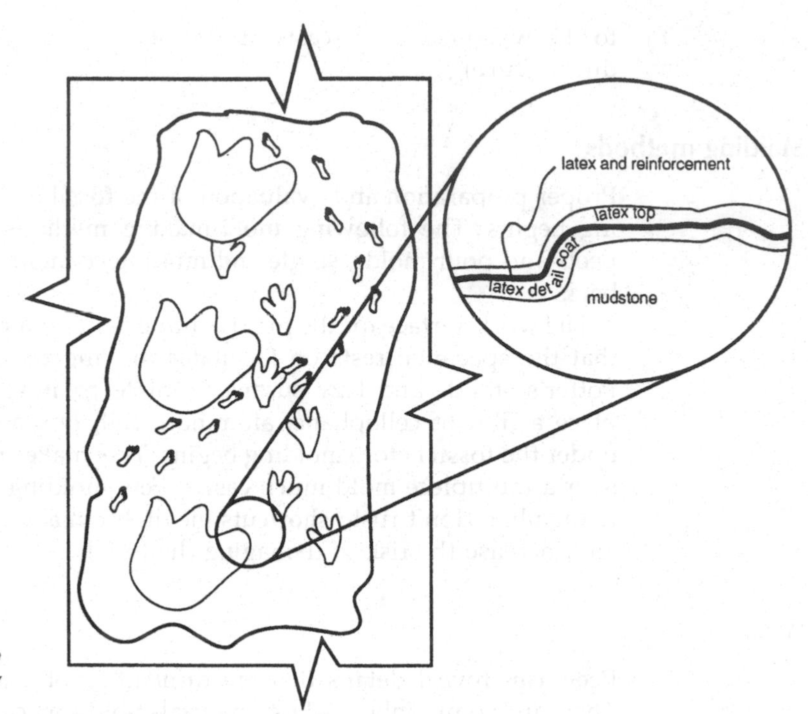

Figure 10.1. Steps for the construction of peel molds. Example is a trackway in the field.

After the detail and secondary coats of latex are completed, apply gauze or cheesecloth to strengthen the rubber peel. Cut the material into strips and dampen with water. To help the cloth lie down, paint another coat of latex on the specimen, and before it completely dries, apply the cloth. Cover the surface and overlap along the seams. Apply a "thinned" coat of latex to saturate and bond the cloth to the underlying rubber. Coat the specimen with additional layers of latex until the peel is built up to a suitable thickness, usually about 2–3 mm. Delicate specimens may require a thin peel for easier removal. Large plaques or specimens with natural molds or casts may require thicker peels.

The peel may be removed after the final coat has dried. A fan or hot air dryer will accelerate drying of the latex. Rixon (1976) suggests applying a 5% ammonium hydroxide, a colorless solution of ammonia gas in water, to the specimen before applying the rubber to facilitate the penetration into small details and reduce the formation of bubbles. Exposure to this gas is extremely harmful, and this technique should be avoided. A low-pressure stream of compressed air applied to the detail coat will sufficiently drive the latex into small areas and remove most bubbles. Place used latex brushes in kerosene. Kerosene softens and expands the dried latex; making it easier to remove between coats.

The use of silicone for making peels, with some minor variation, is similar to the latex procedure (Figure 10.1). A low viscosity silicone (40,000–60,000 cps) is mixed according to manufacturer's instructions. The detail coat is brushed over the specimen. A low pressure stream of compressed air will eliminate most bubbles on the surface. Apply the silicone as a thin, even coat to avoid trapping air and creating bubbles. Silicone is "self-leveling" and will fill most voids as it "bleeds" over the specimen. If runoff is a problem, construct a dike of clay or wax around the edges of the specimen. A layer of a thicker silicone is applied next with a brush or spatula. Use a high-viscosity silicone (500,000 cps) or a thin compound made thixotropic by the addition of fumed silica fibers. Add the fumed silica after the silicone base is combined with the activator. While the mold surface is still tacky, apply a layer of nylon or fiberglass onto the peel. Depending on the size of the peel, it is best to cut the fabric into workable strips or squares and apply in an overlapping patchwork. A stiff brush helps the material lie flat over the silicone. Cover the surface of the peel completely. The layer of fabric is saturated with another layer of silicone. A final coat of silicone paste "sandwiches" the fabric into the peel. If desired, a mother mold is useful for holding or storing the mold in its proper form.

Pour molds

For purposes of discussion, pour molds are defined as simple one- or two-piece molds made by pouring rather than brushing individual layers of a molding compound over a specimen. Silicone rubber (40,000–60,000 cps) is the recommended molding compound; polyurethane can be substituted but should be used with caution on fossils (see materials section for more information on polyurethane elastomers). Pour molds include one-piece (plaque) molds, molds of microvertebrates and small fossils, and standard two piece molds. Multipiece "flexible" molds are discussed in a separate section.

One-piece molds

A one-piece, or plaque, mold is the simplest to make. The specimen is placed on a base of clay. A dike (plastilene clay, cardboard sheets joined together with hot glue) is built up around the specimen like a frame to contain the molding compound. A groove along the inside of this wall will act to form a lip on the mold. This small detail sounds insignificant, but when the mold is removed, this groove is now a raised wall or dike along outside edge of the mold. This dike will keep the casting compound from overflowing the mold. The silicone rubber, mixed and evacuated, is poured over the specimen. Care should be taken not to trap any air by pouring the rubber slowly and evenly over the surface of the specimen. Plaque molds of suitable thickness (1 cm or more) may not need a mother mold for support.

Slit molds

Slit molds are fabricated in one piece around a specimen (see Figure 10.2). The molding rubber is poured or brushed onto the specimen. After the rubber is cured, the mold is cut or "slit." This slit allows the specimen and later casts to be removed from the mold. The slit in latex rubber molds is temporarily joined together with rubber cement. With silicone rubber molds, the slit is held together with masking tape or pressure from the mother mold. Pour holes are provided for the casting compound. Use a scalpel to cut a sharp, even slit in the mold. Slit

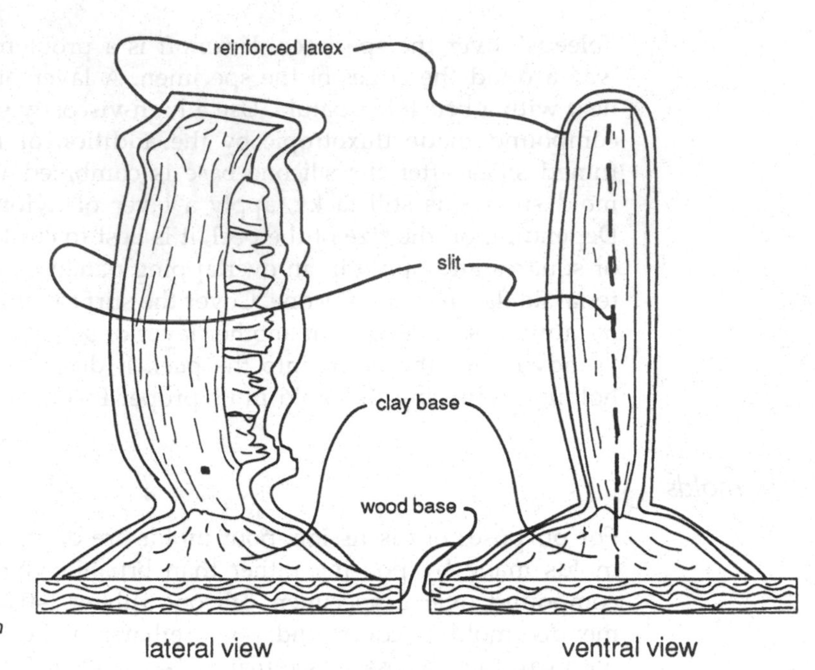

Figure 10.2. Steps for the construction of slit molds.

molds are used where a minimum of seam lines is desired on the cast. Slit molds are generally not suitable for lay-up casting.

Microvertebrate molds

Molding and casting microvertebrate fossils (and other small fossils) is a specialty unto itself. Microvertebrate fossils are often delicate and need special preparation before molding. The techniques were first described by Waters and Savage (1971) and later reviewed by Reser (1981) and Chaney (1989).

The use of silicone rubber for molds of microvertebrate fossils dictates close attention of the specimen before molding begins. For illustrations, we will consider a one piece mold of a small mammal jaw (see Figure 10.3). Silicone rubber is invasive, flowing into small crevasses and cracks. Particularly with jaws, it is wise to look for loose teeth, deep holes, and small cracks. Harden any loose teeth and fill in any openings along the gum line with cotton sealed with melted polyethylene glycol (Carbowax). Clay is also used but can be difficult to remove. Loose surfaces and cracks can be hardened with a consolidant (see Chapter 2).

After the specimen is prepared for molding, it is placed on a clay base. The base is labeled with appropriate museum numbers, element, genus, and so on by "dotting" the information in the clay with a pin vise. Label makers or lettering sets also work well. This provides a record of the specimen with each cast.

A clay wall is constructed around the specimen, high enough to hold the proper amount of silicone. A groove or moat is tooled along the margin of the base where it meets the inside wall. This seals the base to the wall and makes a lip on the silicone mold, which acts as a dike for the casting compound. RTV is mixed, deaired, and applied to the specimen. Use a small spatula and apply the

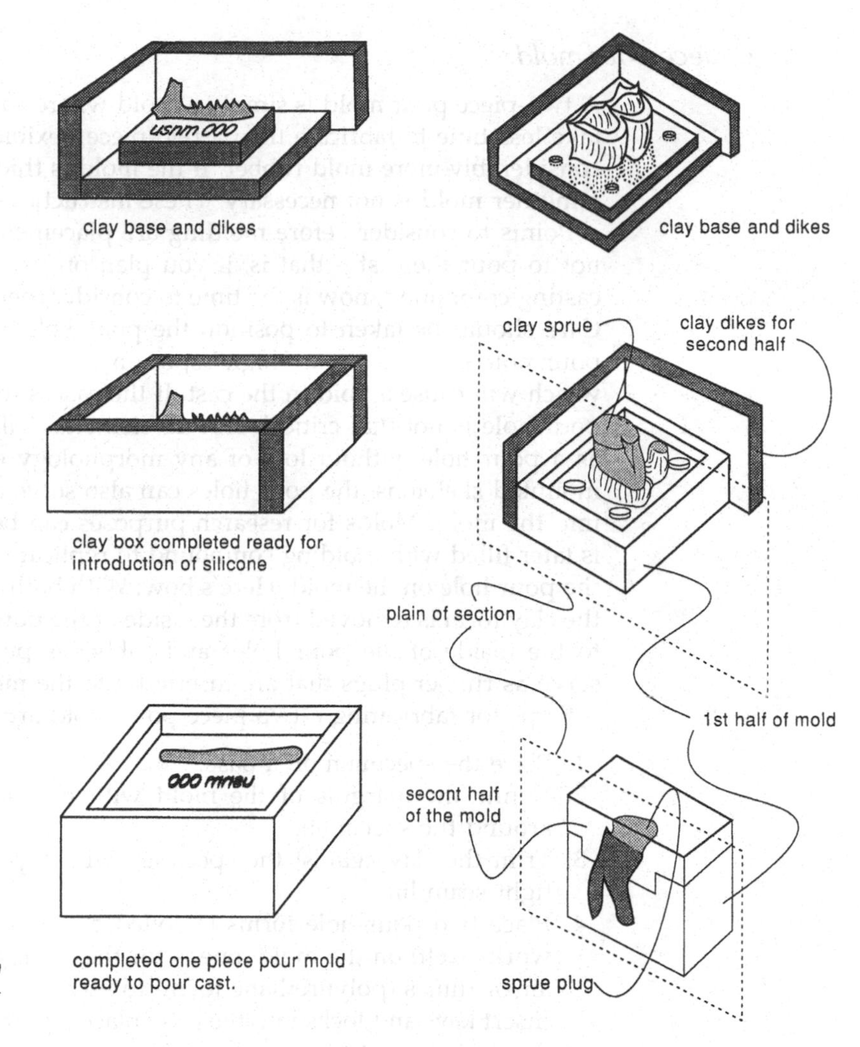

Figure 10.3 Steps for the construction of pour molds both one- and two-piece.

silicone to the inside surface of the clay wall; the silicone will run down the wall to the base and over the specimen. Prevent the formation of air bubbles by applying the silicone slowly. Do not let the silicone entrap air by overlapping on itself. Gently pull the silicone away from the specimen if it is building up too quickly. Once the fossil is completely covered, the remaining silicone is applied.

Most microvertebrate molds are made in one piece. For two piece microvertebrate molds, a key or lock is tooled in the clay base before the first pour of silicone is applied (see Figure 10.3). After the silicone cures, the mold is turned over and the clay is removed. Be particularly careful not to pull the mold away from the specimen, as this breaks the tight fit, leaving small gaps that ultimately result in wide seam lines. A separator is applied to the silicone surface and a clay wall is built up around the mold. The keys appear as raised surfaces on the mold. Mix and apply the silicone to this side. After curing, the silicone is parted along the seam line and the specimen is removed.

Two-piece pour mold

A two-piece pour mold is simply a mold where each half is poured. Pour molds take less time to fabricate than a two-piece flexible mold, but their use calls for considerably more mold rubber. If the mold is thick enough to resist deforming, a mother mold is not necessary. These instructions are for RTV silicone.

Points to consider before molding are placement of seam line and to pour or not to pour the cast – that is, if you plan on using pour holes (sprues) for the casting compound, now is the time to consider their placement in the mold. *Note*: Care should be taken to position the pour holes away from critical areas. The pour hole is placed on a "high" spot on the specimen in order not to trap air, which will cause a void in the cast. If the cast is for exhibition, placement of the pour hole is not that critical. Not all elements will have areas that can be used as a pour hole without loss of any morphology on the cast. For casts used in mounted skeletons, the pour holes can also serve as openings for bent steel cast into the mold. Molds for research purposes can be fabricated with a sprue that is later filled with molding compound to replicate the morphology that is lost to the pour hole on the mold. Here's how: With both halves of the mold completed, the clay form is removed from the inside of the pour holes. A separator is applied to the inside of the pour holes and rubber is poured into the opening. These serve as rubber plugs that are inserted into the mold during casting.

Steps for fabricating a two piece pour mold are the following:

1. Place the specimen on a base.
2. Define the margins of the mold with clay and build up the clay evenly around the specimen.
3. Trim the clay against the specimen with a pin vise or fine edge tool for a tight seam line.
4. Place two pour hole forms (a rolled cylinder of clay flattened on one side works well) on the mold; one is for the casting compound and the other for air or fumes (polyurethane foam) to escape.
5. Insert keys and locks into the clay; place a U-shape groove in the clay around the perimeter of the specimen.
6. Build a dike around the margins of the base.
7. Pour molding compound over the specimen.
8. Turn the entire mold over, leaving the specimen in it (remember, this is a two piece mold!).
9. Remove the clay and clean the surface of the mold of any debris.
10. Check the seam line for any extra rubber or irregular surfaces; with a scalpel or sharp tweezers remove any debris along the seam line.
11. Build up another dike around the mold.
12. Apply separator to the mold surface.
13. Reapply molding rubber.

After the rubber is set, gently split the mold along the seam and allow the two parts to separate. Work slowly and evenly to avoid tearing the mold or damaging the fossil. If some surfaces of the mold were not sufficiently coated with a separator, they may bind together. Stretch the mold gently at this point and, with a sharp scalpel, cut the bonded areas apart. Often just touching the rubber with the edge of the scalpel blade will be enough to release the two sides of the mold. Check for any fragments of bone that remain in the mold. Wash the mold in warm, soapy water to remove any clay or parting agent residue. Allow the finished mold to fully cure for a few days before pouring the first cast.

Flexible molds

Flexible molds are defined as two or more multipiece molds built up to a suitable thickness by applying layers of the molding rubber to the specimen. Silicone and latex rubber are the recommended molding compounds. Flexible molds are the most complex to build. Specimen preparation and the choice of materials are critical for a successful mold and protection of the specimen. Undercuts on a specimen and fragile or complex morphology demand a high degree of planning before fabricating the mold. Major points to consider are choice of molding materials; required number of pieces for the mold; position of seam lines; areas of rubber-on-rubber contact in the mold (any large fossa, area inside zygomatic arch); placement of plugs and pour holes on the specimen; and position of plugs (removable or permanent) in the mold.

The following steps outline the procedure to make a multipiece mold in silicone or latex rubber. A multipiece mold of a skull is illustrated in Figure 10.4. Latex may be substituted with minor variations, as noted in the text. For additional information on physical properties of the rubber and the handling and compatibility of molding compounds, see the materials list at the beginning of this chapter. Before molding any fossil, study the specimen. Look for areas that may cause problems: undercuts, openings, cracks, old glue joints, loose or cracked teeth. Complete the necessary preparation.

Follow these steps for fabricating a flexible mold:

1. Place the specimen on a base.

2. *Position of seam lines:* Determine the placement of the seam lines on the specimen. If the seam line is too thick on the cast, it will have to be tooled. Seam lines can follow a natural edge or ridge on the specimen. If the specimen is for research, placing the seam along a suture or other important antatomical features should be avoided, as later tooling of the seam on the cast can alter or obliterate these areas. The specimen is placed on a base. Clay is applied over the base and brought up to the edge of the specimen as a continuous, clean sharp line.

3. *Pour holes:* If desired in the mold, they are placed on the specimen. Position the pour holes in an area that is not morphologically significant, if possible.

4. *Plugs:* Some areas of a specimen need to be shut off from molding: unprepared matrix-filled cavities; areas that reveal morphology too fragile to mold, deep foramen, undercuts or deep recesses in the specimen. Other openings may need to be preserved in the mold. Both situations can be accommodated. A temporary plug made of clay or cotton capped with melted polyethylene glycol (PEG) is used to fill deep holes or undercuts on a specimen. Preserving a fenestra or arch is slightly more complicated. The opening is partitioned with a clay, wax or acetate barrier. When the molding rubber is applied to the specimen, one side of this partition is painted. After the mold is built up, the partition is removed and a parting agent is painted on the rubber that once was against the partition.

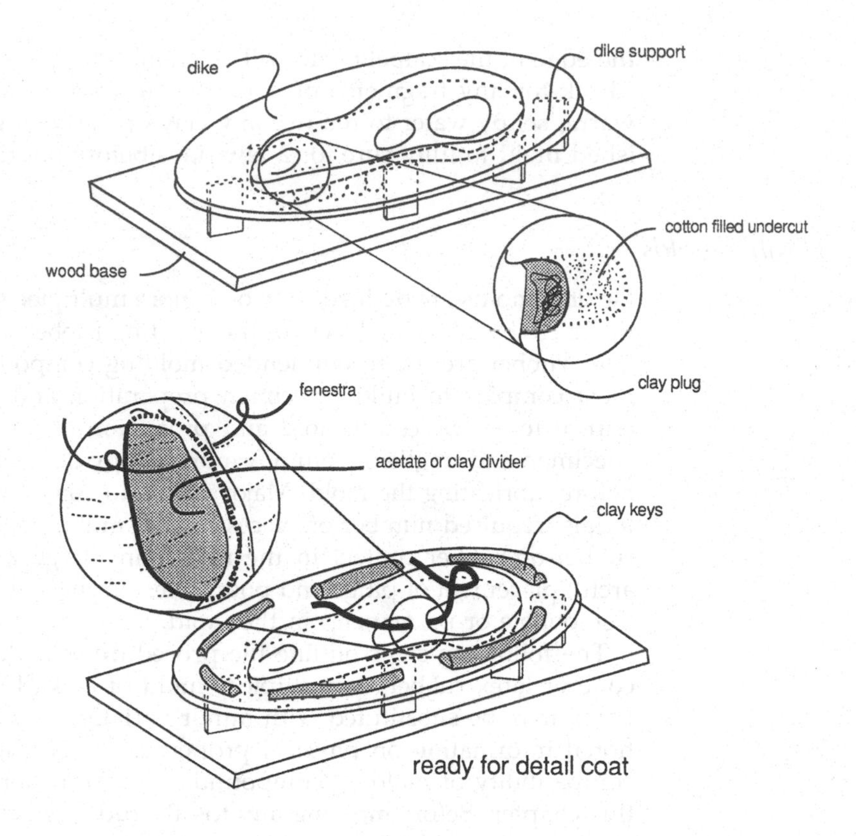

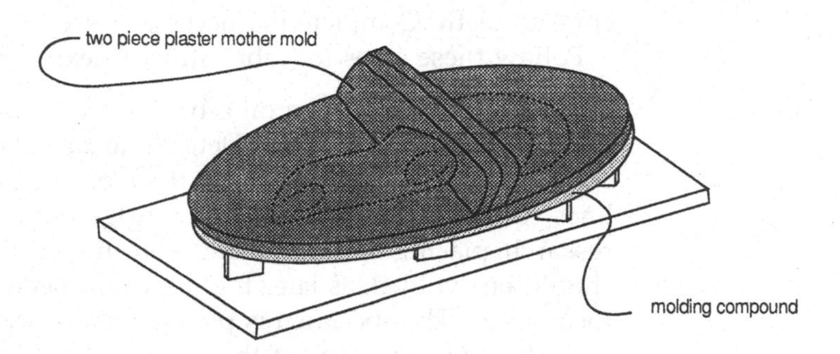

Figure 10.4 Steps for the construction of multipiece flexible molds.

Rubber is applied to the specimen and the barrier, forming rubber-on-rubber contact in the opening. If the partition is left in place, and it is often less trouble to do so, this seam line between the two areas of rubber filling the opening will come out relatively thicker (basically the thickness of the barrier). Acetate plastic sheets about 1 mm are recommended for this. They are stiff and resistant to deforming. For some specimens it is preferable to have rubber on rubber contact.

5. *Keys and locks placed in mold:* Impressions are made in the clay to act as guides for proper alignment of the mold (see Figure 10.4). Hex stock, heads of

machine bolts, ends of tool handles, and clay sculpture tools are all good pattern makers. A shallow trough or groove (a wire "noose" mounted on a wooden handle works best), 1–2 cm from the edge of the specimen in the clay, is tooled around the specimen. In some molds, excess resin contributes to an extremely thick seam line on the cast. An extra key made from a half round coil of clay is placed around the specimen about 0.5 cm from the edge of the fossil. This clay snake is covered with the rest of the mold in rubber. When that side is turned over, and all the clay is removed, leave this clay "snake" in the mold. After the mold is completed, and the pieces are separated, the clay snake is removed. This leaves a D shaped hollow groove in the mold that now serves as a trough for any extra resin.

6. *Applying molding rubber:* The detail coat of silicone or latex is brushed on one part of the specimen. Repeated layers are applied until the specimen is no longer visible through the rubber. Each coat is about 1 mm thick. Cloth is added and sandwiched into the mold with additional rubber. After one piece is completed, the clay wall that divides the separate pieces of the mold is removed. Separator is applied to the exposed rubber "flange." The process of applying the rubber is repeated until the individual pieces of the mold on the visible side are finished.

7. *Before applying the mother mold:* The surface of the mold is inspected for thin areas and undercuts that require more rubber or in-filling. With *silicone molds,* undercuts are filled with a paste of 100% silicone caulk and water. Use a separator if you want the plugs to be removable. If the opening is large, cut a piece of flexible foam rubber that is slightly smaller than the undercut. Coat the entire piece of foam with the caulk mixture and place in the undercut; fill any voids with additional caulk. Vermiculite mixed with RTV or latex rubber also works well as a plug compound. It conforms to irregular shapes well. Use plastic wrap as a separator. Removable plugs can also be made from thixotropic polyurethane elastomer. No separator is required with most, but test to be sure. Use a flexible polyurethane (TAP stretchy) thickened with fumed silica. For *latex molds,* removable plugs are made of plaster (coat the mold with soap first), silicone or polyurethane. Permanent plugs can be fashioned out of cut pieces of foam rubber. Paint with thin latex or dip the pieces of foam into thin latex and push them into the undercut. Before they are completely dry, cover the plug with latex soaked gauze. The flexible but firm plug is now incorporated into the mold. Use permanent plugs in areas where the mold is easily removed from the specimen to prevent unnecessary breakage or damage to the fossil.

8. *Applying the mother mold:* The mother mold or jacket is applied to the outside surface of the mold; FGR plaster is recommended. Be sure that the mother mold can come off the mold before you begin to fabricate it. By design, multipiece molds have multipiece jackets. On the other hand, because rubber is stretchy and flexible, and plaster is rigid, a one- or two-piece mold can also have a multipiece jacket because of undercuts or projections on the specimen. Follow ridges when making multipiece molds.

A few undercuts on the skull are filled in. The mother mold is constructed in four pieces for easy removal. A barrier of dishwashing detergent applied to the plaster lip before making the next piece acts as a separator. For large jackets, cellophane film placed on the flange helps to separate the individual plaster pieces of the mother mold. Pilot holes drilled through the plaster jacket and mold help keep the separate pieces together. Hand-tighten machine bolts with wing nuts to secure plaster pieces in place. Clamps distributed evenly around the mold keep even pressure on the seams without distorting the mold. Do not overtighten the bolts or clamps. If the mother mold is constructed of resin and fiberglass, a layer of aluminum foil strips applied to the outer surface of the rubber mold serves as a surface separator and sealer. It will adhere to the underside of the fiberglass mother mold when it is removed. The foil works to seal the inside surface of the jacket and facilitates easy removal of the jacket from the mold.

9. The mold is turned over and the base of clay is removed. Steps 6 through 8 are repeated.

10. Now it is time to take apart the mold, piece by piece, that took many hours to construct. The mold is disassembled in reverse order of construction. Remove the mother mold and slowly, without undue stress on the specimen, gently part the edges of the mold. Sometimes, a stubborn area that will not separate just needs some gentle persuasion. If you are not in the mood, do it another day. Many hours of labor are invested in the mold, so think about the situation before rushing into it and risk destroying the fossil. Here are some suggestions for difficulties that may arise when removing the mold (a jet of compressed air or soapy water may also help):

The mother mold will not come off: Chances are the jacket is hung up on an undercut. Forcing the jacket off will probably break the specimen. If the area can be located, grinding away that part of the jacket may help. In some cases, the jacket is too big and will need to be cut off the mold. Use a flexible shaft grinder (Foredom) with an abrasive wheel to score the jacket.

The mold will not part along the seams: Silicone sometimes leaks through the seam line and onto the specimen (along the internal surface of the mold; now you know what those curious lines are on your cast that are not on the specimen). Separate the rubber until you are close to the specimen. With a sharp scalpel, gently slice at the fused seam. It usually does not take much to separate the thin film of silicone and it is preferable to tearing the mold. A more serious situation occurs when a rubber-on-rubber contact fails to separate in the mold. Try to avoid tearing the mold apart. What usually happens when brute force is applied to a bonded rubber surface is that the specimen starts breaking apart inside the mold. In these circumstances, it is better to cut the specimen out of the mold. The mold may still be usable as a slit mold.

The mold separates, but the specimen adheres to the mold: This is serious. The tensile strength of the molding rubber is far greater than the fossil inside. Pulling on the mold will undoubtedly damage the fossil, potentially beyond reasonable repair. If the mold is latex, soaking it a kerosene bath might soften the mold as it enlarges in order to facilitate its removal. Solutions for a polyurethane or silicone molds are slim. Solvents might work to soften the surface; heat can be applied, but do not use open flames. Industrial hot air dryers work well. The mold is probably a total loss at this point.

Broken specimen in the mold: Harden as many fragments of bone as possible in the mold before removing them. The mold is an accurate record of where the pieces go; don't rush to take them all out. If only a thin veneer of bone is preserved in the mold, don't try to remove it. It can be cast out of the mold in plastic (silicone mold) or plaster (latex mold). Coat the bone sliver in polyethylene glycol and grind down the cast to the surface of the bone. Reattach the fragment to the specimen. Warm water will remove the PEG.

Casting

Casting of fossils has been discussed by numerous authors over the years. In particular, see Rigby and Clark (1965), Rennie (1969), Rixon (1976), Converse (1984), Smith (1989), Smith and Latimer (1989), and Chaney (1989). Also, see Chase (1979) for an excellent review on materials and techniques pertinent to the paleontologist.

Casting materials

The type of mold and the material it is made of often dictate the choice of casting compound. Usually more than one casting compound may be suitable for the job. Other considerations are purpose of the cast (teaching, research, or exhibit), complexity of the mold, and compatibility of materials.

Casting techniques vary, depending on the casting agent used. Use these instructions as a guide only. They are not a substitution for thorough testing of compatibility and physical properties of all materials. Beware that companies often change the chemistry of their products or suppliers of various ingredients without notifying the consumer. What may be considered only a "slight" change by the manufacturer can have a serious effect on a fossil specimen or alter physical properties of a compound.

Compatibility of the mold material and the cast material is extremely important. Full knowledge of incompatibility is necessary to avoid a loss of a great deal of time, effort, material, and money. In the worst case, complacency in this area can end in the total destruction of the mold. This may result in the undesirable need once again to subject a valuable specimen to the molding process. If the desired casting material is not compatible with the molding compound, a separating agent must be applied to the mold to function as a barrier in prohibiting chemical reactions between the two materials. If the casting material will physically or chemically bond to the mold, a release agent must be applied to the mold. Keep in mind that any additional substance on the mold surface reduces the replication of details in the cast.

The complexity of the mold can limit the suitability of some casting compounds, particularly in a hollow cast. Improperly applied casting agents may result in voids and bubbles in the cast. The larger the cast the stronger the material must be. Most often this is accomplished by the addition of reinforcing materials.

Fiberglass is the most common reinforcement material used in cast production. It is produced by drawing single fibers of special glass into very small diameter strands. For use with resins, the glass is generally given a surface treatment to reduce self-abrasion, to increase compatibility with resins, and to adhere the resin better to the glass. Some glass fibers are specially treated to saturate with polyester, epoxy, or plaster compounds. Fiberglass is available in chopped mats, woven mats, roving, woven strips, and chopped loose fibers. These materials are available in a variety of strand thicknesses and compositions. Most chopped and woven mats are composed of bundles of very fine strands. These are formulated for use with polyester resin in the plastics industry. The other type of glass cloth is a random weave, thick strand monofilament used for added strength in injection molding. This monofilament mat (or continuous strand cloth) is recommended for use with Hydrocal gypsum cements. Be sure to use a fiberglass cloth compatible with the resin or gypsum cement chosen. Some glass fibers formulated for use with gypsum cement will not saturate or "wet out" in resin compounds; others do not saturate when used with a plaster mixture. Do a compatibility test between materials to be certain.

Casting compounds

Gypsum plaster and thermosetting plastics are the two most commonly used materials for making casts in vertebrate paleontology. There are numerous plaster and plastic compounds available for casting. We discuss the most commonly used materials for applications in paleontology.

Plaster

Plaster is a safe, economical, easy to use, and widely available casting compound. It has a variable working time, very low expansion coefficient, and range of hardness for tooling and casting applications. It is widely used in the preparation lab as a casting compound and in mother mold fabrication. In the field, plaster is used to make casts of natural molds and, in the time-honored tradition burlap and plaster field jackets.

Plaster of Paris usually refers to general purpose molding and/or casting plasters whose source historically was the gypsum deposits of Paris, France. Plaster or gypsum (the names are used interchangeably) compounds are formulated from pulverized gypsum that is dehydrated (calcined) in order to remove most of the chemically bound water. Gypsum cements consist of a finer ground gypsum with nearly all the water removed. Alum or other salts may be added to the compound. In general, the gypsum cements are four to five times harder than general-purpose industrial plasters. Although any plaster needs only 19% water to rehydrate, industrial plasters require more water per volume (70 parts water to 100 parts plaster) than gypsum cements (25–45 parts water to 100 parts plaster) to make a proper mix. Gypsum products use lesser amounts of water, giving them the special characteristic of being harder, stronger, and denser when cured. These properties, along with their finer grain size, make them preferable for casting. In contrast to other casting materials, gypsum cements and most industrial plasters do not shrink on setting but have a positive expansion. As the cast plaster sets up, the mass begins to heat and expands. This expansion coefficient varies among the different plaster compounds. Plasters differ in their hardness, density, and strength, as well as in their working and setup times. Note that some plasters remain fluid for a longer period of time after mixing; others have greater plasticity and gradual setting action for tooling, or even very rapid stiffening after curing begins. Additives are available to change the physical properties of plaster. A premixed solution of equal parts potassium alum and borax to 20 parts water to industrial molding plaster produces a plaster comparable to White Hydrocal (Chase 1979). These chemicals make molding plaster harder and denser, but speed up the setting time (as will the the addition of salt to old plaster). Retarders work to slow the cure time of plaster. Chase (1979) mentions the use of carpenter's glue and casein or citric, boric, and phosphoric acids and their salts as retarding agents. He also offers a recipe for a water-soluble plaster: combining 1 part (by volume) corn or potato starch to 3 parts plaster. This produces a plaster that dissolves in hot water.

Types of U.S. Gypsum plasters

The following plasters are useful as casting compounds in paleontology labs. Additional information on the more than 30 different U.S. Gypsum plaster and gypsum cements is available from the U.S. Gypsum Company (101 S. Wacker Dr., Chicago, IL 60606-4385). Literature is also available from the company's Tooling and Casting Division (800-621-9536).

1. *Molding plaster:* Also referred to as plaster of paris and casting plaster; these are the least expensive grades of plaster. Recommended for quick casts, mother molds, and plaster jackets. Setup time between 20 and 30 minutes. Casts made from this grade of gypsum should be hardened by brushing or soaking in a solution of polyvinyl acetate or acrylic resin to prevent chipping and scratching of the surface. A good utility plaster.
2. *Dental plaster:* Similar to fine molding plasters, but often has a faster setup time.
3. *White Hydrocal:* A U.S. Gypsum product. A good casting plaster, with a normal set time of 25 minutes. Expansion is about twice that of molding or casting plaster with a higher dry strength.
4. *Ultracal 30:* A U.S. Gypsum product. A hard, strong gypsum cement recommended for its extreme accuracy and good surface hardness. Low expansion, gradual setup and long period of plasticity desirable in casting. A good tooling medium.
5. *Hydrocal FGR Gypsum Cement:* A relatively new product available from U.S. Gypsum. Hydrocal FGR-95 (= fiberglass-reinforced) gypsum cement is a specially formulated plaster used in combination with continuous strand fiberglass mat, chopped fibers, and woven fiberglass cloth to produce lightweight, strong, and resilient glass-reinforced plaster casts. Continuous strand mat (0.75–1.0 oz.) is recommended. Information regarding local suppliers of fiberglass for this purpose is available from Nicofibers, Inc. (Box 26, Shawnee, OH 43782, 614-394-2491).

We discuss FGR-95 in detail: FGR-95 is denser and harder than standard molding and casting plasters. It has a working time of up to an hour (mix with cold water) and stays "creamy" or plastic until just before setting up. This allows complete saturation of the fiberglass and adequate working time. A bag of FGR-95 is about twice the cost of standard plaster; however, less is used per cast when combined with the fiberglass. For most applications; it can be used as a substitute for resin in composite fiberglass resin cast and mother mold fabrication.

The main type of cast produced with FGR-95 are lay-up casts, using similar procedures employed when using polyester or epoxy resin (Chaney 1987). It works well for a wide range of cast sizes, from small to very large (Chaney et al. 1991), plaques, and three-dimensional specimens. Small casts can be poured. The many advantages of FGR-95 over plastic casting resins include the following:

1. Safety: gypsum cement is nontoxic, nonallergenic, and odorless; toxic gases are not emitted when subjected to heat or during production.
2. Fire resistance.
3. Paintability: surface accepts most oil-based, alkyl or latex paints and most stains, lacquers, varnishes, or shellacs.
4. Accurate reproduction: releases or separators are not required with any of the common molding materials.
5. Dependable results: plaster does not chemically react with the mold resulting, in longer mold life.
6. Economical: less expensive than most thermosetting plastics.

Mixing plaster

Water used in mixing plaster should be clean. High concentrations of soluble salts or other chemicals can migrate to the surface of the mold and produce efflorescence on the cast. Potable water should be fine. Try distilled water if results are unsatisfactory. Variations in water to plaster ratio will affect the absorption, strength, and quality of the finished cast. Plaster of Paris is mixed 70 parts water to 100 parts plaster. Refer to manufacturer's instructions for the recommended ratio of other products.

Use a clean plastic or rubber bowl filled with slightly less water than the desired amount of plaster required for the job. Avoid adding water to dry plaster or large amounts of plaster to water. This produces lumpy plaster and air bubbles that cause imperfections in the cast. A flexible bowl makes it easier to remove any cured plaster or residue. Mixing plaster in a bowl lined with plaster residue or plaster chips causes the mixed plaster to set faster. Use water at room temperature. Colder water retards setting, warmer temperature accelerates it. Sift or strew the plaster into the water slowly and evenly without stirring until "mudcracks" appear in the mixing bowl. Let it soak for a few minutes. Soaking removes the air surrounding each gypsum particle, allowing complete saturation to occur. Mix the plaster gently by hand or with a spoon or spatula. Keep the mixing device or your hand in the plaster. Raising it above the surface introduces air into the plaster. Gently tapping the mixing bowl up and down on a countertop will cause trapped air bubbles to rise to the surface, where they can be skimmed off. Mix large volumes with a rubber disc mounted on a shaft attached to an electric drill or a commercial plaster mixer. Placing mixture under a vacuum helps remove trapped air. Plaster becomes warm or even hot during curing. Do not remove the cast from the mold until the set plaster is cool.

Casting with plaster

A solid plaque cast from a one-piece, or "open-faced," mold is the easiest to make. First fill the mold with water and drain, allowing some water to remain in small indents and undercuts. Slowly pour the plaster mixture plaster into the mold. The plaster displaces the water as it flows onto the surface and into detailed areas. To reduce bubbles forming on the surface, allow the plaster to flow down the side of the spoon or spatula before it enters the mold. Pour the plaster at one position on the mold and allow the mixture to flow slowly and evenly over the surface. Vibrate the mold to disperse any trapped air bubbles to the surface. A centrifuge may be used with open molds to eliminate air bubbles.

When casting in plaster using a multipiece mold, the cast can be poured solid, hollow, or layed up with glass-reinforced gypsum, depending on the size and complexity of the mold. Solid plaster casts are easily made from small to medium-sized piece molds. Follow the above directions for mixing and pouring the plaster. Before the plaster fully cures, piece the molds together. Add a little extra plaster before closing up the mold to facilitate sealing the seams if needed.

Hollow casting is done by wetting the mold, pouring in the plaster, sloshing it around to coat the inside of the mold, and pouring out the excess. Hollow plaster casts are fragile and not recommended for most purposes. If a hollow cast is desired, plastic resins used with fiberglass and polyurethane foams produce a stronger and more durable cast. (This is discussed in the section on casting with thermoplastics.) A rotating casting machine removes the tedious hand turning of molds. They work the best with equidimensional specimens such as human skulls. Detailed plans for a rotating casting machine are available in McCrady (1978) and Smith and Latimer (1989).

Fabrication with gypsum cement and fiberglass is recommended for plaque casts and medium to large casts from piece molds. Paint or pour a thin layer of plaster (FGR-95) in the mold to capture the surface details. A low pressure stream of compressed air or bulb syringe will help to eliminate any bubbles on the surface. Next apply a layer of a lightweight (0.25–0.50 oz.) surfacing mat (also know as angel hair cloth) saturated with plaster to the mold. Follow this with a layer of plaster-saturated 0.75–1.0 oz. weight continuous strand mat. This completes the plaque cast. After the last layers of fiberglass and plaster are applied in a piece mold, run a bead of plaster along the seams that join the individual pieces of the mold together and close the mold. The cast may be left hollow or filled with an expanding polyurethane foam for added strength and durability. Plaster casts may be pulled from latex, polyurethane, polysulfide, or alginate molds without a release or separator.

Thermosetting plastics

Thermosetting plastics become solids when cured by application of heat or chemical means. These materials cannot be reheated or reshaped after curing. The long chains of molecules polymerize and form cross links during curing. These bonds permanently lock the molecules in position. This results in a plastic that is heat and chemical resistant and can only be destroyed by extreme heat and ultraviolet light. Once set, they will not melt or flow. During the polymerization process the material shrinks, thus introducing linear distortion. How much of this occurs depends on the material, rate of polymerization, temperature, and volume of material. Because all plastics are made in many formulations, their physical and chemical properties are variable. For casting, a resin with these properties is desirable: minimal shrinkage, relatively low viscosity, low exothermic cure, and adequate working time.

A valuable reference for plastics and related materials is the *Encyclopedia of Plastic*, published yearly by McGraw-Hill. This book contains information on plastics, manufacturers, and suppliers. It lists addresses and phone numbers.

Thermosetting plastics include several compounds used extensively for molding and casting in paleontology: epoxies, polyesters, silicones, and urethanes.

Epoxies

Epoxy resins are formed by the reaction of epichlorohydrin with a condensation product of phenol and acetone known as bisphenol-A. Epoxies that cure at room temperature are generally the easiest to use. Silicone rubber is the recommended molding compound for use with epoxy.

Epoxy offers many advantages as a casting compound: low degree of shrinkage; high fidelity; chemically very inert; cures at room temperatures; and hard, strong cure. Fumed colloideal silica and microballoons are fillers used to increase viscosity. Pigments are available to opaque and color the normally clear-yellow epoxies. These additives may change the cure time and physical properties of the resin. Always test for compatibility. Disadvantages to consider when using epoxy are: higher cost; its good adhesion that makes it difficult to separate from most mold materials except silicone rubber; and slightly more hazardous to use than polyester resins.

Epoxy is the material of choice for casting small specimens in silicone molds. This combination produces the greatest fidelity, and by using extremely slow cure (two or more days) they are the best for dimensional stability. Epoxies are primarily used in RTV pour molds with no separator required. They may be thickened as necessary with fumed colloidal silica (Cab-O-Sil) or reinforced with fiberglass in hand lay-ups of larger casts.

Epoxy casts have been shown to work well for use in scanning electron microscopy in the study of bone surfaces, teeth, and fossils. Epoxy casts routinely give resolutions of 0.1–0.25 μm at magnifications of 1,500×–2,000× (Rose 1983). For visual examination of an epoxy cast, it is necessary to add opaque epoxy colors prior to mixing the resin and catalyst; otherwise the light plays off the inner surfaces of the cast and obscures surface detail. Opaque coloring materials are not necessary for casts used in SEM.

Microvertebrate epoxy casts

Duplicating small vertebrate fossils in epoxy resin was first introduced to paleontologists by Waters and Savage in 1971. Prior to their publication, there was a lack of accurate casts of delicate and rare microvertebrate fossils. Now almost every preparation lab is equipped to produce epoxy duplicates of small fossils. Researchers, teaching collections, and exhibit museums have benefited tremendously from Savage and Water's important contribution to molding and casting.

The technique for making epoxy casts of microvertebrate fossils has changed little since its introduction by Savage and Waters (1971). First, make a silicone mold of the specimen (see the first part of this chapter for instructions on mold making). The epoxy is mixed according to manufacturer's instructions. Tap 4-1 epoxy resin works well for duplicating microvertebrate fossils. It has an adequate working time, cures relatively quickly for repeated castings, accepts pigments well, is water and chemical resistant, and maintains a hard surface cure with excellent fidelity.

Follow these steps for making an open pour microvertebrate cast from a silicone rubber mold:

1. Mix the color pigment into the epoxy resin before the catalyst or hardener is added; usually a few drops of pigment per ounce of resin is satisfactory. Use white or gray to opaque the resin first and then add the desired tinting colors (i.e., brown, amber). Use caution, as too much pigment may affect the cure.
2. Add the hardener and mix thoroughly.
3. Pour or brush the resin into the mold.
4. Place the mold in a hand-driven centrifuge for 15–20 revolutions to help force the uncured epoxy into the smallest details of the mold and remove the air incorporated into the epoxy during mixing; or place the epoxy-filled mold in a vacuum chamber at 32 in./Hg for about 15 minutes to eliminate any trapped air bubbles. The epoxy will expand out of the mold under a vacuum, so be sure to place the mold in a tray or container to catch the excess resin.
5. Cure and remove from mold.

Epoxy resin usually cures in 24 hours. Accelerate the cure by placing the cast in a vented oven at 55° C or next to a sunny window. The life of a mold can be prolonged by removing the cast while it is still "elastic." Epoxy has a memory and will continue to cure after it is removed from the mold.

Epoxy generates heat when curing. The larger the volume of resin, the greater the heat. Use caution when handling large volumes of mixed resin. When the resin undergoes curing, large amounts of fumes are given off. The heat may be intense enough to burn the mold, melt mixing containers, or damage the cast. Check to see that the epoxy you choose is compatible in this regard for casting purposes.

Medium and large epoxy casts

Epoxy resin is also an excellent casting compound for medium and large molds. Refrain from making these casts solid. The exothermic reaction of the epoxy generates great heat that can destroy a mold. Instead, use a lay-up method to fabricate the cast. Shell Epon Resin 815 and V-40 hardener are recommended for casts of this size. This compound has a long working time; it slowly cures in 24 to 36 hours and generates little heat; it accepts pigment well; it is of a medium viscosity, so it paints onto and coats a silicone mold evenly (some epoxies create a spiderweb effect when painted on a mold due to surface tension of the resin); and it cures with a hard, paintable finish. When cured, this resin may be drilled, sanded, and routed. It can be re-epoxied and patched easily when incorporating an internal steel armature for casts used in mounted displays. It adheres well to steel surfaces for casting the armature directly in the mold.

Most formulations of epoxy resins and auxiliary materials can have adverse health effects, from minor skin irritations to serious systemic effects. Observe proper precautions and handling recommendations.

Follow these steps for making an epoxy cast from a piece mold or multi-piece pour mold:

1. Thicken the epoxy resin with a suitable filler such as Cab-O-Sil. Carefully paint this thixotropic "detail" coat into the mold. Use a stream of low pressure compressed air or bulb syringe to help eliminate any bubbles on the surface.
2. While the first coat of resin is still tacky, apply a second coat with a brush or spatula, filling in undercuts and thin areas of the mold. For a small two- or more piece mold, run a small bead of resin along the seam line and close the mold together. A thickness of 1/16–1/8 inch is suitable and the cast can remain hollow. If shallow undercuts are a concern, they can be filled solid before the mold is closed up. Small specimens can also be filled with resin up to the seam lines and closed up. Very little heat is generated to damage the mold. Any heat dissipates through the thixotropic resin without noticeable effects.
3. For larger casts, a third coat of resin can be applied with fiberglass cloth to reinforce the cast. Do not thicken this resin, as it will not properly saturate the cloth. Use a lightweight fiberglass cloth or medium weight surfacing mat for best results.
4. Run a bead of resin along the seam line, close the mold, and let cure overnight. If the mold has pour holes, proceed to step 5, otherwise allow the cast to cure for 24 hours before demolding.
5. Pour expanding polyurethane foam into the mold through the pour holes. Be sure to do this in a well-ventilated room or outdoors. A 2 1/2-pound density foam expands quickly to fill the mold and seal the individual pieces of the cast within. If the bead of resin between pieces was not applied or not cured prior to addition of the foam, the foam may bleed into the seam lines and deform the cast. Care must be taken that excessive foam is not trapped in a cast, as its pressure over time (weeks) will split the cast.

RTV silicone rubber is recommended for use with epoxy casts. Silicone molds require no separator or release agent when used with epoxy resin. The use of a separator will extend the life of the mold, but it also reduces the fidelity of the cast. For exhibit casts (i.e., large skulls or reconstructed specimens with limited research purposes) a release agent or protective barrier coat may be useful. Epoxy resin is not recommended for use with latex, polyurethane, and polysulfide molds without a release or separator. Do not use with alginate molds.

Polyester resin

Polyester resins are thermosetting, unsaturated alkyd resins dissolved in styrene and other monomers. They are catalyzed at room temperature by the addition of an organic peroxide, usually methyl ethyl ketone peroxide (MEKP). MEKP is especially dangerous to the eyes. Slight exposure to one's eyes without an immediate rinsing out can result in permanent blindness.

The cure time is accelerated by the addition of a promoter such as cobalt napthenate (CoNap), a dark purplish liquid, which reacts with the catalyst to produce heat. The cure is also affected by room temperature and humidity. MEKP will react violently with the accelerator and may result in fire or explosion.

Never mix MEKP directly with CoNap. Thoroughly mix each additive separately with the resin and in the proper sequence.

Polyester resins are available in a wide variety of formulations. For molding and casting purposes, use a resin that contains a premixed accelerator. Adding MEKP is all that is needed for the resin to cure at room temperature. Mix 0.25–1 g of MEKP to 100 g of resin (or about 2–3 drops per ounce). Use less catalyst for thicker casts, more for thin surfaces. The amount of MEKP used will vary the cure time of the resin. Overcatalyzed polyester resin may undergo increased shrinkage, crazing, cracking, or even burning during the final curing state because of the greater heat generated in the exothermic reaction. An undercatalyzed mixture remains tacky, and may be difficult to remove from the mold. After measuring the correct amount of MEKP, mix thoroughly with the resin. Scrape the sides and bottom of the container. Use a Jiffy mixer for larger volumes. The volume of resin, amount of MEKP used, temperature, humidity, laminate thickness, and type of molding rubber will affect cure time and quality of the finished cast.

Certain resins remain tacky after curing because oxygen in the air inhibits their complete cure. These laminating resins are useful for bonding individual layers of fiberglass. They should not be used for the "gel coat" (a top or surface coat that is the first layer of resin painted into the mold; this layer becomes the surface of the finished cast). Surface cure resins have oils or waxes dissolved in the mixture. The wax gradually rises to the surface during polymerization, forms a barrier, and prevents air from contacting the uncured resin. Because polyester resin shrinks to a modest degree in the mold, some casts remain tacky for quite a long time if laminating resin is used for the gel coat. This is easily avoided by using a surfacing resin for the gel coat. This way you are guaranteed a cast with a hard, durable surface. It has been noted by some workers and in technical manuals that a surfacing resin does not bond as well to a laminating resin as laminating resin does to itself. This situation is improved by applying the laminating resin while the surfacing resin gel coat is still tacky.

Polyester resin casts will experience surface chalking, discoloring, and crazing after long exposure to ultraviolet (UV) radiation. These effects may be reduced or eliminated by adding a UV absorber into the resin. Mix the powder 0.25%–1% by weight into the resin. Some UV absorbers will inhibit the cure.

Polyester resin is not recommended for pouring solid casts because of the heat generated during curing. Cold set or low exothermic epoxy resins are preferable. If polyester resin is used, casts greater than 2 cm thick should be poured in layers. Allow each layer to cure between coats. Be careful not to introduce air bubbles into the mixture.

Larger casts are made by laminating polyester resin and fiberglass into the mold. The technique is similar to that outlined for epoxy resin and fiberglass casting. Prepare a batch of gel coat resin by first adding pigment to resin. The pigment makes it easier to see that even layers are painted into the mold. Mix in the proper amount of MEKP and add fumed colloidal silica to thicken the batch. The thixotropic gel coat will not run off vertical surfaces and pool less in undercuts. After building up the gel coat to a suitable thickness, apply a layer of fiberglass and resin. Use surfacing mat or cloth. Cut the fiberglass into workable pieces. Fiberglass tape, though expensive, is easily cut into convenient strips.

Saturate the cloth with unthickened resin and apply over the gel coat. Tamp the cloth down with a brush to remove any trapped air bubbles. If the cast is in a piece mold, run a bead of resin along the seams before closing the mold up. Allow the cast to cure overnight before removing it from the mold. If pour holes were constructed in the mold, a hollow cast can be filled with a lightweight (2–9 lb. density) expanding polyurethane foam.

Polyester resin used with silicone, latex, or polyurethane rubber requires a separator in the mold. Polyvinyl alcohol is recommended. For latex molds, apply two to three coats of caranuba wax, then brush or spray on an even coat on polyvinyl alcohol.

Water-extended polyester resin

Madsen (1974) introduced water-extended polyester resin (WEP) to paleontologists; see Chase (1979) for a good review on the uses of this product. WEP resin is an emulsion of resin and 50%–60% water catalyzed with hydrogen peroxide. It was considered an economical alternative to the standard polyester resins and thought preferable to heavier casting plasters. Used almost exclusively in solid pour molds, this material produces strong but heavy casts. They are easy to paint once the external water evaporates (however, this can take months). Minute cracks in the surface of the cast cause water-based acrylic and latex paints to blister and peel over time. Colored casts are made by dispersing dry pigment in the resin before producing the emulsion. Due to the entrained water, a separator is generally not necessary for use with silicone, latex, or urethane rubber. Polysulfide molds cannot be used with WEP. Compatibility tests should be run prior to use of this material. Polyvinyl alcohol will not work as a separator because it is water soluble and is quickly dissolved by the water in the WEP. Despite its relatively quick set up time (less than 6 minutes) and cure time of about 1 hour, WEP is complicated to mix properly and exhibits shrinkage of 2%–5%. This is unacceptable for research casts. Although good exhibit quality casts have been produced using WEP, there are lighter, safer, and less expensive materials available for casting (see FGR plaster).

Polyurethane

Polyurethane resin and rigid casting foams are useful for casting. They usually come as a two part system, one part resin and one part hardener. Chase (1979) and Smith and Latimer (1989) discuss casting with polyurethane resin. Jensen (1961) introduced paleontologists to expanding polyurethane foam as a casting compound for fossil vertebrates.

The suggestions for handling polyurethane elastomers discussed in the molding section of this chapter also apply to polyurethane resins. One disadvantage of these materials is their sensitivity to moisture. Moisture in the hardener produces bubbles and reduces its curing ability, resulting in an inconsistent cure and soft spots on the cast. Solids form in resin exposed to moisture, causing crystallization and waste of material. If an open container is not used in a short period of time, crystallization may occur. Covers can bond so tightly to their containers that they must be cut off. Besides being extremely sensitive to moisture, some polyurethanes react to petroleum products. Avoid using wax mixing cups and parting agents and barrier coats that are petroleum based.

Polyurethane resin produces hard, strong, inexpensive casts. Casting techniques with polyurethane resin are very similar to those discussed using epoxy resin. However, these materials are much more sensitive to exact measuring and mixing than most other resins. Thorough mixing is necessary to achieve properly cured casts. Polyurethane resin has a relatively short working and cure time. For larger lay ups and solid casts, its exothermic cure may be problematic. Polyurethane is used in many anthropology labs for making hollow casts of skulls. To prevent excessive heat from damaging the mold as the resin cures, the polyurethane is applied in layers (also known as "shelling") to the mold.

Polyurethane foam

Polyurethane foam (also known as rigid foam) is available as a two part system in a variety of densities. Mixing equal amounts by volume of the prepolymer and catalyst together produces an expanding foaming reaction. Air is trapped within the foam as it expands and skins over inside a mold. Fluorocarbons or carbon dioxide is added to the resin to produce this foaming action. (Most manufacturers have eliminated fluorocarbons from their foams.) Lower density foams (2–9 lb./cu. ft.) are recommended for casting. They are lightweight and strong and expand with enough velocity to fill a mold uniformly. The higher the density, the heavier and harder the cured foam. Lower density foams expand faster, but are lighter and softer. Some are available in aerosol cans. For casting, the two-part mix-as-you-go works best because it is easier to control the amount of material needed. Most canned foams do not expand with the velocity needed to fill a mold uniformly.

The foam is mixed in equal proportions by volume. Mixing small amounts by hand with a wooden tongue depressor is sufficient. Amounts greater than 16 ounces should be mixed with a Jiffy mixer attached to a power drill. When the mixture is "creamy," it is ready to be poured into the mold. This requires the mold to have pour spouts, and two are recommended: one for pouring and the other for release of fumes. Be sure the pour holes direct the foam solution to the bottom of the mold. Areas with undercuts can be cut off if the foam does not reach the bottom of the mold before expanding. In the event a resin-foam cast is not filled to satisfaction, drill a small hole in the problem area and refill it after the cast is pulled.

Let the foam cure overnight in a piece mold. Some casts can be pulled earlier, but problems develop along weak seam lines if the cast is pulled prematurely. Without adequate support of the mother mold, the cast can distort under the pressure of the expanding foam. Before pouring, be sure the mother mold is securely bolted or lashed together with rubber inner-tube strips. Failure to do so can result in blowing the mold and mother mold apart.

The foaming reaction produces harmful levels of carbon dioxide gas and diisocyanate monomer (Chase 1979). Work in a well ventilated lab, fume hood, or outdoors. Follow all recommended handling procedures. The components are mildly alkaline, so rubber gloves are recommended.

Polyurethane foam can be used by itself as a casting compound or with a plastic resin. The expanding foam used in casting can have irregular results, depending on the shape of the mold. The foam expands more in areas of less resistance and incompletely fills larger voids. Two suggestions to counter these problems: (1) As the foam expands out of the vents, temporarily block the holes. Try blocking one at a time and then briefly together. This will help fill large voids. (2) Paint a gel coat of resin in the mold before the foam is used. For piece molds, a few coats built up to 1/16–1/8 inch is usually sufficient. This may be reinforced with a layer of resin and fiberglass cloth. Fill undercuts and areas of detail solid with resin to ensure they are properly covered. Reinforce the cast with a layer of resin and fiberglass if desired. This is recommended with polyester resin because of its brittleness but may not be needed with epoxy and polyurethane gel coats. Use of a resin gel coat eliminates any concern about the skinning over of the foam in the cast. Foam adheres very well to the resin.

Uncoated polyurethane foam degrades rapidly when exposed to ultraviolet light or unfiltered fluorescent lighting. This causes the surface to become powdery.

Use the recommended release agents if applying a resin gel coat to the mold before pouring the foam. When used by itself, polyurethane foam requires a separator or release agent with most cold set rubbers. Most RTV silicone rubbers (Silicone, Inc.'s, GI-1000 e.g.) do not require a separator. Do not use with alginates. Foam sticks to most surfaces when fully cured. Applying some petroleum jelly to the outside of the mother mold in the area around the pour holes will make removal of excess foam much easier.

Materials and suppliers

This is a list of suppliers of products mentioned in the discussion on molding and casting. The sales office of major manufacturers can give you the name of a local distributor in your area. Check to see if they have a toll free 800 telephone number. See Chase (1979:262–267) for a large list of suppliers.

General suppliers of molding and casting supplies, materials, and equipment

Douglas & Sturgess, 730 Bryant St., San Francisco, CA 94107, 415-421-4456. FAX 415-896-6379, catalog available.

TAP Plastics, 3011 Alvarado St., San Leandro, CA 94577, 415-357-3755, catalog available.

BJB Enterprises, 6350 Industry Way, Westminster, CA 92683, 213-598-7777, catalog available.

Acryloid

Rohm-Haas Co., Independence Mall West, Philadelphia, PA 19105, 215-592-3000.

Alginate

Douglas & Sturgess. See general suppliers.

Butvar

Monsanto, Bircham Bend Plant, Indian Orchard, MA 01151

Cab-O-Sil

Cabot Corp., 125 High St., Boston, MA 02110

TAP Plastics. See general suppliers.

Douglas & Sturgess. See general suppliers.

Circle K Products (west coast supplier to Silicones, Inc.), 20814 S. Normandy Ave., Torrance, CA 61953, 213-320-4218.

Dow Corning Type E Silicone

Dow Corning, Midland, MI 48640, 800-248-2345, Product information: 800-248-2481

Dow Corning Regional Sales Offices Fremont, CA, 510-490-9302, and Mount Olive, NJ, 201-661-0702.

K. R. Anderson, 2800 Bowers, Santa Clara, CA 95051, 800-672-1858

Fiberglass. See also Surmat

TAP Plastics. See general suppliers.

Douglas and Sturgess. See general suppliers.

Fumed colloidal silica, See Cab-O-Sil.

Latex rubber

Douglas & Sturgess (Latex No. 74). See general suppliers.

Chicago Latex Products, 3019 W. Montrose Ave., Chicago, IL 60613

Meguiers paste wax

TAP Plastics. See general suppliers.

Plaster, gypsum cements, FGR-95

U.S. Gypsum Co., Industrial Gypsum Division, 101 S. Wacker Dr., Chicago, IL 60606-4385, 312-606-4000, Order center: 800-621-9523

Polyester resin

TAP Plastics. See general suppliers.

Douglas & Sturgess. See general suppliers.

Polyethylene glycol (Carbowax)

Union Carbide Corp., 270 Park Ave., New York, NY 10017

Fisher Scientific, 711 Forbes Ave., Pittsburgh, PA 15219, Sales offices located statewide, 800-766-7000

Polysulfide rubber

Smooth-On, Inc., 1000 Valley Rd., Gillette, NJ 07933, 201-647-5800

Polyurethane elastomers

BJB Enterprises. See general suppliers.

Smooth-On, Inc., 1000 Valley Rd., Gillette, NJ 07933, 201-647-5800

Polyurethane foam
TAP Plastics. See general suppliers.
Douglas & Sturgess. See general suppliers.

Polyurethane casting resin
BJB Enterprises. See general suppliers.
Kindt-Collins Co., 12651 Elmwood Ave., Cleveland, OH 44111, 216-252-4122

Polyvinyl alcohol
TAP Plastics. See general suppliers.
Douglas & Sturgess. See general suppliers.

Release agents
TAP Plastics. See general suppliers.
Douglas & Sturgess. See general suppliers.
BJB Enterprises. See general suppliers.
Silicones, Inc., 211 Woodbine St., Box 363, High Point, NC 27261, 818-886-5018
Circle K Products (west coast supplier to Silicones, Inc.) 20814 S. Normandy Ave., Torrance, CA 61953, 213-320-4218

Shell Epon Resin
Shell Chemical Co., 1415 W. 22 St., Oak Brook, IL 60522, 800-323-3405
E. V. Roberts and Associates, Inc., 2624 Barrington Ct., Hayward, CA 94545, 510-784-0414. FAX 510-784-0326

Surmat
Nicofibers, Inc., Box 26, Shawnee, OH 43782

Vinyl polysiloxane
President putty: Colteen Co., Carlsbad, CA, 800-228-0470
Express impression material system: 3M Dental Products Division, St. Paul, MN 55144-1000, 800-634-2249

Water-extended polyester resin (WEP)
Ashland Chemical Co., 8 E. Long St., Box 2219, Columbus, Ohio 43216

References

Baird, D. 1951. Latex molds in paleontology. *Compass of Sigma Gamma Epsilon 28*(4):339–345.

Chaney, D. S. 1985. PNC-724 urethane elastomer (molding compound). *Society of Vertebrate Paleontology News Bulletin No. 134*, pp. 57–58.

———. 1987. Fiber glass reinforced gypsum cement: Applications and methods for use in the laboratory, field, and museum. *Journal of Vertebrate Paleontology*, (Supplement to No. 3) 7:13A–14A.

———. 1989. Mold making with room temperature vulcanzing silicone rubber (pp. 284–304). In: R. M. Feldmann, R. E. Chapman, and J. T. Hannibal (Eds.), *Paleotechniques*. Paleontological Society Special Publication No. 4.

———, and M. B. Goodwin. 1989. R.T.V. silicon rubber compounds used for molding fossil vertebrate specimens: A comparison. *Journal of Vertebrate Paleontology* 9(4):471–473.

Chaney, D. S., P. Kroehler, and A. D. Lewis. 1991. Molding and casting the world's largest turtle, *Stupendemys geographicus. Journal of Vertebrate Paleontology* (Supplement to No. 3) *11*:21a.

Chase, T. L. 1979. Methods for the preparation of palaeontological models. *Special Papers in Palaeontology No. 22*, pp. 225–267.

Cherry, R. 1967. *General plastics*. Bloomington, IL: McKnight & McKnight.

Converse, H. H. 1984. *Handbook of paleo-preparation techniques*. Gainseville: Florida State Museum.

Heaton, M. J. 1980. New advances in latex casting techniques. *Curator 23*(2):95–100.

Jensen, J. A. 1961. A new casting medium for use in flexible and rigid molds. *Curator 4*(1): 76–90.

Leiggi, P. 1989. The use of vinyl polysiloxanes and quick setting epoxies in paleontology. *Journal of Paleontology 63*(2):256.

McCrady, A. D. 1978. Casting machine plans. *The Chiseler 1*(2):17–21.

Madsen, J. H., Jr. 1974. Derkane vinyl ester resins – an alternative to plaster-of-Paris. *Curator 17*(no. 1):64–75.

Rennie, G. S. III. 1969. Reproduction of a skeleton. *Discovery 5*(1):17–22.

Reser, P. K. 1981. Precision casting of small fossils: An update. *Curator 24*:157–180.

Rigby, J. K., and D. L. Clark. 1965. Casting and molding (pp. 389–413). In B. Kummel and D. Raup (Eds.), *Handbook of Paleontological Techniques*. San Francisco: Freeman.

Rixon, A. E. 1976. *Fossil animal remains: Their preparation and conservation*. London: Athlone/University of London.

Rose, J. J. 1983. A replication technique for scanning electron microscopy: Applications for anthropologists. *American Journal of Physical Anthropology 62*:255–261.

Smith, J. and B. Latimer. 1989. Casting with resins (pp. 326–330). In: R. M. Feldmann, R. E. Chapman, and J. T. Hannibal (Eds.), *Paleotechniques*. Paleontological Society Special Publication No. 4.

———. 1989. Preparation of hollow casts (pp. 331–335). In R. M. Feldmann, R. E. Chapman, and J. T. Hannibal (Eds.), *Paleotechniques*. Paleontological Society Special Publication No. 4.

Von Fuehrer, O. F. 1939. Liquid rubber as an enlarging medium. *Museum News of Washington 16*(14):8.

Walker, A. 1980. Functional anatomy and taphonomy (pp. 182–196). In A. K. Behrensmeyer and A. Hill (Eds.), *Fossils in the making*. Chicago: University of Chicago Press.

Waters, B. T., and D. E. Savage. 1971. Making duplicates of small vertebrate fossils for teaching and for research collections. *Curator 14*:123–132.

Wilson, M. A. 1989. Enlarging latex molds and casts (pp. 282–283). In: R. M. Feldmann, R. E. Chapman, and J. T. Hannibal (Eds.), *Paleotechniques*. Paleontological Society Special Publication No. 4.

Cast painting

Michael Tiffany and Brian Iwama

In the process of molding and casting, painting the casts is the last step. A good paint job can make a poor cast look good, but a bad paint job can make a good cast look terrible. Not painting the casts is usually not an option either. The surface of any unpainted cast will appear flat and without detail, but with the addition of properly applied colors, surface textures and details will be revealed.

Paint can even be used to duplicate various subtle textures present in the original that are visual rather than physical in nature. The "grain" visible on the surfaces of some teeth and bones is an example of this. Sometimes the painted cast will look better than the original object because the paint has highlighted surface details and textures so as to make them more visible.

One asset of working with a cast is that, unlike a flat piece of canvas, there is often a good deal of surface texture to work with. It only remains for you to bring it out with a little paint. Before setting out to paint any cast though, it is helpful to determine the use to which it will be put. Will it be used as a study or reference specimen – and if so, what features are going to be studied? Will it become part of a public display, close to the public view? Will it form the central part of the display, or will it just be a part of a whole (i.e., a single bone in a complete skeleton)? What will the illumination of the background be like?

If the cast is to be used as a study or reference specimen, the purpose of the paint job is to reveal surface texture and detail. The paint scheme used need not exactly duplicate the colors of the original. The color of the original object may not even show up surface details – bones from tar pits, for example, are often so black or brown that it is not easy to distinguish minute features. The paint scheme used on the cast can be relatively simple. One or two applications of a thin wash of a dark colored paint (brown or black, e.g.) on a light colored cast will deposit pigment in low spots to reveal surface irregularities. If the cast is also to be used as a teaching specimen, the paint scheme can more closely approximate the colors of the original if the duplication helps to provide more information.

Casts forming part of a public display do not require an elaborate paint scheme. For instance, a mounted composite cast skeleton of a dinosaur can have a very simple scheme composed of a single base color for the bone with a couple of thin washes to reveal surface texture. Exact reproduction of each bone's color scheme is not necessary, as it would likely be unnoticed. It might also make the skeleton resemble a patchwork quilt, as some of the original bones might be radically different in color. If casts of less grandiose objects are to be displayed close to the viewing public, a more elaborate paint scheme may be desirable (especially if the original material is in relatively close proximity). If the illumination will be subdued, a lighter paint scheme can improve its appearance on display.

If your efforts produce a finished product that is lighter or darker in color than the original, do not be discouraged. Variation in nature is the rule rather than the exception, and the natural color of an original can change because of aging or treatment with different consolidants. If you set out to duplicate the original color scheme to the letter, you may fail more often than you succeed. However, if you try to reproduce the flavor of the original, you may succeed more often.

Basic equipment

The list of necessary equipment is not extensive: brushes, sponges, and low-lint rags, mixing palattes, spatulas, small glass jars, paper towels, an airbrush or spray gun (optional), and a loose-leaf notebook. The only expensive items are the brushes and the airbrush or spray gun.

Brushes are available from a large number of manufacturers in a range of qualities. It is better to purchase good quality brushes because they will last longer (when treated properly) and will be less likely to shed their bristles, marring an otherwise good paint job. Brushes are available in a number of bristle materials, including sable, camel hair, and nylon. Each bristle type is often available in two shapes, flat or pointed, and a range of internationally coded sizes (000, 00, 0, 1, 2, etc.). Large rounds, fan-bristle, and angle-bristle brushes are useful for many applications. It is recommended to keep a wide variety on hand.

Sable brushes are a good choice for use with either water-based paints or oils. They are easier to load with pigment, allowing the color to be applied more evenly, and do not leave brushmarks. Nylon bristle brushes are stiffer and usually leave brushmarks, which can sometimes be useful when a graining effect is desired. Nylon brushes are also good for dry-brushing techniques. It is generally best to use a single brush with only one painting medium, as both water- and oil-based media leave residues even after cleaning. In the case of oil-based paints, for instance, the residual oils left behind are incompatible with water-based paints.

Sponges and low-lint rags are good for use in both applying paints and removing excess paint. Natural sponges are preferred to the common household sponges, especially when using them to produce mottling effects on a cast.

Mixing palattes can be made from almost any handy item: clean pie tins and lids from plastic yogurt or ice cream containers (cheap and disposable), old dinner plates, scrap pieces of plexiglass, and the like. Small glass plates are very desirable because they are easy to mix on and require very little effort for cleanup. Dried paint can be scraped off with a razor blade and a fresh surface is ready again.

It is best to use spatulas for dispensing paint from jars and for mixing paint on your palatte. Also very useful are tongue depressors or coffee stir-sticks. Both are made of wood and can be thrown away after one use or cleaned and reused. Paint brushes are not good for mixing paint, as they are easily ruined and can contaminate jars of paint (either with other colors or, in the case of acrylics, introduce mold spores, which can ruin the paint).

Small glass jars are excellent for storing paint colors that are used often or must be kept to touch up damaged casts. It is much easier to mix a large volume of color when painting many casts of the same specimen than trying to duplicate the color each time. It is a good idea to check paints stored in jars periodically to make sure they have not dried beyond the point of usability. Addition of water or other thinner will extend them indefinitely.

Paper towels are a must, as one always has to clean up. Towels that have been used to dry cleaned brushes or wet hands can be saved, dried, and reused (in the interest of conservation).

A loose-leaf notebook is important for keeping a variety of notes: good color mixes, mistakes that you don't want to repeat, suppliers, techniques. A written record makes these easier to access. You can even make your own color mixing charts in a notebook, painting small samples of a color mix directly on the page to refer to when needed.

Airbrushes and spray guns are optional. They are expensive to purchase, require a source of clean compressed air, and must be kept clean. They are indispensable for covering large areas rapidly with thin, even layers of paint (does not obscure surface texture or detail) and can be used to achieve effects not possible with a brush. Merging different colors with a feathered edge (transition of one color to another by fading out over varied distances without mixing the colors) or shadowing can be easily achieved with practice.

Spray guns are best for covering large areas. The paint is emitted in a conical spray of small droplets in a relatively large area of coverage. Airbrushes, however, expel the paint in a very fine mist with a narrow area of coverage. Although this is best for achieving effects like feathering and shadowing and for painting small objects, paints used in an air brush must be highly thinned. Most airbrushes are designed to be used with specially formulated inks and paints. Problems arise when using other paints that have their pigments in the form of larger particles that can clog the airbrush tip or will not flow evenly.

There are many brands of airbrushes, such as Badger, Paasche, and X-acto, available in either single- or double-action models (the latter are preferable). Highly recommended is the X-acto model 8010 Dual Action airbrush, the cheapest double-action airbrush available. It is rugged, spare parts are easily obtained, and because it is designed for use with enamel paints, it will handle the large-particle pigments in various other paints as well.

In order to obtain the best results from your equipment, as well as to enjoy a long service life, it is important to use it properly and keep it clean. Paint brushes should be cleaned immediately after use, repeatedly dipping into cleaner/thinner and stroking gently across a lint-free rag or paper towel until clean. They should then be re-formed and stored bristle-end up in a jar or holder. If the brush came with a plastic protective tube for the bristle end, save it and replace it after each use. Don't clean brushes by allowing them to stand bristle-end down in thinner or cleaner – it is the quickest way to ruin a good brush.

Airbrushes should always be kept clean. When changing colors, or at the end of a painting session, attach a bottle of the appropriate thinner and spray until no more color is visible and clear thinner is all that is emitted by the airbrush. When you are finished, the piece of equipment should be disassembled, thoroughly cleaned, reassembled, and carefully packed away.

Paints

Painting media and techniques for a particular project are often determined by the casting media you will be working with. Some casting media are receptive to a variety of paints, whereas others are problematic. With this in mind, we discuss the various painting media along with their characteristics, and advantages and disadvantages. Most of the readily available paints commonly used in cast painting are either water- or oil-based paints.

Water-based

Water-based paints are perhaps the best type to use when considering ease of cleanup, safety, and drying time. When dealing with polyester resin and epoxy casts, however, adhesion problems arise. An undercoating must be applied to these materials; matte enamel spray-on is superior for this purpose (more on this in the section on oil paints).

Acrylics

Acrylic paints are commonly used on casts. They are water-soluble acrylic polymer emulsion paints that dry to a waterproof matte or satin finish within minutes. In this family are two types of acrylics: Cryla/Flow Formula Cryla and Acrylic Gouache. Cryla is described as being a heavy, medium-based paint that requires considerable thinning before application. Flow Formula Cryla requires less thinning or can be used straight from the container. It dries with a slight sheen to the finish. Acrylic Gouache enjoys more popular use and has a wide selection of brands available. Examples include Grumbacher, Hunt, Liquitex, Newton, Shiva, and Windsor. Like Flow Formula Cryla, it requires less thinning or can be used straight from the container. It also dries rapidly, but to a matte finish. Both types of acrylic paints – the Crylas and Acrylic Gouaches – can be mixed together easily.

Acrylic paints are easily worked by the novice. Because they dry to a flexible waterproof finish, they will not lift with the application of subsequent layers of paint. Thinning is accomplished with water. Thick, out-of-the-container paint can be brushed on thin enough to produce an opaque layer without obscuring surface details. The speed with which acrylics dry cuts down on the drying time necessary between layers, but it can create problems when trying to blend or shade colors directly on the cast.

Cleanup of brushes is easily accomplished with warm water and a mild soap. Brushes must be cleaned immediately after use because of the short drying time; if paint cakes in the bristles, a small amount of alcohol will remove the particles.

Acrylics work well on all plasters and WEP (water-extended polyester). They will also adhere well to polyester (fiberglass) casts if the surface is gel coated. Acrylics are economical to use and therefore often employed, especially on large cast specimens. The durability of the paint job is enhanced by the application of a proper fixative.

Inks

There are a variety of inks useful in the painting of casts, and probably as many techniques for using them. No longer limited to the standard black inks, we now have colored opaque, transparent, and autographic inks. Brand names most commonly encountered include Pelikan, Speedball, and Osmiroid, though others undoubtedly exist. Korns makes a very good quality autographic ink.

When painting plaster casts, inks can be used in several ways. One must always be careful in applying ink to dry plaster – the pigment will disperse into the plaster and become almost unnoticeable when dry. To avoid this, a general thin wash of diluted ink can be applied over the entire surface. Building up several of these thin washes will impart enough color to darken the plaster and give it a good base for more detailed painting. If a thin coating of sealer or hardener is applied to the plaster before painting on the ink (or after base-coating the cast with several washes), one can then work with the ink as a highlighter. Butvar thinned with acetone is a good hardener and sealer, although there are many polyvinyl acetate alternatives to choose from. Once the hardener/sealer is dry, ink can be applied with a brush to highlight surface detail. The ink is generally thinned with water (at least in the case of opaque inks) in a ratio of one part water to three parts ink. When brushed carefully onto a cast where there is much surface detail, the pigment will be drawn into the pits, cracks, and depressions in the surface. To feather these darker areas into the lighter, smoother surface areas, a dry brush can be used while the ink solution is still relatively wet. A finger also works wonderfully for this application (use quick strokes and be careful not to let the finger rest on the surface, leaving fingerprints). Any number of color inks can be used, though reds and browns are especially useful. Autographic ink, which is normally used for making lithographs, has a wonderful brown color that dries relatively uneven – perfect for that fossil "flavor" we are trying to achieve.

It is important when using inks to let each successive application dry thoroughly before covering it with another layer. Though they are water-resistant when dry, inks will bleed terribly if still slightly wet when another layer of ink is brushed on. In some cases this bleeding effect can be desirable, but in general it is a nuisance to be avoided.

With acrylics, it is best to apply some sort of fixative or sealer to the finished product to enhance its durability.

Water colors

Traditional watercolors come in two basic varieties: tube colors and dry cakes of color (those found most commonly in watercolor sets). Both work well for the purposes of painting casts. They differ from acrylics in that generally they are not as opaque and they do not dry as rapidly as acrylics do. The pigments in watercolors act much the same way as those in ink, but there is generally a wider color range to choose from. It is difficult to build up a quick opaque layer with watercolors, but they are very good when applying washes of color. The pigments tend to be drawn into cracks, pits, and depressions in the cast surface, bringing out minute surface details.

The tube-type watercolors can be used in much the same way as acrylic paints – straight from the tube or thinned with water. Of course, when used directly from the tube, they will produce a more opaque finish than if they are thinned. The dry cake watercolors must be used with a wet brush. In this case, however, one is not necessarily restricted to the use of water as a thinner. They can be effectively used with lacquer thinner, acetone, or alcohol, depending on the surface to which they are applied. The colored lacquer, when applied to a cast coated with a lacquer base, will fuse to the base and "float" the pigment. Similarly, the colored acetone or alcohol will fuse with compatible undercoatings and produce similar results. Mottling effects can be easily obtained using this method, though it does take some practice and experimentation. The color cakes must be cleaned with a dry, lint-free cloth before using water or other thinners again.

The dry cake watercolors are among the cheapest pigments available to the cast painter, and their utility in painting applications makes them a good investment.

Oil-based

Enamels

Enamels are not commonly used to paint casts. However, they can be useful in cast painting, as a foundation for other paints. The solvents in enamel paints have a slight etching action on certain materials, such as polyester resin, forming a surface that the paint will bond to upon drying.

Matte enamels work best as undercoatings because they provide a surface with sufficient "tooth" for the adhesion of other paints without beading upon application or flaking after drying. They are available in a variety of colors, so finding a suitable base coat color is quite likely. Enamel paints range considerably in price, quality, and color, depending on the purpose they are intended for (i.e., hobby vs. automotive use). They are available in small bottles, cans, or spray cans. Airbrushing is the best method of application, but careful brushing or working with a spray can will work almost as well.

The enamels in bottles or cans usually require considerable thinning before they can be applied with an airbrush. Those that come prethinned for use in an airbrush are not suitable for application with a paint brush, as there is too much solvent for direct application and they can etch the surface severely. When using an airbrush the excess solvent is lost as vapor between the nozzle and the surface of the object.

Oils

Oil paints generally give the best results. The colors are richer than those of other types of paints and they have excellent mixing properties which allow a fantastic range of color mixes. Various quality grades and color ranges are available from a number of manufacturers. Avoid the least expensive grades (usually labeled "student's"), as the color permanence and smoothness of flow are not as good as those of the other grades. Those labeled "artist" are the highest quality and will give the best results.

Oils are slow-drying paints with varying rates of drying time. Certain colors, mostly the lighter shades and colors produced from dyes (phthalo green, ultramarine blue, amaralyzinine crimson), tend to require more time. The speed at which a color dries can be varied with the addition of thinners or mediums. Oil paints are thickly pigmented and require considerable thinning in order to avoid obscuring surface detail.

Because of their relative expense, oils are primarily used on smaller cast specimens and where a very good quality paint job is required. Their greatest utility comes in the painting of epoxy resin casts, to which most other paints will not adhere well. Both oil paints and dyes for epoxy resins can be mixed with the uncatalyzed resin to produce casts with the basic color of the original. In this way, casts also possessing the basic appearance of the original object material – opaque, translucent, or transparent – can be produced.

Opaque dyes for epoxy resins can be combined with oil paints and fumed silica or powdered fiberglass to produce opaque casts. To produce a translucent or transparent cast, a minimum of fillers is used and only oil paint is mixed in for color. A smoky appearance is produced by mixing separate cups of clear resin (with minimal filler) and oil paint–colored resin. The clear resin is first painted into the mold halves. The colored epoxy is then slowly added and carefully swirled around until the desired smoky appearance is achieved. A light sprinkling of sand or other particulate matter, even oil paint that is not thoroughly mixed, can be added in order to obtain a speckled appearance. (Translucent or transparent casts are more commonly encountered in archaeology and anthropology than paleontology, but the information is given in the event that you are called upon to produce one.)

Final painting of the surface with thinned oil paints is done as with any other medium. Thinning should be done with either turpentine or lacquer thinner so the paint will stain evenly and adhere to the surface. If lacquer thinner is used, drying time is minimal and subsequent layers can be added soon after painting.

Oil paints are compatible with, and will adhere well, to all the casting materials discussed in this chapter.

Lacquers

Lacquers are not the paint of choice in most cases, but they do have some applications in cast painting. Notably, where other paints tend to flake and peel, lacquers will adhere to epoxy resin casts and fiberglass casts with a smooth surface. When using lacquers, it is not necessary to buy a spectrum of different colors. It is, in fact, not necessary to buy colored lacquers at all – a clear matte lacquer primer and a clear matte lacquer paint will suffice. With some lacquer thinner and either dry cake watercolors or powdered dry colors, one is set for painting.

As described in the section on watercolors, lacquer thinner can be used as a vehicle instead of water (clear lacquer paint works, too). Once the cast is coated with a layer of matte lacquer primer (it can either be sprayed on with an airbrush or painted on carefully with a fine-bristle brush), you can add colors to it by floating the pigment onto the sealer with color mixed in lacquer thinner, or paint more color on with pigment in clear lacquer paint. It is necessary to let each layer dry thoroughly, as the addition of more paint or lacquer thinner will soften the primer layer briefly. This allows for some bleeding of the pigment (flotation) and subsequently a nice mottled effect.

Lacquers can also be obtained in opaque colors, which are handy for painting large areas of base color or painting out areas of matrix. Once the base color is added to the primer, the same pigment flotation techniques can be used to create a mottled effect.

A fixative is usually not required, as lacquers are quite durable, but sometimes the cast surface will become shiny after the addition of several layers of paint. If this is the case, you can very carefully brush on a final coating of the matte lacquer primer or spray on a clear matte lacquer fixative. Be especially careful at this stage: Too much brushing will smear the paint job and too much spray can make it run. It is best to brush on the final coating in small areas and let each application dry thoroughly before continuing on to another section. It sounds like a complicated process, and it is, but the final results can be amazing and well worth the effort.

Dry colors

Dry colors, for coloring cement, are useful for several applications in painting casts. Besides the obvious ability to color the casting material – be it plaster, fiberglass, WEP, or epoxy – they can be mixed with just about all media previously described without ill effects.

When mixed with enough water to put the pigment into suspension, the colors can be added to acrylics, inks, and water colors. Mixed with a little lacquer thinner they can be combined with lacquers or oils. Of course, they do not have to be added to anything – they color quite nicely on their own. When using them alone, it is important to fix each layer of color because the powder does not adhere well. Sometimes it is best to use the colors in a thinned solution of hardener or sealer (Alvar, Butvar, etc.), which can be painted onto the cast without concern about adhesion.

Powdered colors can be mixed into the casting material or painted onto the surface of the mold before addition of the casting material. It requires a lot of pigment to color plaster, and the strength of the final casts may be questionable if too much pigment is added. The color can be added directly to the plaster and mixed well before adding the plaster to water, or stirred into the water before adding the untinted plaster. If the cast is made with tinted plaster, damage from chipping or scratching will not be quite so obvious as in untinted casts.

With fiberglass, the dry colors can be mixed with resin, filler, and catalyst and then painted onto the surface of the mold as a gel coat. The same applies to WEP and epoxy casts. This cuts down immensely the time required to paint the cast after demolding. However, always test all dry pigment colors that will be used to make sure there is compatibility. Sometimes a pigment can interfere with the action of the catalyst, requiring more catalyst than normally used or preventing proper reaction and curing of the resin.

On plaster casts, dry colors can be sprinkled onto the surface dry and then rubbed in with a finger to produce fine shading. This works well to bring out surface features on relatively regular surfaces with fine features. The technique also works well on WEP casts. After rubbing, the surface should be fixed with an appropriate sealer.

Though not exclusively "dry" colors, pastels should also be mentioned briefly here. They come in three types – chalk, oil, and water-based – and can be used much the same way as other dry colors. Powdered by gently rubbing on coarse sandpaper, they can be applied to painted casts by rubbing. This will give a dusty, abraded appearance to a cast, which can be desirable in some cases. Powdered oil pastels can be added to paint thinner and brushed on; similarly, water-based pastels can be added to water for thinner color solutions. Pastels adhere best to matte surfaces, and must be fixed to the surface with spray fixative.

Solvents and medium

Only solvents (thinners) appropriate for each painting medium should be used. *Do not* use solvents meant for another painting medium – you can end up with a serious mess on your hands. Sometimes thinners are even brand-specific, formulated for use exclusively with a particular brand or line of paint. This situation is usually encountered when dealing with hobbyists' enamel and acrylic paints. Where you have a choice of thinners, it sometimes pays to experiment on a scrap piece of casting material (e.g., a reject cast) in order to see if the paint behaves in the manner expected or gives the desired effect.

Acrylic gloss medium and acrylic matte medium are suitable for use with poster and watercolors as well as Cryla and Acrylic Gouache. Both are milky white acrylic emulsions that can be used for either thinning or glazing, and both dry to a colorless, waterproof finish. The matte medium is added to the paint where a matte finish is desired. Both media can be combined in differing ratios to achieve a range of gloss.

Turpentine is available in either household or distilled grade. Household turpentine is suitable only as a brush-cleaning fluid. The distilled grade is used for thinning oil paints and to reduce their drying time. However, be aware that painting over previous layers with turpentine-thinned paint will reactivate the underlying layer. The reactivated layer may either lift or blend with the new layer being added. More experienced painters use this phenomenon to achieve shading and mottling effects. Mixing turpentine with oil paint will make it flow better and reduce its high-gloss finish, though if too much is added the richness of color will be lost.

Linseed oil is the medium upon which many oil paints are based. It slows down drying time and produces a glossy finish. Copal medium or copal oil produces a gloss finish but speeds up drying. Lacquer thinner can be used in oil paints to dramatically reduce their drying time.

Sealers

In order to prevent damage to surface details on softer casting media due to chipping and scratching, consolidation of the cast is recommended. Impregnation of well-dried plaster casts with any of the polyvinyl acetate consolidants is suggested (water-thinned Elmer's glue or acetone-thinned fiberglass resin are also options). Vacuum impregnation with a polyvinyl acetate (e.g., VINAC B-15) dissolved in ethyl alcohol is probably the best method.

Once the cast has been painted to satisfaction, the durability of the paint can be enhanced by the use of a sealing overspray. This forms a colorless, transparent layer that binds the underlying paint medium, as in dry colors or pastels, or adds a final plastic coating to protect against wear and tear. Sealing oversprays are available for both water- and oil-based media. They are usually available in clear gloss, semi-matte (satin), or matte formulas. They tend to act optically in such a way as to accentuate the color scheme, allowing the shades and tints of the many thin washes of color to show through with an impression of depth. One should always use a sealer that is compatible with the final painted medium. Clear semi-matte or clear matte finishes are usually most desirable, except for teeth, which are often glossy in fossil specimens. Clear gloss varathane works well on teeth, tusks, or any specimen that requires a gloss finish.

A variety of commercial art fixatives are available. Acrylic, lacquer, or enamel base formulas are usually sold in spray form and often contain strong solvents. The final paint layer should be thoroughly dry before addition of the sealer, and it should always be applied in two or three layers. Sprayed on too thick, they will tend to run and pool, where they may lift the underlying paint layers and cause bleeding. If you are unsure of the suitability or effect of a fixative on a particular painting medium, by all means try it first on a sample piece.

Following is a summary table showing the compatibility of the painting media discussed with the most commonly used casting media. It is not meant to be all-inclusive but merely a rough outline to work from.

	Acrylic	Ink	Watercolor	Enamel	Oils	Dry colors	Lacquer
Plaster	E	G	E	E	E	E	E
WEP	E	G	G	E	E	G	E
Fiberglass	G–F	P	P	G	G	P (+G)	G
Epoxy	F	F–P	P	E	E	P (+G)	G

E = excellent, G = good, F = fair, P = poor, + = in combination with clear lacquer. Note that those media with "poor" compatibility ratings can be used with addition of under-coating.

Colors

Following is a table of the most commonly used colors for painting casts, accompanied by an accessory list of useful colors. Remember that these are colors preferred by the authors; others are available that you may find useful. Personal preference plays a role in this as well!

Basic palette	Accessory palette
White	Raw sienna
Raw umber	Burnt sienna
Burnt umber	Yellow oxide
Yellow ochre	Paynes grey
Cadmium yellow	Cobalt blue
Cadmium red	Mars yellow
Ultramarine blue	Indian red
	Naples yellow
	Black

Notice that the basic palette has relatively few colors (most of them are earth tones) and that black is missing altogether. This is the basic group of colors for producing just about any color you will ever run across when painting casts. Burnt umber is used where you would normally think to reach for the tube of black (in most cases black paint is really just a very dark, dark blue, so when you mix it with yellow you end up with green). Raw umber keeps your palette in the range of earth tones when mixing, though. Ultramarine blue is useful for making colors darker (approaching black). Mixed with burnt umber or burnt sienna, blue will make a very rich dark brown, which is less harsh and more visually pleasing than black. It is always best to use titanium white when working with oils and acrylics, because zinc and flake white do not carry pigment well.

Techniques

As stated before, there are probably as many techniques for painting casts as there are colors and types of media. Here we present the basic technique of painting transparent or semitransparent layers over a base coat to achieve the desired results. Of course, it is always easier to paint a dark color over a lighter one, and some transparency can be maintained between the layers. If you are trying to paint a lighter color over a darker one, total coverage is usually necessary, obscuring the darker color below (there are exceptions, but you won't run across them too often).

The first thing you want to do, before applying any paint to the cast, is to make sure that all separators and dust and dirt are removed from the cast. If the cast is epoxy or fiberglass, warm soapy water and a soft-bristle brush (like a toothbrush) are good for cleaning the surface. WEP casts can be cleaned in this fashion as well, although you will have to let the cast dry again thoroughly before painting, because the surface will absorb some water. Plaster is usually workable as it emerges from the mold. Letting it dry thoroughly is usually the only requirement before beginning the paint job.

Once the cast is dry and clean and you are ready to paint, you must decide on the material. Remember that your base coat – whether it be enamel or lacquer or a sealer like Butvar – can be tinted as well, though your best bet is to use something safe like acrylics or dry colors at this stage. Your entire paint job can depend on the quality of the base coat, and any unusual reactions caused by mixing of incompatible paints can cause flaking or bubbling.

With a good base coat in place, you can proceed to painting thin washes. It is usually best to pick a color that predominates on the actual specimen you are reproducing and paint the entire cast that color. You can then add more colors to re-create the scheme of the original.

The size of the cast will dictate somewhat the tool you use to apply the paint. If it is a small cast, like a small dicynodont skull, for instance, you will probably do just fine with a #2 sable brush. The brush will allow you to cover large areas as well as work small places with relative ease. Larger specimens, like a dinosaur skull or femur, will require larger brushes (1"–3") and sponges. You will find the natural sponges to be a great resource in painting very large specimens: They can put large amounts of paint onto a large surface rapidly, relatively evenly, and without brushmarks. By tamping them across the surface lightly you can even mottle the surface. A low-lint rag will work in much the same way, although paint seems to be absorbed and held much more readily by a rag than a sponge.

With the base color applied and dry, you are ready to work up the rest of the color scheme. Dry brushing – using a color that has been minimally thinned with solvent or thinner – will add color and grain to the cast. Often the uneven surface of a bone can be brought out with the addition of paint with a dry brush or rag. Texture that does not already exist can be added in the same way – use a sponge to mottle paint onto the surface or dry-brush with a nylon bristle brush to produce "grain."

Fossils are generally covered with irregular splotches of color, sometimes dark and sometimes light. You can use a stiff bristle brush and thick paint to spatter paint on the cast. By changing the distance between your brush and the cast, you can regulate the density of the splatters and the relative size.

Most of the painting media can be combined on any single cast, though you must always be careful to let each layer of a particular medium dry thoroughly before switching to another. Sometimes it is necessary to apply a sealer before switching to another medium that will have a more desirable effect. For instance, you may begin by painting a large plaster cast of a dinosaur femur with thinned tube-type acrylics. Once it is dry, you decide to dry-brush the more rugose areas of muscle attachment, also using acrylics. Once these are dry, you coat the entire cast with two layers of thin Butvar, floating the pigment onto selected areas of the articulate surfaces to give them a more mottled look. Another layer of Butvar (carefully applied) preserves this addition. Ink is then used to highlight cracks in the bone surface, and a dark enamel is finally splattered in wide areas over the entire cast to add a granular appearance. Several more coats of Butvar are carefully applied to seal the whole, and a matte acrylic fixative is sprayed on for a final finish.

Summary

We have gone over the essential elements of cast painting – equipment, painting media, and some basic techniques. Doubtless there are points we have missed; there may be techniques and processes not mentioned that some of you practice yourself or have heard about.

Equipment necessary for successful cast painting can range from the very inexpensive to the relatively costly. Good brushes are probably the most basic tool you will encounter, followed by a good finger and an eye for duplication. After these come the sponges, mixing palattes, spatulas, and airbrushes and spray guns. Painting media are varied but, again, the most simple can end up being the most valuable. Dry cakes (or semimoist) of watercolor or dry powdered pigments can be extremely useful with a wide range of thinners, solvents, and casting materials. The color palette you will be using most frequently is relatively simple as well. A few well-chosen colors will work wonders on almost any painting job you encounter.

The techniques outlined here are relatively few and simple. There is no standard right-or-wrong formula to use, but there are a few basic rules to remember that will keep you out of trouble every time. Start your paint job using lighter colors and work into the darker tones. You can always cover up a poorly chosen color if it is lighter; making a black cast white again usually requires drastic measures. Remember that you can always let the previous layer of paint dry and then make corrections. After the base coat, keep the layers of paint thinned and don't worry about adding too many layers (as when correcting a not-so-good choice of color). The more paint you add, generally the better "flavor" you will achieve. Fossils are usually rich with color, sometimes undiscerned, and the more colors your cast contains the more fossillike it will appear. Work with the surface textures of the cast, these are your road markers and will help you in painting. Textures pick up color readily and help to convey a sense of structure and form.

Perhaps the most important thing to remember about cast painting and all its aspects is *keep it simple*. That applies down the line – with equipment, media, colors, and techniques. The less complicated your procedure, the less complicated the solutions required to rectify. Painting casts should not create a sense of panic but, rather should allow you to explore and create.

Acknowledgments

The authors wish to thank the following people (in alphabetical order) for sharing their knowledge and for their support and assistance: Maggie Dickson, Dr. G. Edmund, Mark Goodwin, Catherine Hollett, Anne MacLaughlin, Peter May, Dr. C. McGowan, and Ian Morrison.

11

Mounting of fossil vertebrate skeletons

Kenneth Carpenter, James H. Madsen, and Arnold Lewis

There is little doubt that the "new golden age of vertebrate paleontology" is due in large part to the public's fascination with dinosaurs. Most natural history museums have been slow to capitalize on this interest, but in recent years we have seen many new fossil displays and fossil exhibit renovations. As a result, the number of fossil skeletons mounted for display has increased significantly. Too often, however, the quality of the mounts is poor because of unrealistic deadlines, anatomical mistakes, incomplete specimens, and inadequate training of those mounting the skeleton.

Brief history of mounting skeletons

In 1806, the first mounted fossil skeleton, a mastodon, went on display at Charles Peale's museum in Philadelphia (see Simpson 1942 and Alexander 1983 for a history of the collecting of this unique specimen). The legs of the skeleton may be seen in Peale's 1822 painting *The Artist in His Museum* (Alexander 1983 fig. 3). The techniques used in the mount, unfortunately, were not recorded. This exhibition of a prehistoric animal skeleton made a tremendous impression on the public, as evidenced in the following years when several more mounted skeletons were placed on display. They included a *Megatherium* in Madrid, Spain (cited in Flower and Lydekker 1891); a composite mastodon with an exaggerated length of 10 m (32 ft.) and a height of 4.5 m (15 ft) (Simpson 1942), displayed at various locales in New York State; and a composite archaeocete whale with an exaggerated length of 35 m (114 ft) (Kellogg 1936). The latter mount was rather crude, with boards used to hold the skeleton together. This innovation did, however, make it easier to disassemble and move the specimen from city to city for repeated exhibition.

The greatest impact on the general public and natural scientists, however, was the first mounted dinosaur skeleton, *Hadrosaurus foulkii*, exhibited at the Academy of Natural Sciences in Philadelphia (see Figure 11.1). This specimen was

Figure 11.1. Workshop of Waterhouse Hawkins showing (A) the mounted skeleton of Hadrosaurus foulkii *(fossil bone dark), (B) a partially mounted cast of the skeleton of* H. foulkii, *(C) an articulated segment of plaster of Paris casts for the skeleton being mounted, and (D) modern skeletons used for comparative purposes. (Courtesy of the Academy of Natural Sciences of Philadelphia Library.)*

mounted by Waterhouse Hawkins under the direction of Joseph Leidy in 1868. The specimen was not complete and Hawkins had to rely on his own imagination in reconstructing the missing bones. He gave the skeleton seven cervicals, a mammalian scapula, and an iguana-like skull. Hawkins did, nevertheless, anticipate many of the techniques used today: Missing limb bones from one side of the body were modeled as mirror images of their counterparts preserved on the other side; vertebrae were drilled and supported by an iron rod; limb bones were strapped to the supporting armature with metal bands; and missing parts were reconstructed after those of a modern mammal or lizard.

In selecting the stance, Hawkins relied on Leidy's observation that "the enormous disproportion between the fore and hind parts of the skeleton of *Hadrosaurus* has led me to suspect that this great herbivorous Lizard sustained itself in a semi-erect position on the huge hinder extremities and tail" (1865:97). The completed skeletal mount became the forerunner of the traditional, tripodal pose for bipedal dinosaurs. The mammalian characters of the mount, seven cervicals and the shape of the scapula, may also have been influenced by the kangaroo model. Leidy distributed plaster of Paris casts of the *Hadrosaurus* skeleton to the U.S. National Museum and the Field Museum of Natural History. These were some of the first casts of a dinosaur skeleton ever displayed.

Hawkins's mounting techniques were later improved upon by Louis Dollo and his assistants at the Musee Royal d'Histoire Naturelle in Brussels. Dollo and his crew mounted a group of *Iguanodon* skeletons during the late 1870s and early 1880s. Before mounting the skeletons, life-size drawings of the skeletons were prepared. Individual bones or sections of bone were then suspended from scaffolding in the position they were to assume. These were then attached by metal bands to an iron armature (Colbert 1968, pl. 14; Norman 1985:12, 28).

While mounting the skeletons, Dollo noted anatomical similarities between birds and *Iguanodon* (Dollo 1883). For reference in mounting the skeletons, he utilized the skeleton of an emu, as well as that of a kangaroo (Norman 1985:28, skeletons near knee). This ensured that the bones were articulated as accurately as possible. Unlike Hawkins, Dollo and his crew did not attempt the reconstruction of missing bones (Colbert 1968, pl. 16, e.g., missing sacral spines on one skeleton). This tradition of not reconstructing missing parts is continued today by some preparators (e.g., note missing cervical ribs and chevrons on *Lufengosaurus* skeleton in Dong 1988:14).

One of the first descriptions for techniques used in mounting fossil skeletons was given in 1909 by A. Hermann. He presented detailed discussions, supplemented with illustrations, of many techniques used at the time. Some of the techniques were those developed earlier by Hawkins and Dollo; others, such as the temporary wood cradle for supporting the vertebrae, were apparently developed by Hermann or his associates. Since the publication of Hermann's handbook, several short descriptions on mounting skeletons have appeared – for example, Drevermann (1930), Schultz and Reider (1943), Gilpin (1959), Bessom (1963), Bathel (1966), Majumdar (1974), Rixon (1976), and Converse (1984). Detailed discussions on mounting are presented by Madsen (1973), Panofsky (1973), and Burke et al. (1983).

Mounting skeletons: What makes a good or bad mount

The standard for a good skeletal mount has been refined over the years. For a long time it was traditional to exhibit a specimen standing statically like a trophy skeleton. All that was important was having the bones correctly articulated and the armature not too obvious (Hermann 1909). Today, however, most mounts are planned and constructed in an attempt to "breathe" life into the fossil bones. Skeletons are dynamic, standing in midstride, inviting the visitor to imagine the living animal walking, running, or fighting (see Figure 11.2). Unfortunately, some of the newer "action" mounts suffer from incorrectly articulated bones, poor workmanship, and the lack of a natural and realistic pose. These problems are elaborated upon here.

Except for casts of a single specimen, no two skeletons can be mounted the same way because each will present its own problems to overcome in design and construction. Therefore, what follows is meant as a guide for those who mount skeletons, with the provision that some modifications and departure from what is presented are often necessary.

Posture: Some fundamentals of tetrapod anatomy

Hermann once lamented, "Going through our exhibition halls, I often feel a strong temptation to make changes in our former mounts; changes, which I am sure, would improve the naturalness of those skeletons immensely" (1909:308). The reason for this attitude was that improvements in photography allowed the motion of animals to be "frozen." Many of these photographs, showing animal locomotion in sequence, were published in Muybridge's 1887 classic *Motion in Animals* (reprinted in 1957). In our opinion, this work remains a standard reference for determining limb positions and angles in mounting vertebrate skeletons.

Many recently completed skeletal mounts show anatomical liberties that would give Hermann cause for concern. These errors suggest inadequate knowledge of basic vertebrate anatomy and include incorrectly articulated vertebrae, ribs mounted perpendicular to the long axis of the skeleton, radius not crossing over the ulna, overextension of the appendicular joints, and feet placed too far apart (Carpenter 1989).

Some useful anatomical references for those mounting skeletons include Sisson and Grossman (1938), Gilbert (1968), Gray and Goss (1973), Hildebrand (1974), Romer and Parsons (1977), and Hildebrand et al. (1985).

Vertebrae

In mounting vertebrae, we must consider the limitations imposed by the shape of the anterior and posterior faces of the centra and by the articulation of the pre- and postzygapophyses (see Figure 11.3). In many dinosaurs the articular faces of the centra may be oblique to one another, so when the neck is correctly articulated, it has a sigmoid curve (see Figure 11.4). The amount and direction of movement between the adjacent vertebrae are limited by the position of the zygapophyses and thickness of the cartilaginous intervertebral disks. The posteriorly projecting postzygapophyses overlap the anteriorly extending prezygapophyses of the following vertebra. The articulating surfaces of the centra and zygapophyses must be in contact with each other because in living animals they are encased in a cartilage capsule. The amount of movement between any two adjacent vertebrae is limited; therefore the wide range of movement in the vertebral column is the cumulative result of movement involving many vertebrae. The lifelike, graceful curves in a mount are attainable only after taking this into account.

We must understand that the vertebral column in most tetrapods acts like a suspension bridge to support the gut, lungs, and other internal organs. For this reason, the vertebral column is slightly bowed or curved upward. When an animal runs, the back flexes or arches considerably (see Figure 11.5). A side-to-side bending of the vertebral column also occurs when an animal walks or runs. This

Figure 11.2. Example of a dynamic mount, the small carnivorous dinosaur Troodon *sp. (right) and the small herbivorous* Orodromeus makelai *(left). (Museum of the Rockies.)*

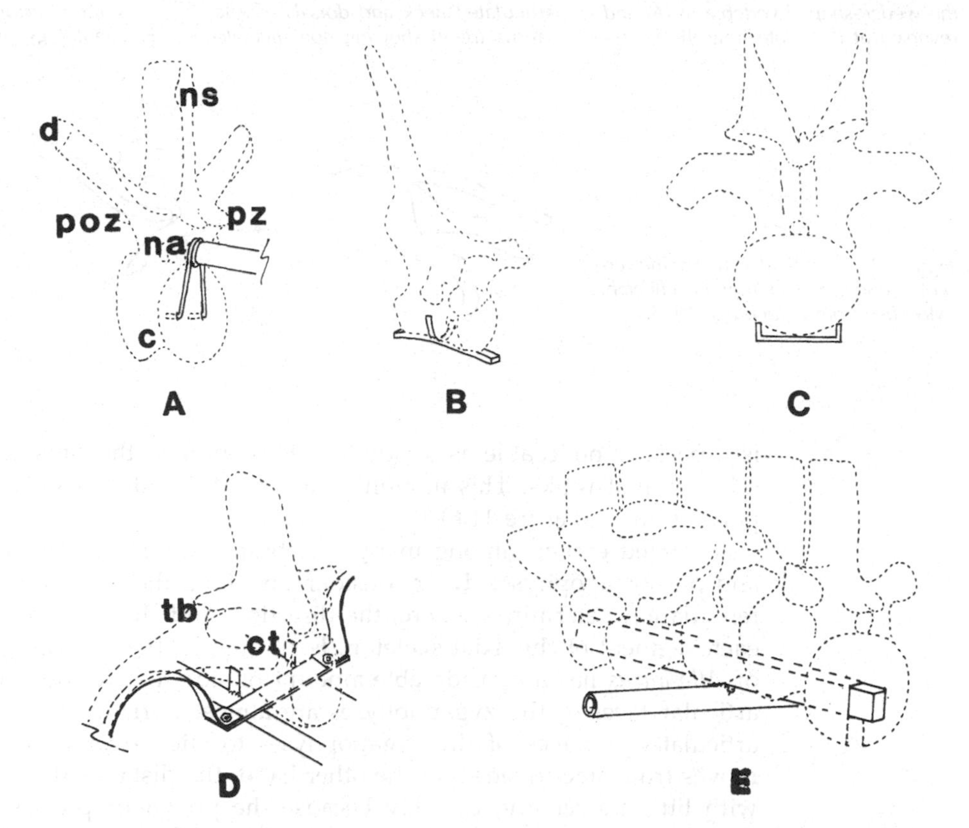

Figure 11.3. Examples of mounting vertebrae and the sacrum. (A) Bar through neural canal with heavy wire support. (B) Bar and clamp. (C) Cradle method using channel iron. (D) Modified clamp method with rib straps. (E) Notched sacrum with steel bar and leg cross-support. Vertebra parts: c, centrum; d, diapophysis; na, neural arch; ns, neural spine; pz, prezygapophysis; poz, postzygapophysis. Zygapophysis refers to both the pre- and postzygapophyses. Rib parts: ct, capitulum; tb, tuberculum.

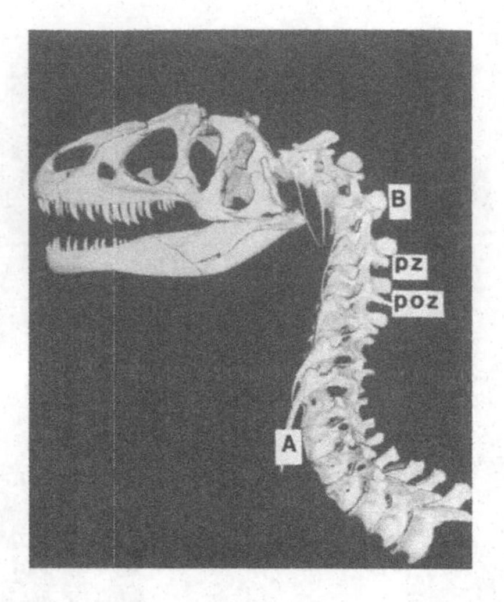

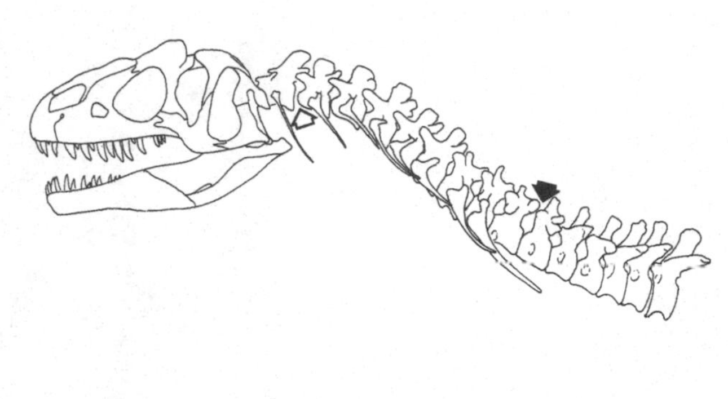

Figure 11.4. Neck and first three dorsals of Allosaurus fragilis *as correctly mounted (left). The sigmoid or S-shaped curvature of the neck is due to the wedge-shaped vertebra at (A) and reversed at (B). Note how the prezygapophysis (pz) reaches forward to the postzygapophysis (poz) of the preceding vertebra. In this manner the vertebrae are interconnected. Incorrectly articulated neck and dorsals of* Allosaurus *(right) showing nonarticulated centra (open arrow) and nonarticulating zygapophyses (closed arrow). (Based on specimens mounted at the Royal Ontario Museum, Utah Museum of Natural History, Tyrrell Museum of Palaeontology, and others.)*

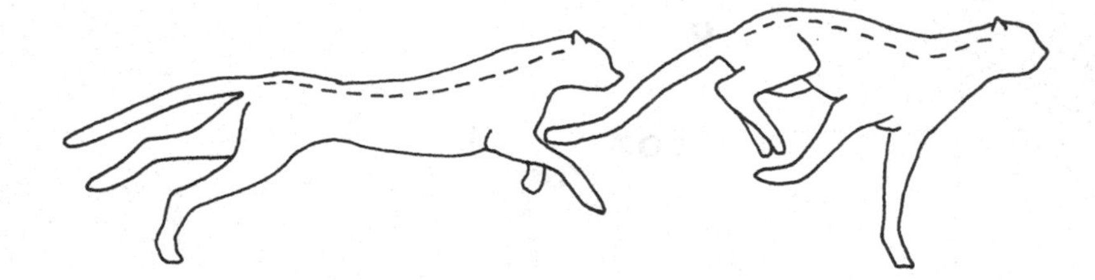

Figure 11.5. Vertical flexion of the vertebral column in a running cheetah. (Modified from Hildebrand 1985.)

is especially noticeable as a side-to-side motion of the hips when a person or other animal walks. This motion maintains the center of gravity over the supporting limbs (Figure 11.6).

As stated earlier, among many vertebrates, the shape and angle of the pre- and postzygapophyses differ considerably along the vertebral column. This differentiation determines where, the direction, and the amount of movement for each segment of the axial skeleton. For example, the anterior section of the tail of *Allosaurus* has a considerable amount of side-to-side movement because the articular faces of the zygapophyses are almost horizontal. This allows the flat articulated surfaces of the zygapophyses to slide over each other as the tail moves from side to side. On the other hand, the distal portion of the tail is stiff, with little movement, possibly because the prezygapophyses overlap most of the length of the vertebra in front of it. These differences in the zygapophyses allow the tail to be used as a counterbalance, because the tail can move from side to side to maintain balance alternately over the supporting leg during each step.

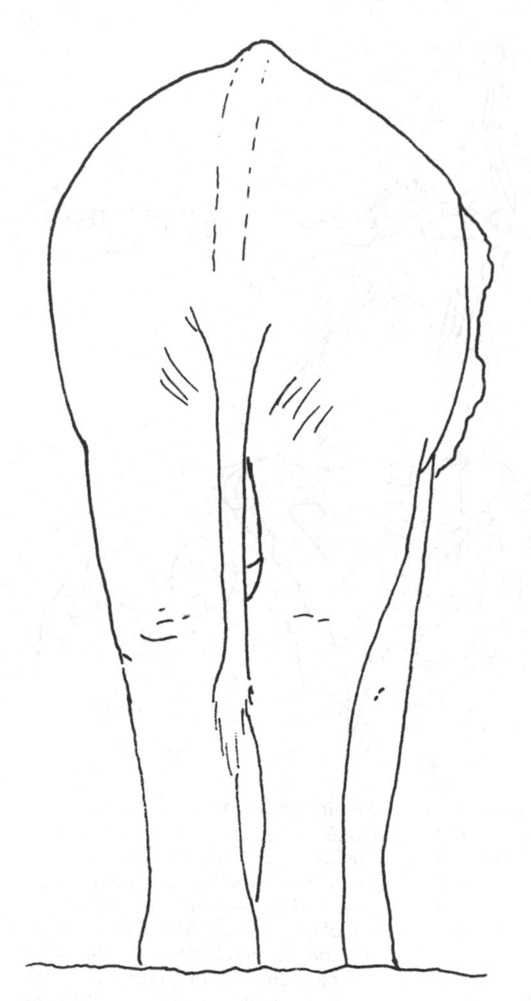

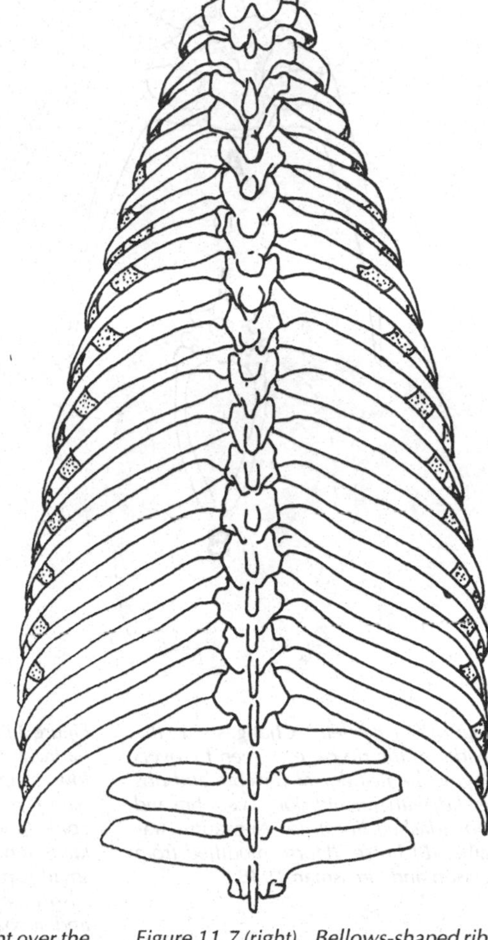

Figure 11.6 (left). Rear view of a walking elephant showing the narrow gait (feet almost in front of each other) and the movement of the hips toward the left to transfer the body weight over the left rear leg. Right rear leg has just stepped forward. (Redrawn from a photograph in Kunkel 1981.)

Figure 11.7 (right). Bellows-shaped rib cage of a horse formed by posteriorly angled ribs; seen in dorsal view. (Modified from Ellenberger 1956.)

Ribs

Ribs attach loosely to vertebrae, and in the living animal function as bellows to inflate and deflate the lungs (in mammals the diaphragm also aids in this). In most vertebrates the rib attaches to the vertebra at two points called the capitulum (head) and tuberculum (handlelike process below the head) (see Figure 11.3D). The capitulum attaches to a winglike process of the neural arch, called a diapophysis, that bears a facet for the rib head. The tuberculum attaches at a facet near the anterior edge of the centrum. In the thoracic region of mammals, these facets may lie at the junction between adjacent vertebrae, so the tuberculum is shared by two vertebrae. Because the diapophysis is posterior to the facet for the capitulum, the rib must angle posteriorly. Viewed dorsally, this angle is typ-

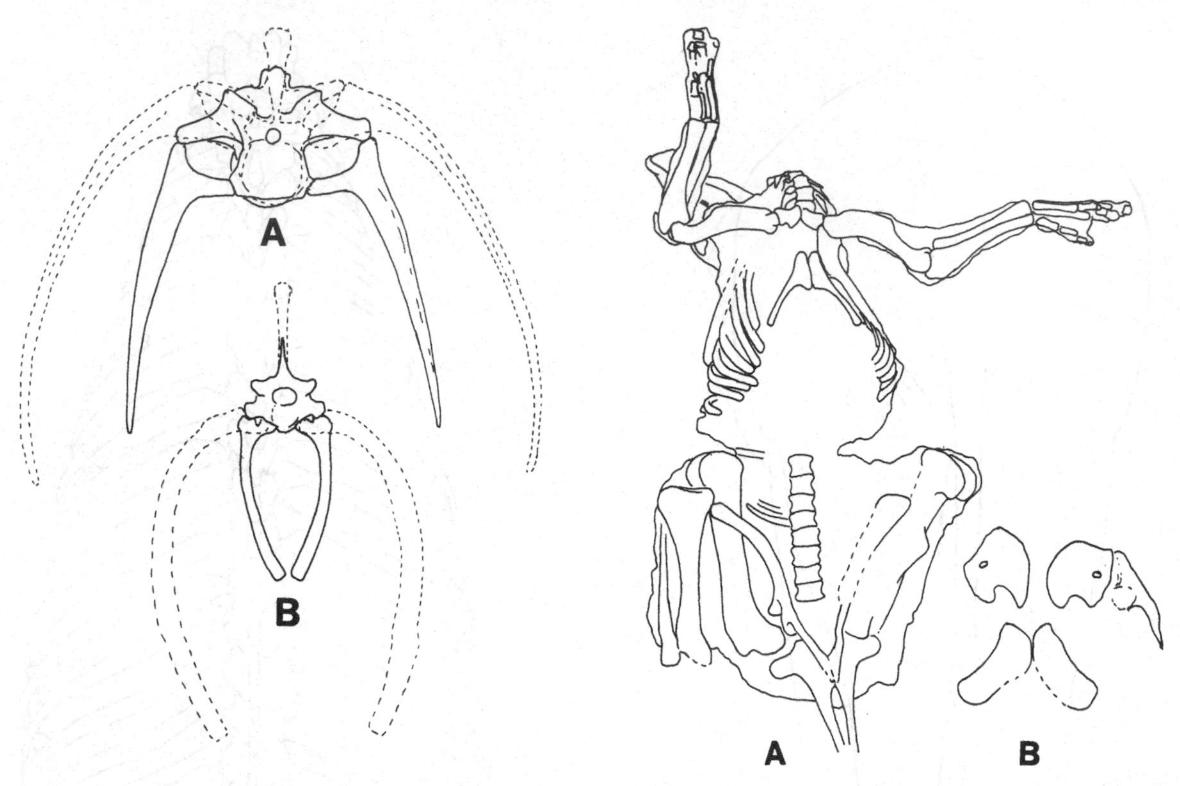

Figure 11.8 (left). Changes in the width of the rib cage as seen by overlaying the first dorsal vertebra and ribs (solid) with the mid-dorsal vertebra and ribs (dashed lines). (A) Allosaurus fragilis; *(B) horse. (Horse modified from Sisson and Grossman 1938.)*

Figure 11.9 (right). Three-dimensional hadrosaur "mummy" showing the bellows-shaped rib cage (A) (compare with Figure 11.7). Note that the coracoids touch, although in other articulated dinosaur specimens there is a small gap (see Norman 1980, fig. 56). Also note the gap between coracoids and sternals, a condition also seen in other articulated dinosaur skeletons. (Modified from Osborn 1912.) (B) Coracoids and sternal plates of the ceratopsian Leptoceratops gracilis *as found. Note small gap between coracoids and between coracoids and sternal plates. Sternal plates diverge as they do in the hadrosaur. See also* Chasmosaurus *sp. in Sternberg 1951, pl. 55, and* Protoceratops *in Lull 1933, pl. 4.*

ically 10 to 40 degrees from the perpendicular (see Figure 11.7). This angulation and an increase in the length and curvature of the ribs posteriorly to the middle of the thorax (see Figure 11.8) give the rib cage a bellows shape. An example of a bellows-shaped rib cage from the fossil record is the uncrushed *Edmontosaurus annectens* "mummy" at the American Museum of Natural History (see Figure 11.9A).

Forelimb

The position of the scapula in fossil mammals is relatively easy to determine because their modern counterparts provide accurate models for comparison. The position in dinosaurs is often less easily determined. However, articulated specimens usually retain the scapula in its correct position. Such examples show that the scapula is usually parallel or at a slight angle to the vertebral column (see

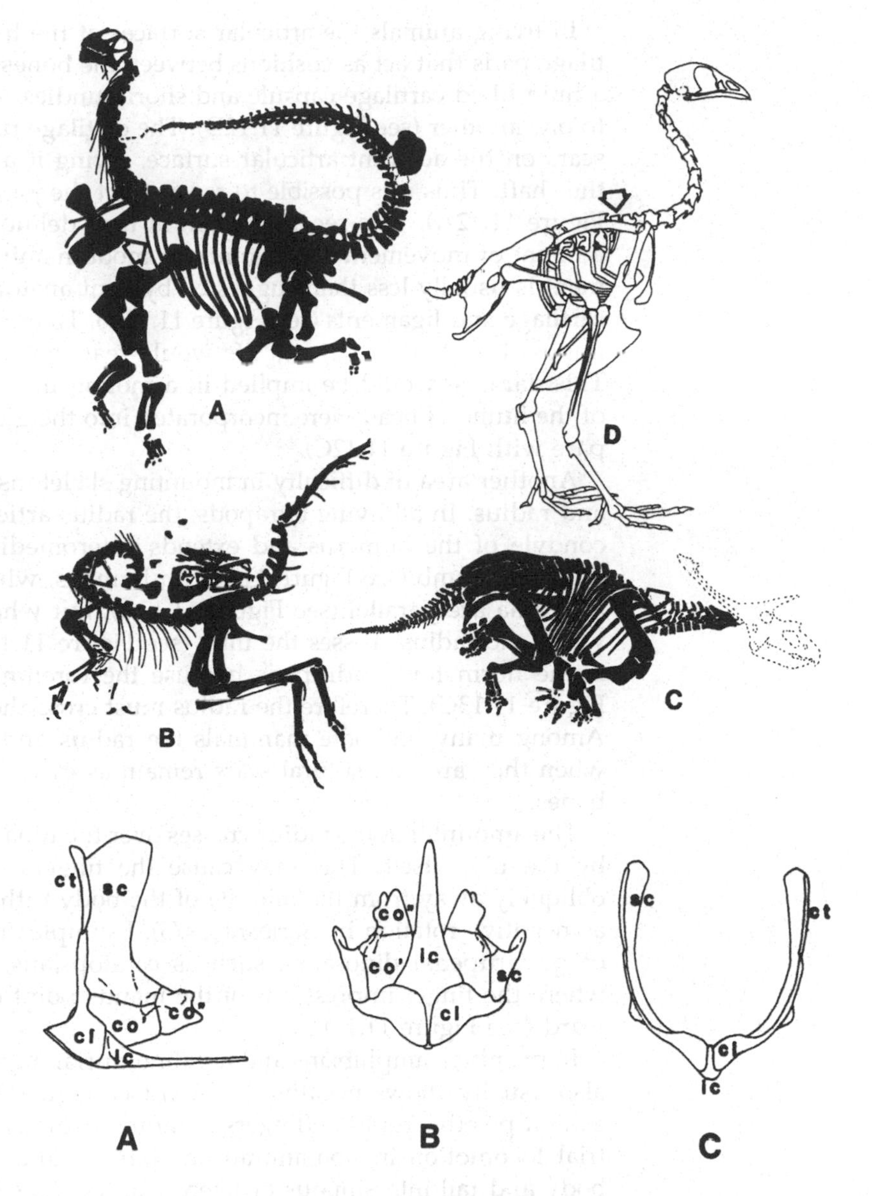

Figure 11.10. Scapula position relative to the first dorsal rib (arrow) as found in the articulated dinosaur skeletons (A) Camarasaurus lentus, *(B)* Compsognathus longipes, *and (C)* Anchiceratops longirostris. *In all of these the first dorsal rib almost bisects the scapula, a condition similar to that seen in a chicken (D). [(A) modified from Gilmore 1925; (B) modified from Ostrom 1978; (C) modified from Lull 1933; (D) modified from Chamberlain 1943.]*

Figure 11.11. Example of a primitive type shoulder girdle (of Dimetrodon*) showing the various bones that widely separate the coracoids (compare with Figure 11.9). Seen in (A) left lateral, (B) ventral, and (C) anterior views. cl, clavicle; ct, cleithrum; co', anterior coracoid; co", posterior coracoid; ic, interclavicle; sc, scapula.*

Figure 11.10A–C). Furthermore, the scapula often extends half its length in front of the first dorsal rib, as is observed in birds (Figure 11.10D). The forward extension of the scapula and the anterior narrowing of the chest bring together the coracoids so that they almost meet at the midline (see Figure 11.9); the sternal plates (when present) are separated from the coracoids by a gap of variable width.

In nondinosaurian reptiles and amphibians, the scapula is almost vertical and the coracoid nearly horizontal (see Figure 11.11). Unlike dinosaurs, the coracoids in these animals (e.g., crocodiles, amphibians, plesiosaurs) do not meet but are separated by several bones, including the interclavicle.

In living animals the articular surfaces of the limb bones are covered by cartilage pads that act as cushions between the bones. The entire joint is encased in a fluid-filled cartilage capsule and short bundles of ligaments connect the bones to one another (see Figure 11.12D). The cartilage pad and encasing capsule leave scars on the adjacent articular surface, giving it a different texture than that of the shaft. Thus it is possible to reconstruct the joint cartilage of fossil bones (see Figure 11.12A). This reconstruction in turn delineates the theoretical maximum amount of movement at the joints. In modern animals, however, inferred movement is usually less than suggested by joint anatomy because of the presence of cartilage and ligaments (see Figure 11.12D). To move the limbs beyond the limits imposed by the cartilage in life would tear the cartilage and damage the joint. This damage would be implied in a mount, if, for example, the cartilage scars of the humeral head were incorporated into the glenoid (see Figure 11.12B; compare with Figure 11.12C).

Another area of difficulty in mounting skeletons is the articulation of the ulna and radius. In all living tetrapods, the radius articulates with the outer (lateral) condyle of the humerus and extends anteromedially to the carpal of the first digit, or thumb (see Figure 11.13). In humans, when the palm is up, the radius and ulna are parallel (see Figure 11.13A), but when the palm is rotated downward, the radius crosses the ulna (see Figure 11.13B). The palm-down position is the norm for quadrupeds because the forelimb is used in locomotion (see Figure 11.13C). Therefore the radius must cross the ulna in a mounted skeleton. Among many ungulate mammals the radius and ulna are fused together, but when they are not, sutural scars remain as evidence of the crossing of the two bones.

The amount that the radius crosses over the ulna in a quadruped is constrained by the ulna itself. This may cause the fingers of the manus (hand) to face obliquely away from the midline of the body rather than face forward (defined as negative rotation by Sarjeant 1975). Examples may be seen in the trackways of quadrupedal dinosaurs, such as ceratopsians, ankylosaurs and sauropods, where the finger impressions of the manus point obliquely out, not straightforward (see Figure 11.14).

In primitive amphibians and nondinosaurian reptiles, the imprint of the manus also usually shows negative rotation (see Figure 11.15B), but occasionally may exhibit positive rotation (fingers pointing medially inward and forward). Terrestrial locomotion in non-anuran amphibians and reptiles involves moving the body and tail into sinuous or lateral curves. This body movement can result in a relatively narrow trackway because the side-to-side motion brings the feet closer to the midline (see Figure 11.15). Trackway data alone would incorrectly suggest that the animal had an erect or semierect posture. Therefore joint morphology must be considered in selecting the type of posture for mounting the skeleton of one of these animals.

Tail

Trackways of amphibians and reptiles frequently show tail drag marks (see Figure 11.15B), but these traces are rarely observed in bipedal dinosaur trackways. This supports the hypothesis that the tail of bipedal dinosaurs served as a coun-

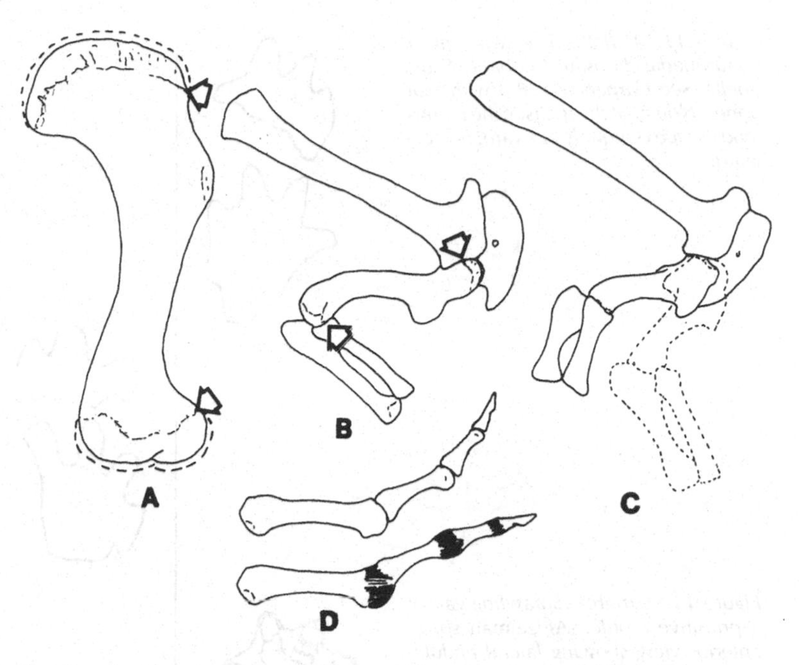

Figure 11.12. Cartilage and ligament constraints in joint movement. (A) Humerus of Allosaurus *showing the rim (arrow) marking the transition from bone (humeral shaft) to cartilage-covered bone (distal and proximal ends). Dashed lines show approximate extent of the cartilage caps. (B) Right forearm of* Allosaurus *with rim of cartilage scars within the glenoid and in articulation with the facet of radial head (arrows). (C) Maximum arm movement of* Allosaurus *(based on articulating casts) keeping the rim of the cartilage scar outside the glenoid and outside the facet of the radial head. Amount of movement is considerably less than in (B). (D) Hypothetical dorsal movement of the human finger elements based on joint morphology (top) and actual possible movement as restricted by ligaments (bottom). This example shows that the actual range of movement based on joint morphology is usually less than what is possible. This implies that the amount of limb movement based on joint morphology in (C) may be greater than was actually possible.*

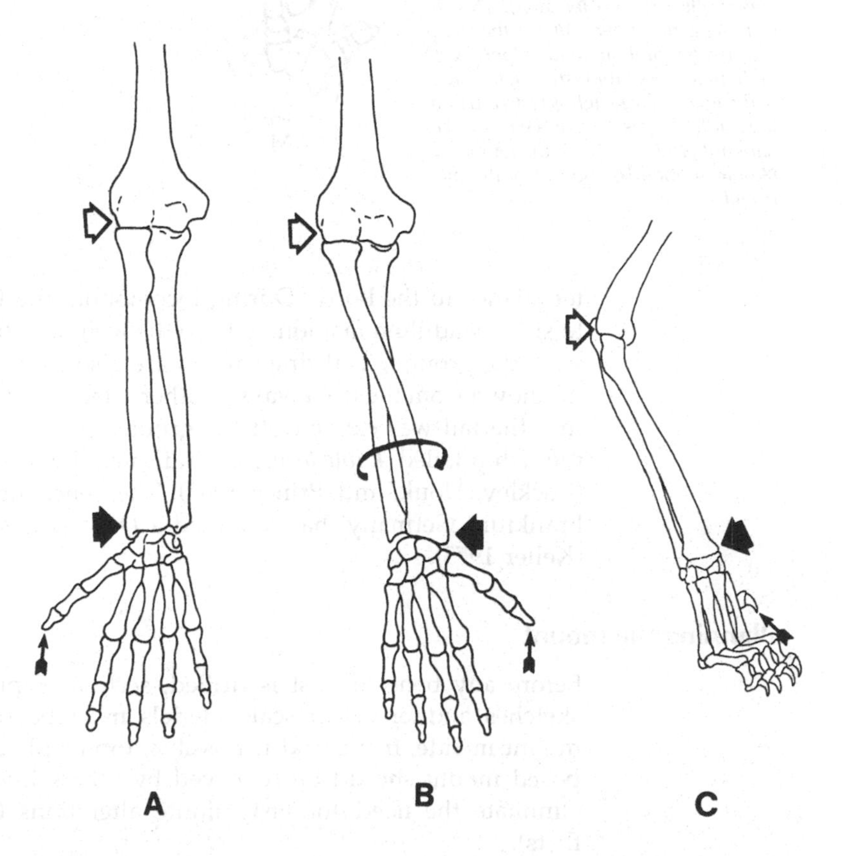

Figure 11.13. Rotation of the radius across the ulna as seen in the right arm of a human. (A) Palm faces forward (or upward), radius and ulna parallel. (B) Radius rotates across the ulna to a palm back (or down) position. The elements assume this position when the arms are used to support the body (e.g., when a person crawls on hands and knees). (C) Oblique view of the right forearm of a cat. Note that the radius (large closed arrows) articulates with the external condyle of the humerus (open arrow), and distally, the radius is on the same side as the first digit (small arrow). [(C) modified from Gilbert 1968.]

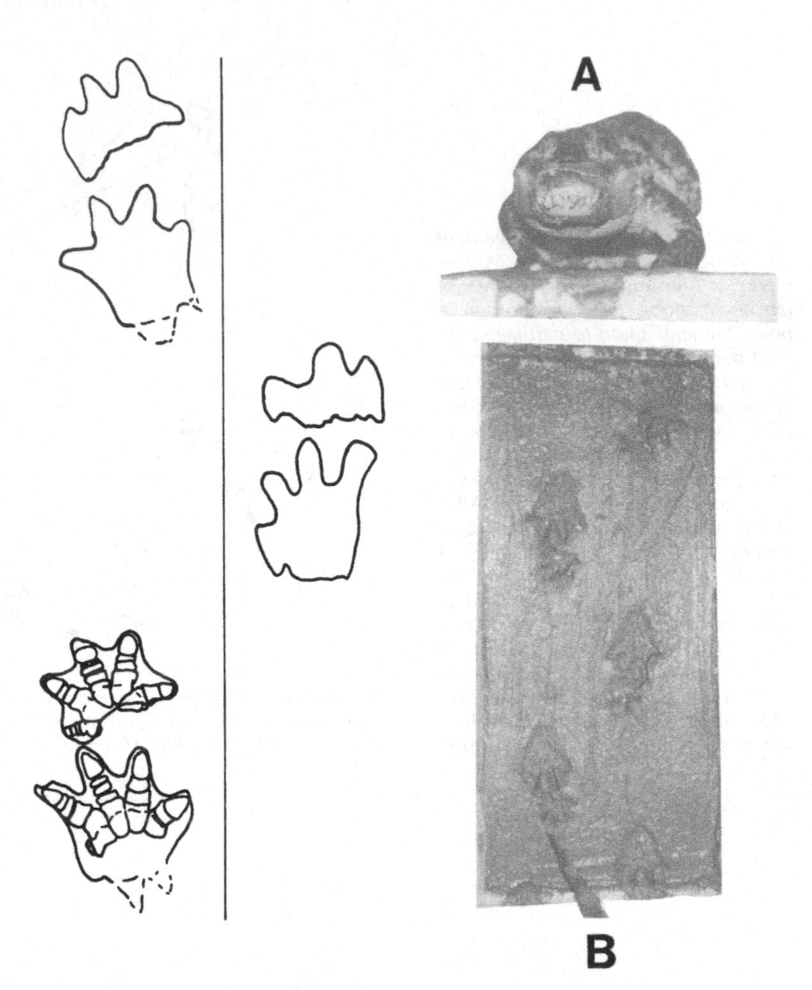

Figure 11.14 (left). Trackway of a quadrupedal dinosaur (probably Sauropelta; see Carpenter 1984) with foot bones. Note that the digits of the manus tend to face obliquely outward, not forward.

Figure 11.15 (right). Sprawling gait of a primitive reptile. (A) Caiman *sp. in anterior view showing lateral undulation of the body during locomotion. The lateral shifts of the body and tail keep the weight over the supporting legs (right fore and left rear). This results in a narrower gait (B) than predicted from limb anatomy alone. Note in (B) that except for first digit, the digits of the manus face slightly outward from the midline. This is negative rotation of Sarjeant (1975). Toes of the caiman in (A) are hidden by the rim of the mud trough.*

terbalance to the body. During locomotion, the tail probably moved from side to side, pendulum fashion, to keep the weight of the body over the foot in contact with the ground. Tail drag marks are also rare in sauropod trackways and are unknown from the trackways of other quadrupedal dinosaurs, further suggesting that the tail was carried off the ground (see Figure 11.14). This also applies to the whip-tailed *Diplodocus*, as indicated by trackways in southern Colorado (Lockley, Houk, and Prince 1986). The Senckenberg Natural History Museum, Frankfurt, Germany, has mounted a *Diplodocus* skeleton with the tail in the air (Keller 1973).

Planning the mount

Before any bone or cast is drilled, section of pipe cut, or length of rod bent, sketches and/or small scale models must be prepared showing the finished mount in side, front, and if possible, top or plan views. The design of the proposed mount should be reviewed by others involved in the exhibit project to eliminate the need for last minute alterations (and possible personality conflicts).

Minor position changes of the bones may be necessary during assembly because fossils can not always be mounted as planned. Ultimately, the muscle scars and articular surfaces on bones define the limits of movement. It is helpful to study existing mounts or photographs of mounts with a critical eye, both for ways of improving the mount and for armature design.

Permanent mounting of fossil bone often limits the accessibility of the specimen for scientific study. However, the mount can be designed to allow the bones to be removed later for study (see below).

Inventory

Before a specimen can be mounted, an inventory of the skeleton is necessary to identify any missing elements. This can be accomplished by checking off the bones on an illustration of the project skeleton (or skeleton of a close relative) or by laying out the skeleton on a table or floor for a visual check. The advantage of the latter is that it allows rough measurements to be made for ordering support steel, fabricating the base, and confirming that the allotted exhibit and ceiling space are adequate.

Missing or damaged bones

Once the missing parts have been identified, casts or actual bones of the correct size can be obtained or the parts reconstructed. This can be done by sculpting the part from wood, plaster of Paris, paper-mâché, or ceramic clay fired in an oven. An alternative is to sculpt the part in Plasticine clay, then make a squeeze mold and plaster or plastic cast(s) of it. It is also possible to modify an existing cast, especially if it is made of water-extended polyester (WEP) or epoxy. This modification can be done with a small electric hand grinders (e.g., Dremmel or Foredom flexible shaft). A proper dust mask should be worn to avoid inhaling the dust, or an exhaust hood or vacuum system should be in place to control the dust.

Reconstructing a skull can be difficult if the thin, delicate bones and openness or fenestrae, such as that of a dinosaur, are required. One successful technique is referred to as the "lost styrofoam method." Sheets of styrofoam are glued together (see Figure 11.16A) using carpenter's glue, Elmer's Glue All, or similar adhesive. An electric knife, serrated steak knife, or X-acto knife is used to carve the styrofoam into the rough shape of the skull (see Figure 11.16B). This should include the palate. Coarse sandpaper or a rasp is used to smooth the contours. The styrofoam is covered, except the cranial fenestrae, with thickened epoxy to a depth of a few millimeters (see Figure 11.16C).

Once the epoxy has hardened, blemishes are removed or repaired, fine detail (sutures, and the like) are carved, and casts of teeth are glued in place. Finally, acetone is poured into the cranial openings to dissolve the styrofoam, leaving a hollow shell. If the braincase is to be visible through the cranial fenestrae, reconstruct it separately and insert inside the skull. The lower jaws can also be sculptured in styrofoam and coated with epoxy. The styrofoam can be dissolved through the mandibular fenestrae. A skull (Figure 11.16D) made by this technique was used to complete a mount of the dinosaur *Tenontosaurus*.

For repairs of damaged bones, plaster of Paris, Durham's Rock Hard Water

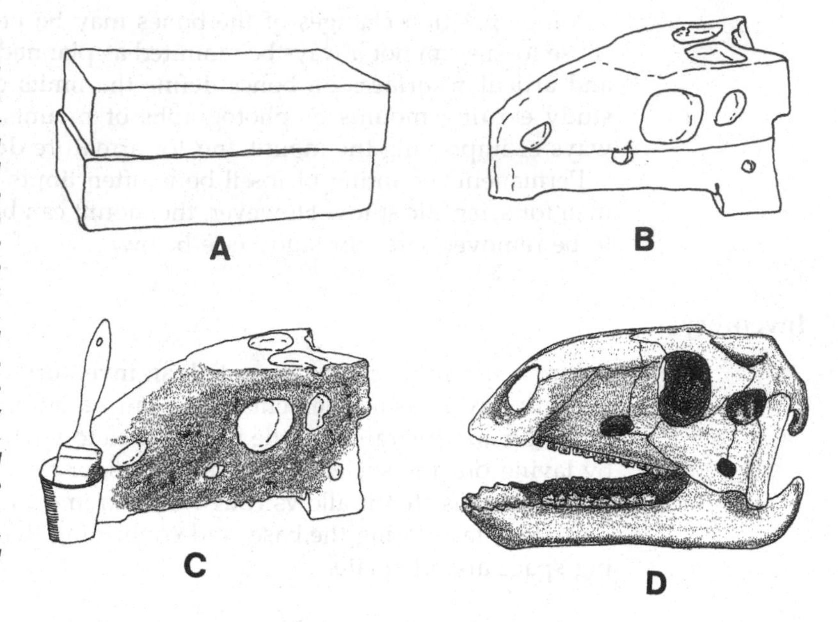

Figure 11.16. Reconstructing a skull using the lost styrofoam technique. (A) Sheets of styrofoam are glued together with a paper or wood glue. (B) A skull is roughed out, with the various cranial openings cut about 1 cm deep. (C) The skull is covered about 1–2 mm with thickened epoxy, except within the cranial openings. When hardened, the surface is sanded smooth, rough areas ground away, and casts of teeth and sutures added. Acetone is poured into all the cranial openings to dissolve the styrofoam, leaving an epoxy shell. The braincase can be sculpted separately and installed after the styrofoam is dissolved. (D) The skull is completed when the lower jaw is sculpted and the skull painted. Skull in (D) is that of Tenontosaurus, *now on display on a skeleton at the Academy of Natural Sciences, Philadelphia.*

Putty, wood putty, paper-mâché, automotive body putty, or thickened epoxy have been used successfully. Plaster of Paris can be carved when still wet. However, once the plaster has begun to set, continued shaping will result in a powdery, weak repair. If the plaster repair is dry, rough shaping can be done with any sharp blade or coarse, abrasive material, such as a knife, small saw blade, rasp or sandpaper. Carving with a knife is made easier if the plaster is dampened with water.

Painting the repair should be done in a contrasting color or shade of the original fossil so that the reconstruction is apparent, however subtle. Acrylic paints, lacquers, or wood stains are good coloring agents.

Adhesives vary widely and include everything from plaster of Paris, acetate (plastic model glues), polyester resins, epoxies, and cyanoacrylates (superglues). Over the years plaster of Paris repairs and restorations endured as long as temperatures and humidity have remained fairly constant and the specimen not subjected to vibration or shock. Acetate or model glues (e.g., Duco) become brittle with age and fail along previous repairs. Some with nitrose (e.g., Ambroid) decompose with time, forming nitric acid, which etches the bone surface and causes the pieces to separate. Many epoxies appear to have a long life, with the possible exception of fast setting types (5- or 10-minute epoxies). Cyanoacrylate is very useful in making fast repairs of small pieces, but the longevity of this adhesive is uncertain.

Tools and equipment

The specific hand tools and equipment needed for mounting a skeleton can vary and depend on the specimen and type of mount. Some or all of the following tools may be needed:

- 16-oz. or 32-oz. ball-peen hammer
- 3-lb. short-handled sledgehammer (used to bend steel)
- Nonswiveling bench vise (used to hold steel for cutting or making sharp bends)
- Vise grip pliers (for clamping pieces of steel together for welding or for bending steel bars)
- Electrician's pipe bender
- Diagonal cutters
- Bolt cutters
- Hacksaw
- Framing square
- Socket wrench set
- Combination wrench set (box-end used to bend steel bars; see Figure 11.36)
- Hex or Allen key set
- Screwdrivers
- Needle-nose pliers
- Wood saw
- Tape measure
- Yellow lumber crayons or welder's chalk
- High-speed steel drill bits
- Bit extensions of various lengths
- Respirator
- Dust masks
- Goggles or full-face shield
- Leather gloves
- Welding face shield and gloves
- Welding apron
- Shop apron
- Tap and die set
- Pipe cutter
- Pipe tristand
- Pipe wrench
- Pipe threader
- C-clamps
- Cold chisels and punches
- Utility pry bar
- Round and flat steel files
- Deep-cut high-speed hole saws of various sizes
- Acrylic paints or wood stain
- Artist brushes of various sizes
- Crescent wrenches
- Diamond tipped coring bit

More elaborate shop equipment can also include any of the following:

- AC/DC arc welder or wire feed welder
- Oxyacetylene torch (for heating and bending thick steel bars)
- Small disc sander/grinder
- Reciprocal saw
- Two-ton hydraulic jacks

- Bench grinder
- 1/4-inch and 1/2-inch electric hand drills
- Electric circular saw
- Chain hoist or ratchet pullers
- Stepladder and/or scaffolding
- Airscribe
- Medium- to high-volume air compressor

Assembling skeletons

Whether to mount fossil bone or casts is a problem without a satisfactory solution. The decision is often not for those mounting the specimen to make, but it includes the perception of what the visitor expects to see, availability and costs of obtaining and preparing a specimen versus cost of obtaining a cast, the exhibit schedule, and the philosophy of the curator concerning the mounting of real bone versus cast. When real bone is mounted, it is preferable that holotype specimens (some paleontologists advocate all bone) should not be damaged (vertebrae drilled and so on). Furthermore, the bones should be available for future study and this may require that they be removable. Techniques for such mounts using external armature are discussed below.

If original specimens are to be mounted, either masonry or diamond coring drill bits with a water swivel and a drill press are used. Large diameter holes (1/2" or larger) in hard bone may be drilled with a diamond coring bit. Softer bone can often be cored with a deep-cut high-speed hole saw or a carbide tipped twist bit. Vibration can be a severe problem when drilling hard, brittle bone. As a result, cracking and shattering may occur. This vibration is probably due in part to differential permineralization of the bone. At present no satisfactory solution has been found to deal with this beyond use of a water-cooled diamond core bit and gentle pressure. Cracking of fossil bone can also occur because of excessive heat buildup. A hollow carbide or diamond drill bit with a water swivel can be used to cool the bone, or the bone can be drilled submerged in a pan of water. This may or may not, however, solve the problem of cracking and shattering due to vibration.

Armature

The choice of using an external or internal armature is sometimes determined by the diameter of the bone. For example, an armature necessary to support a skeleton may be too large in diameter for the bones to be safely drilled. In such a situation the armature must be external or partially recessed. The internal support for a freestanding skeletal mount detracts as little as possible from the exhibit. Casts lend themselves well to internal armature. Seamless tubing, black water pipe, and ungalvanized steel stock are recommended if any welding is required because of toxic gases released by welding galvanized steel and because of the difficulties in welding galvanized steel. All welding should be done with proper ventilation because of the toxic fumes generated by the flux.

Large, yellow plumber's crayons and white alabaster welder's markers are excellent for marking bone for the placement of the external armature. When

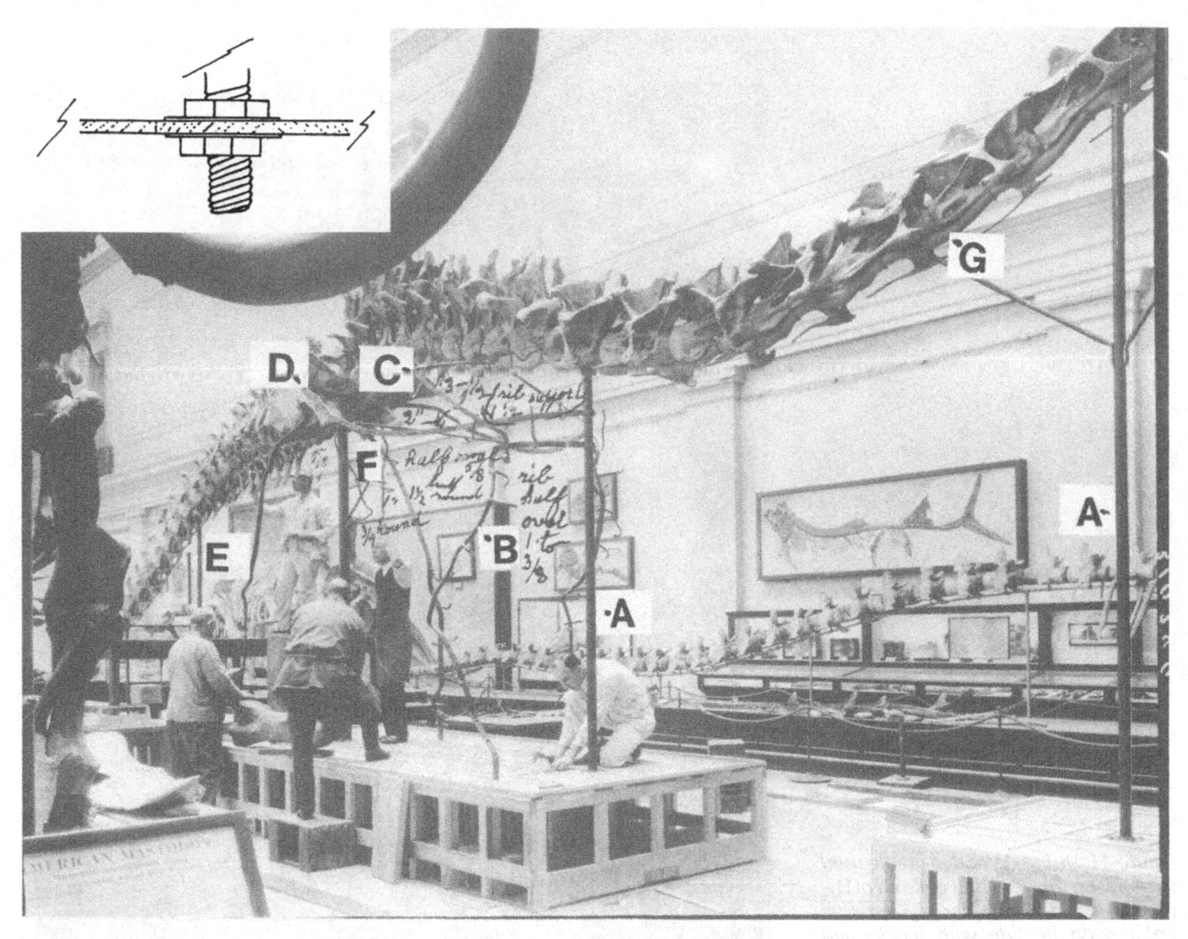

Figure 11.17. Mounting of a Diplodocus longus *at the U.S. National Museum of Natural History. (A) Vertical uprights. (B) Right forelimb armature with clamps to secure elements. (C) Right scapula armature with clamps; note that the armature is attached to the crosspiece connecting the humeral armature to the vertical upright, and to a crosspiece from the vertebral armature. (D) Right rib cage armature and straps for individual ribs. (E) Right rear limb armature with clamps for elements not yet installed. (F) Pubis armature being installed (ischium armature not yet installed). (G) Vertebral armature. Notation on photograph used to aid in the mounting of the* Diplodocus *at the Denver Museum of Natural History. Insert: Adjustable vertical support using nuts and washers. (Courtesy of the Archives Collection, Department of Earth Sciences, Denver Museum of Natural History.)*

selecting steel, we must consider flexibility. To eliminate all flex would require the use of much heavier steel than is practical. The diameter or thickness of the steel to be used will depend on whether the skeleton is to stand by itself, to be supported by cables from the ceiling, or to be attached to rigid supports from a wall or base (see Figure 11.17).

For external armature, half-round steel is ideal. Unfortunately, such steel can only be obtained by special order from a few steel distributors. Half-round can be made by cutting round-bar in half length wise or by grinding away one side. However, the time and expense may not make this a very practical option. An effective half-round can be made by welding bundles of different width flat steel together. However, if the armature must be bent to follow the contours of the

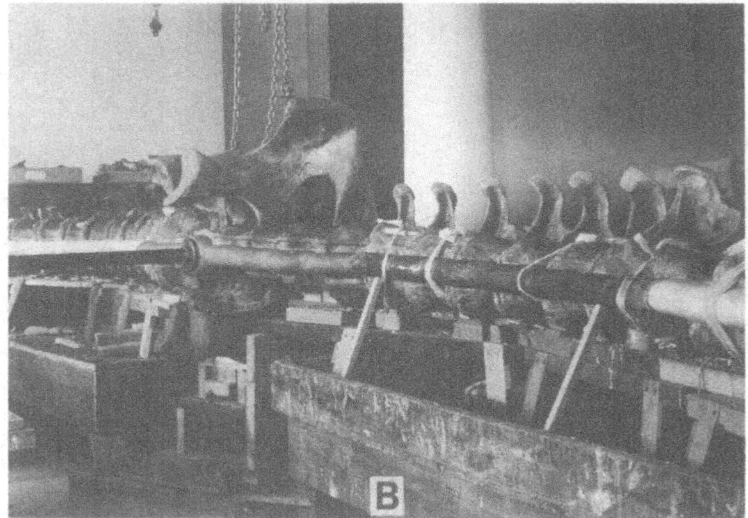

Figure 11.18. (A) Wooden cradle used to mock up the vertebral column of Diplodocus. *(B)* Diplodocus *vertebral column on its side with blocks and wedges of wood supports. (C) Pelvis and hind legs of* Allosaurus *resting on sandbags prior to making the hind leg armature. (D) Front view of the* Allosaurus *pelvis and hind legs on sandbags.*

bone, each piece must be bent before welding them together. Square steel bar can also be used for external support, although it is much more noticeable in the final mount unless the edges are "feathered" to the bone, with plaster or epoxy creating a pseudo half-round.

Mounting

Hermann (1909) recommended that the entire skeleton be mocked up to the pose desired before any armature is prepared. This allows the bones to be realigned to simulate a lifelike pose without violating the maximum movement at the joints. Furthermore, it identifies possible problem areas in mounting and allows exact measurements to be made for the armature segments. Too often this step

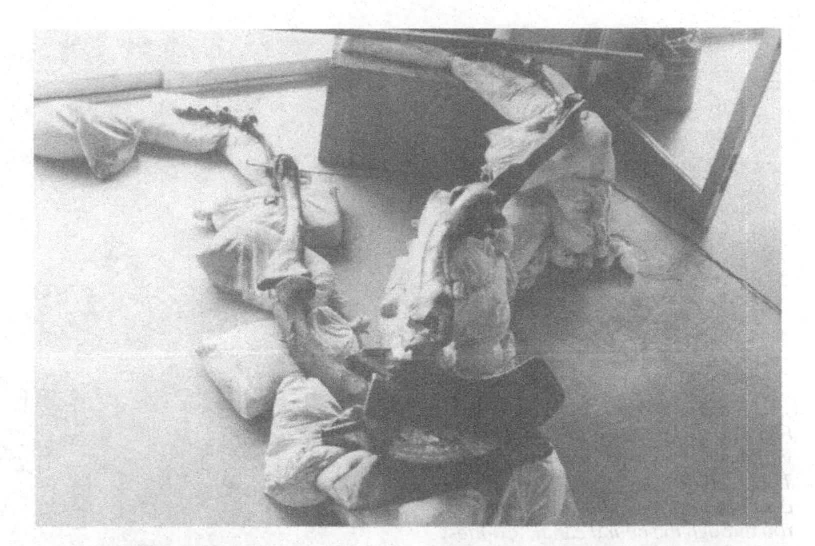

C

D

is neglected, resulting in last minute adjustments to compensate for unforeseen difficulties. The result can be an unnatural or inaccurate pose.

A wooden cradle can be used to support the vertebral column in its desired resting position (see Figure 11.18A). Such a cradle is helpful for determining what processes or neural spines need realignment. Boards, tape, wire, string, and clay can be used to support the rest of the skeleton in the mock-up. If the vertebrae are not too distorted, another method is to lay the vertebrae on their sides in a sandbox or, if the specimen is big, on a ridge of sand or sandbags piled on a table or floor (see Figure 11.18B–11.18D). Next, the rest of the skeletal elements are added to one side and adjusted to the desired pose. If possible, the skeleton should be left mocked up so the armature can be bent correctly at the bone joints.

How the vertebrae and other bones are attached to the armature will depend on whether or not the individual bones are to be removable. Vertebrae can be

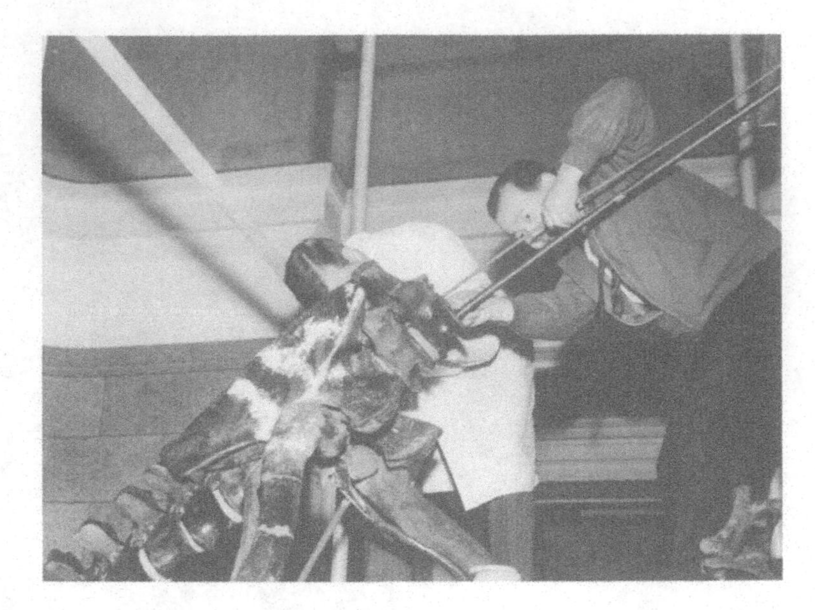

Figure 11.19. Attaching individual vertebrae on the vertebral armature. This method uses a pipe through the centrum and a smaller diameter steel rod through the neural canal. (Courtesy of the National Science Museum of Japan.)

drilled through the centrum, strung like beads onto the armature, and epoxied in place; Durham's Rock Hard Water Putty also works well to hold the bones (see Figure 11.19). Alternatively, the armature can be strung through the neural canal and the vertebrae attached with plaster of Paris, thick epoxy, or a wire (see Figure 11.3A and Figure 11.20B). If the vertebrae are to be removable, they must sit atop the armature and be held with a clamp or peg (see Figures 11.3B and 11.3D and Figure 11.21). Channel rail is ideal because the individual clamps or pegs can sit within the channel. Alternatively, a flat bar can be drilled and/or tapped. Pegs can be made from hex-headed bolts (Figure 11.21A and B) or threaded 1/4 steel rods (Figure 11.21C). Clamps can be made from thin steel straps (Figure 11.4B) or split steel rods (Figure 11.21D). The edges of the centrum can be feathered to the support armature with plaster of Paris or thick epoxy. This makes the armature less noticeable and gives the vertebrae a broader surface to rest upon. A separator should be used to prevent the vertebrae from bonding to the plaster or epoxy.

A poured steel vertebral armature also uses pins to keep the vertebrae in place. Such armatures are expensive but were once used extensively to mount sauropods (see Figure 11.22). These armatures were custom made using a plaster of Paris template of the underside of an articulated segment of vertebrae. Hermann (1909) discusses other methods for making custom designed vertebrae armatures.

If possible, the skeleton may be assembled in segments that are then welded or pinned together. If pinned, a slightly smaller plug of rod or pipe is inserted into the pipe armature. The armature and insert are then drilled and a bolt inserted through them (see Figure 11.23B and Figure 11.24). If the skeleton is to be disassembled, as part of a traveling display, for instance, the hole drilled through the armature may be threaded or a nut tackwelded in place. To avoid getting weld spatter on the threads, thread a bolt into the nut. Universal joints for socket sets can also be used to connect segments of bone, especially in the

Figure 11.20. Armature structure used in the mounting of a titanothere shown on the left in (A). (B) Vertical uprights support the skull, shoulder, and hip region. Close-up shows detail of the vertebral armature passing through a hole in a thin steel bar bolted to the upright. The rib cage armature is also bolted to the upright. In the pelvic region, long bolts attach the pelvic armature to the vertebral armature. (C) Forelimb and scapula armatures. Note that the scapula is held to the proximal end of the humerus by a pin. (D) Posterior view of the forelimb armature to show how it wraps around the olecranon process (detailed in top view to the right). A thin strap (arrow) attaches the forelimb to the body to keep the forelimb in place; it does not bear any weight. (E) Rear limb armature bends around the calcaneum. (F) Armature for pelvic bones in anterior view (top) and ventral view (bottom). This armature supports the pelvis and hangs by bolts from the vertebral armature. (Photograph courtesy of the Denver Museum of Natural History.)

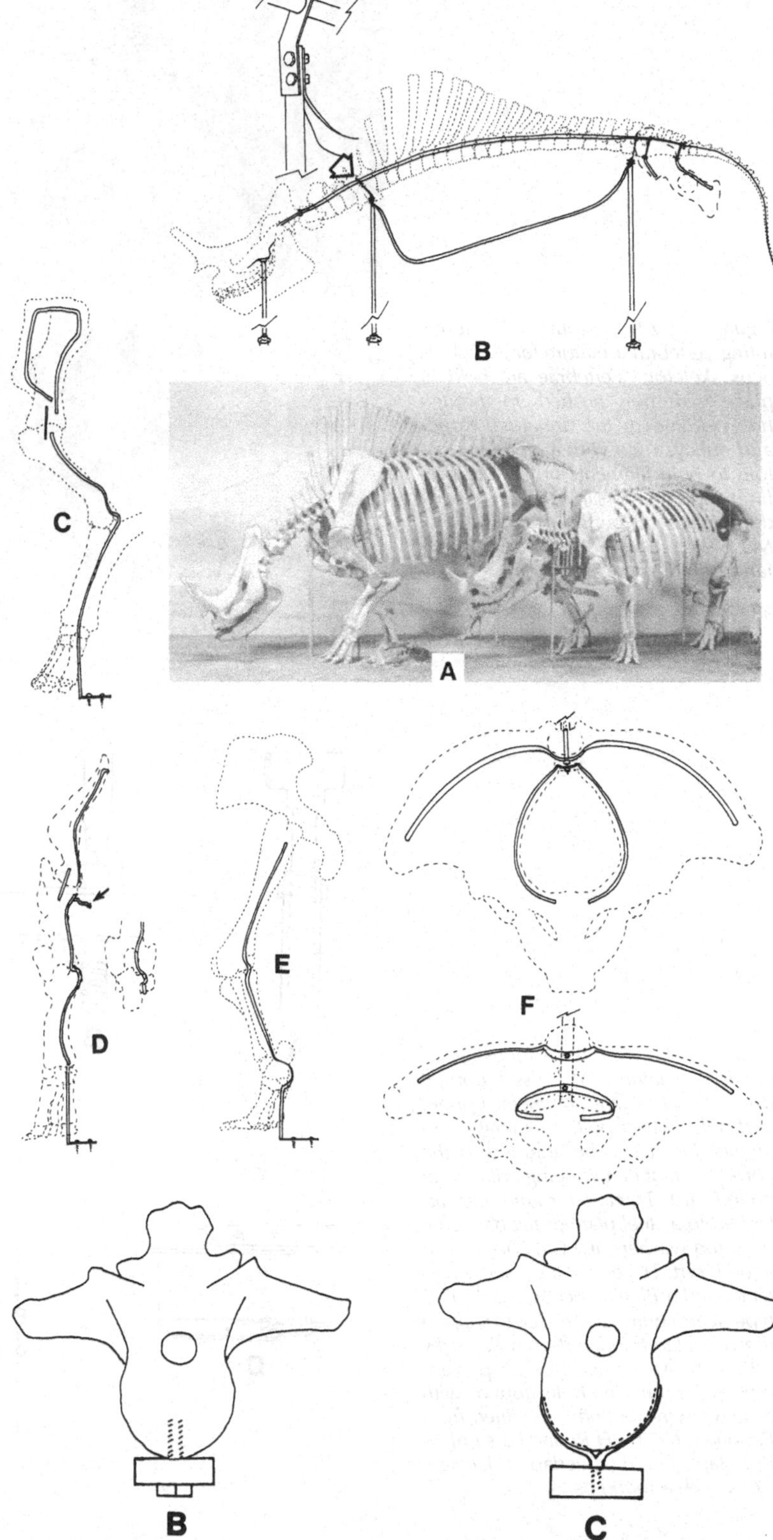

Figure 11.21. Three examples of a removable vertebra. (A) Channel steel using nuts, washers, and a bolt for a pin. (B) Steel bar with a tapped hole and hex bolt for a pin. (C) Clamp made from a split rod.

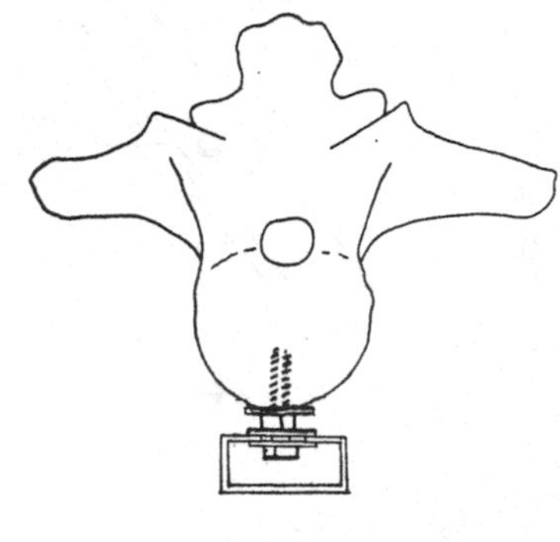

C

Figure 11.22. Custom-made form-fitting vertebral armature for a Diplodocus *skeleton. Vertebrae are held in place with pins pushed up through holes drilled in the armature. Ropes and pulleys used with a wooden scaffold to hold elements of the left forelimb and shoulders in place for making the armature. (Courtesy of the National Museum of Natural History, Smithsonian Institution.)*

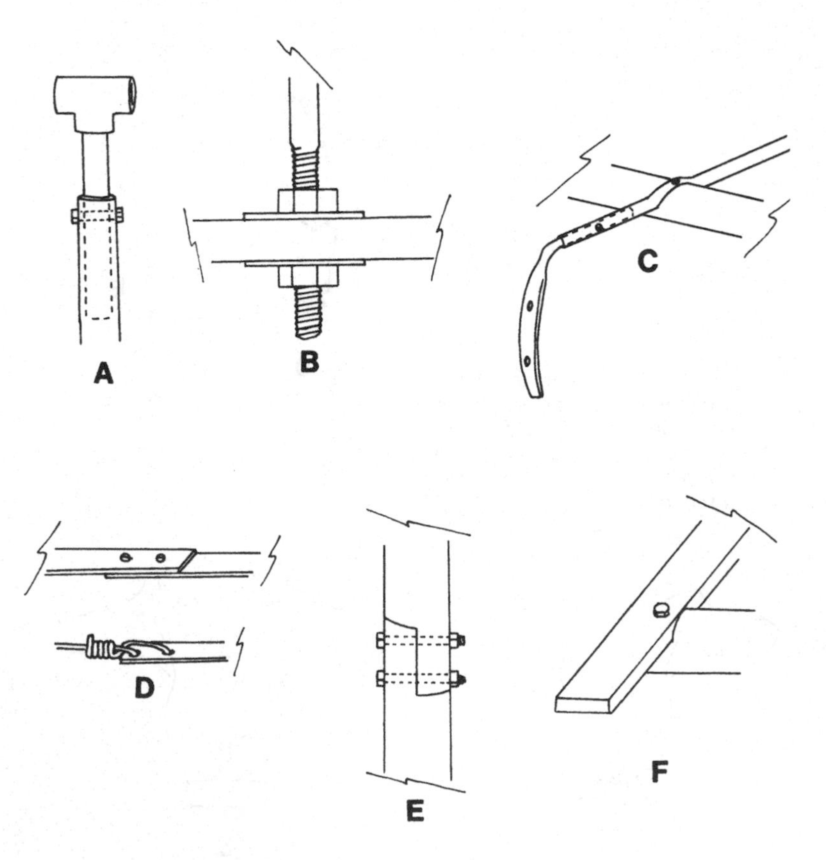

Figure 11.23. Various methods for attaching armature. (A) Cross-T upright inserted into a large diameter upright and held with a bolt. The smaller diameter insert can be adjusted to the correct height before being drilled and bolted. (B) Threaded round bar attached to a steel plate or flat bar using nuts and washers. (C) Crosspiece with a rod insert. The end of the insert is flattened and drilled to receive a bolt. This type of armature can be used to support the scapulae (modified from Panofsky 1973). (D) Attaching a steel strap to another with nuts and bolts (top) or with a heavy wire (bottom) (modified from Panofsky 1973). (E) Round bars united by a lap joint. (F) Overlap of flat steel stock with a round bar.

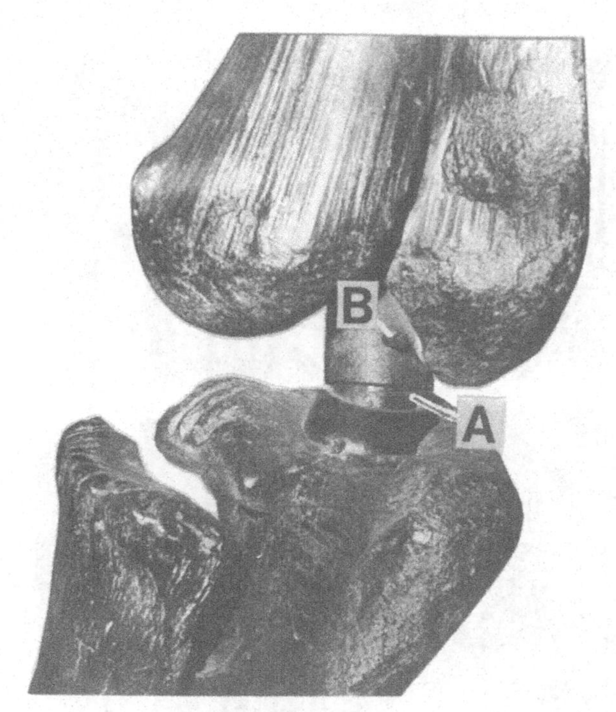

Figure 11.24 (left). Plug (A) insert between two segments of the hind leg. Plug is welded into the lower part and is pinned (B) in the upper part.

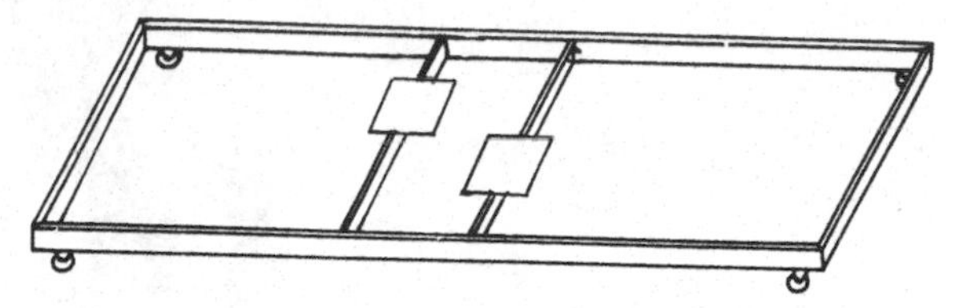

Figure 11.25 (right). Tubular steel base with steel casters and steel plates for the feet.

limbs. Such joints can be purchased individually from hardware stores. Each end of the joint is inserted into the pipe armature of the limb segment and welded or bolted in place. The use of such joints allows the limb bones to be moved after assembly to their final position before welding.

The conventional sequence of mounting is: (1) pelvis or hind legs, (2) vertebral column, (3) ribs, (4) scapulae, (5) forelimbs, (6) skull, (7) chevrons (if present). There are at least two preferred methods of attaching the armature to the base. One is to weld or pin the armature to large steel base plates affixed to the top of a wooden or tubular steel frame (see Figure 11.17 and Figure 11.25). The other method involves threading the end of the armature and using nuts and washers to attach the armature to a base (see Figure 11.17, inset). The advantage of the latter method is that it allows minor adjustments to the height of the armature. If wheels are used to move the skeleton into its display position, they should be made of steel because rubber wheels deform with time. Deformed wheels can vibrate the skeleton when it is being moved, or if deformation is very severe, can prevent any movement.

Begin mounting with either the pelvis or hind legs. Once the armature for them is completed, the position of the rest of the skeleton will fall into place. If the pelvis is mounted first, it is secured at the appropriate height by a floor stand, suspended by cables from scaffolding or with an overhead hoist. The hind legs can then be positioned and the height of the pelvis adjusted as necessary to

Figure 11.26. Mounting the hind legs first allows the feet to be positioned correctly and the pelvis to be at the correct height above the base.

accommodate limb position. If the hind legs are mounted first, the legs can be properly set for the posture and the sacrum will be at the correct height (see Figure 11.26). This is an important consideration if an overhead hoist is not available to support the sacrum at the correct height.

Before mounting the hind legs, it must decided if all the major parts of the limb can be supported by a single piece of steel (see Figure 11.20C–F), or if the leg must be mounted in segments. The following account assumes a large, original fossil skeleton that must be mounted in segments with all but the vertebral armature mounted externally.

To assemble the hind legs, start with the feet because it is difficult to provide enough room for the foot once the lower leg is assembled. The metatarsal and phalanges may be drilled, threaded on thick wire bent at each joint to the correct angle, and epoxied in place. Later, these preassembled digits will be epoxied into the distal ends of the ankle elements. Or the foot bones can be pinned to a thin steel bar extending along the ventral surface of the bones.

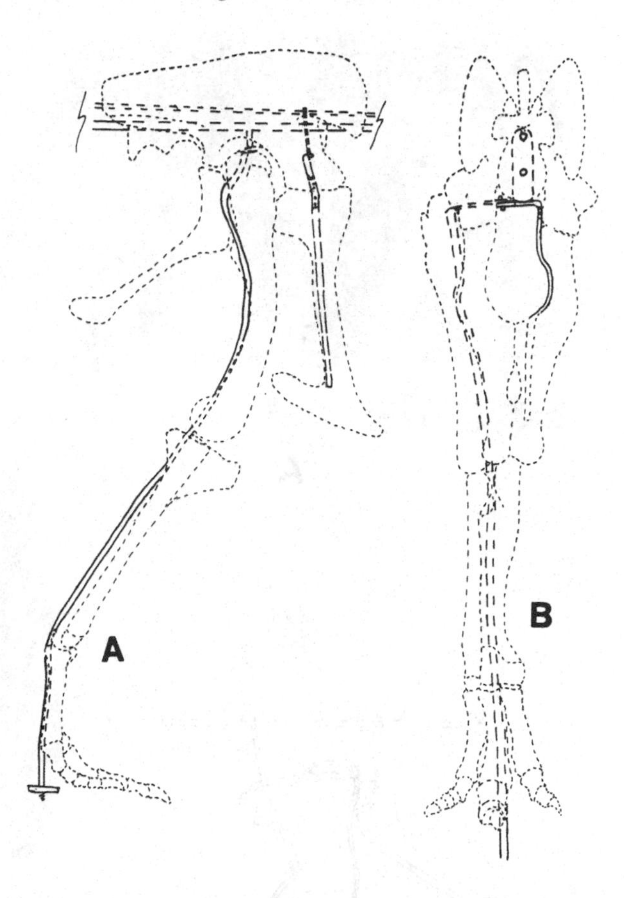

Figure 11.27. Arrangement of the armature for the right hind leg and pelvis of Allosaurus fragilis, *mounted at the National Museum of Natural History (Smithsonian Institution); lateral (A) and anterior (B) views. Sacral armature uses two rods, one through the centrum and the other through the neural canal; both attach to the vertebral armature. The femoral armature attaches to a crosspiece welded to the end of a vertical steel plate extending down through the sacrum. The pubic armature is welded to the end of another vertical steel plate.*

With the foot armature completed, it is temporarily mocked up with the metatarsals on a base. The armature for the lower legs is then cut, allowing for any bend at the ankle, which will require several additional centimeters of armature length. The lower leg support is bent to conform to the posterior surface of the tibia. In mammals, the armature must bend around the calcaneum (see Figure 11.20E). The lower leg armature must be long enough to bend at the knee before attaching to the femoral armature. The patella, or kneecap, of mammals can be glued or pinned in the groove at the distal end of the femur. The femoral armature continues from the lower leg along the posterior surface, then wraps medially below the femur head (see Figure 11.27). The steel must extend beyond the femur head to attach to the sacral armature (see Figure 11.27B and Figure 11.28F–H).

The pelvic armature for mammals differs from that of dinosaurs because the pelvis either consists of paired halves (innominate bones) or a single pelvic complex with the two halves fused together. In addition, the acetabulum is closed in mammals, so connecting the pelvic armature to the femoral armature can be difficult. The armature must extend around the inside of the pelvic canal and attach to the vertebral armature (see Figure 11.20B and F). Attachment to the hind leg armature is crucial only if upright support posts are not used beneath the skeleton. Otherwise the leg armature can end below the femur head (see Figure 11.20E). The pelvic armature hangs from the vertebral armature by long bolts (see Figure 11.20F).

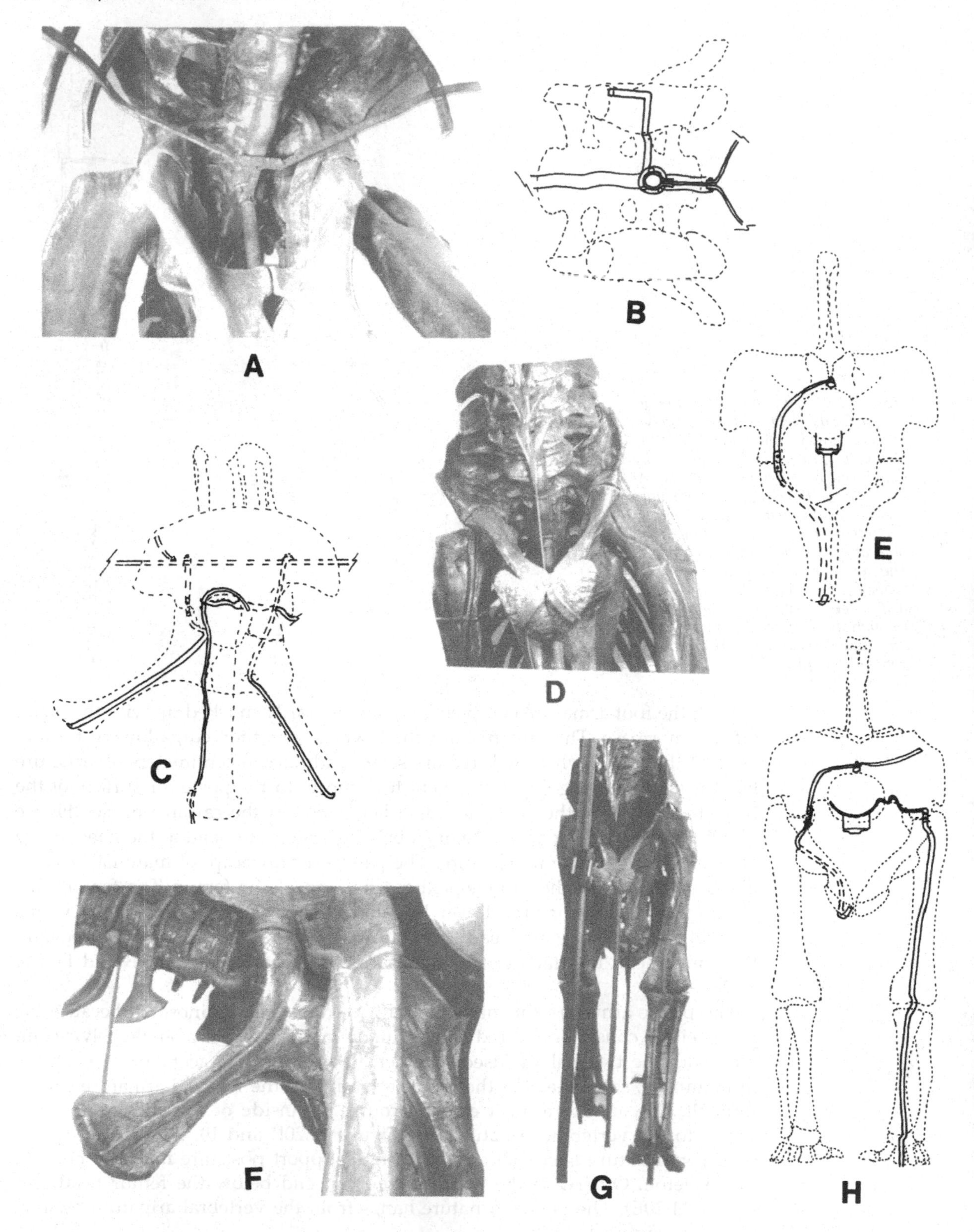
A
B
C
D
E
F
G
H

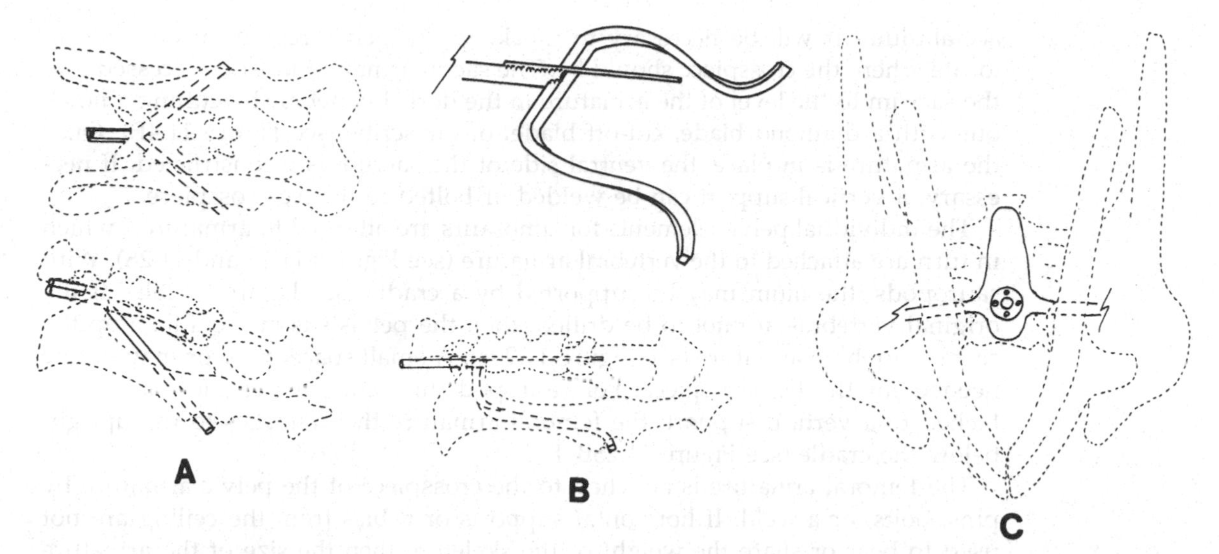

Figure 11.29. Various types of pelvic armature. (A) Y-shaped armature in top view (top) and oblique top view (bottom). The stem, which attaches to the vertebral armature, is two rods or small-diameter pipes welded one atop the other. The ends of the Y exit through a hole drilled in the acetabulum. (B) A lyre-shaped armature that conforms closely to the medial side of the ilia. The stem is a single pipe that attaches by means of a plug to the vertebral armature. The sacrum is supported by a steel rod. The side of the lyre extends to the acetabulum where it exits through a hole. (C) A custom-made sacral support for the pelvis of a Stegosaurus *in ventral view showing the floor flange used to attach the vertical upright. The vertebral column attaches to armature extending through the neural canal. Lateral extensions from the sacral support serve for the attachment of the hind leg armature. Although this support was cast in iron, a similar support could be made with reinforced epoxy resins. [(A) and (B) modified from Panofsky 1973.]*

Another method of mounting a mammalian pelvis uses a Y-shaped or lyre-shaped armature that connects the sacrum, pelvic bones, and hind legs (see Figure 11.29A, B). Such an armature extends through a hole drilled in the acetabulum and attaches to the femoral armature.

In dinosaurs and other lower vertebrates, if the vertebrae are drilled, a metal spacer should be added to the sacral armature so that the crosspiece for the legs is at the level of the acetabulum (see Figure 11.3E and Figure 11.27B). Because the head of the femur fills most of the acetabulum, the crosspiece must be off-centered at a point where the femoral armature follows the femur head into the

Figure 11.28. Armature of the hind legs and pelvis of the Diplodocus *mounted at the Denver Museum of Natural History. (A) Anteroventral view showing upright support attaching to the underside of the vertebral armature beneath the sacrum. (B) Ventral drawing of the same, but also showing the cradle (arrow) to support the ilium at the acetabulum. (C) Lateral view showing the armature for the femur, pubis, and ischium. The pubic and ischial armatures extend up between the sacral ribs to attach to the vertebral armature. (D) Posteroventral view of the ischial armature. (E) Detail of the pubic armature. It passes through the pubic foramen to support the pubis ventrally. (F) Lateroventral view of the ischial armature and the femoral armature where it bends into the acetabulum to connect to the ilium cradle (see C). (G) Posterior view showing the hind leg armature. (H) Armature of the hind leg, ischium, and posterior ilium.*

acetabulum. It will be necessary to mock up the pelvic region on one side to locate where the crosspiece should be. The sacral armature must be recessed into the sacrum to the level of the armature in the dorsal vertebrae by cutting a notch out with a diamond blade, cut-off blade, or air scribe (see Figure 11.4E). Once the armature is in place, the ventral side of the sacrum is reconstructed. If necessary, a vertical support can be welded or bolted to the leg crosspiece.

The individual pelvic elements for dinosaurs are attached to armatures, which in turn are attached to the vertebral armature (see Figures 11.27 and 11.28). With sauropods, the ilium may be supported by a cradle (see Figure 11.28B). If the original vertebrae are not to be drilled, then the pelvis sits in a cradle atop the sacral vertebrae armature (see Figure 11.29c). A small spacer may or may not be needed for the leg crosspiece. For sauropod dinosaurs, the cradle may be attached to a vertical support; the femoral armature then attaches to the upright below the cradle (see Figure 11.28B–H).

The femoral armature is attached to the crosspiece of the pelvic armature by pins, bolts, or a weld. If horizontal supports or cables from the ceiling are not used to bear or share the weight of the skeleton, then the size of the armature used in the hind legs must be large enough to support the entire weight of the rear half of the quadrupedal skeleton or the entire weight of the bipedal skeleton.

Vertebrae may be mounted in segments: cervical (neck), thorax (body), sacrum (pelvis), and caudal (tail). With bipedal dinosaur skeletons, mounting of the vertebrae should proceed with short segments of the tail and back to keep the specimen balanced. A dense polyfoam may be used to simulate cartilage between vertebrae (see Figure 11.30). Madsen (1973) suggests that the foam could be trimmed and molded with a hot soldering iron. However, the production of toxic gases by this technique requires use of an exhaust hood or that a respirator be worn. Alternatively, the foam can be cut to the appropriate shape with scissors and trimmed with a wire brush mounted to a drill, flexible shaft, or die grinder. Eye protection and a dust mask should be worn because of the fine particles produced. If dense polyfoam is used, a band saw with a fine-toothed blade can be used to cut the disks.

Chevrons can be mounted after the caudal vertebrae are mounted. They may be glued into place between adjacent vertebrae, pinned to the underside of one of the vertebrae, or wired into place (see Figure 11.31)

The ribs are mounted after the dorsal vertebrae are in place because they will determine the correct position of the scapulae and hence the articulation of the forelimbs. Ribs traditionally have been mounted to a steel strap or bar that extends along the midlength of the rib cage (see Figure 11.17C and Figure 11.20 A, B). Recent attempts have been made to refine this technique by using steel supports or a polyester reinforcing on the underside of the rib. If a rib cage support is used (perhaps necessary with sauropods or other large skeletons with heavy ribs), it is bowed and is attached to the shoulder/foreleg crosspiece and pelvic armature (see Figures 11.17 and 11.20B). Although the ribs are attached to their diapophyses, the weight is actually borne by the rib strap. For this reason, the strap has to be of the correct thickness without being too obvious. For sauropods, this might require a flat bar 125 mm (1/2") thick and 250 mm (1") wide. The ribs are then drilled and bolted to the bar. If, however, the ribs are too brittle or delicate, each may be mounted to a thin metal strap. This strap extends down the inner side of the rib and is pinned to the rib cage support.

Figure 11.30. Ventral view of the cervicals and skull of Allosaurus showing intervertebral discs (e.g., at arrow) and cradle for supporting the skull.

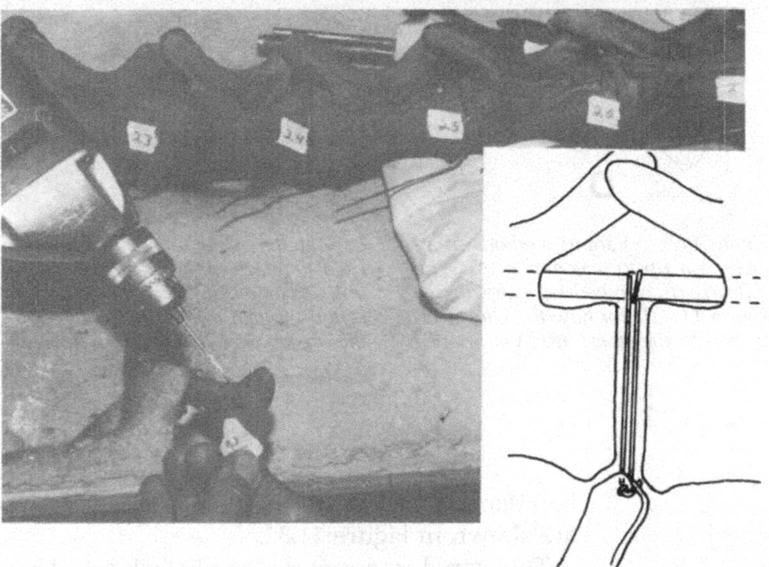

Figure 11.31. One method of attaching chevrons involves drilling the two articular faces and gluing this to wires. Alternatively, the drill holes can extend through the proximal end of the chevron. A wire wrapped around the armature between vertebrae is looped through the holes in the chevron and bent up against the bone (insert).

If no rib strap is used, the ribs must be supported entirely by steel rods or bars bent to conform to the posterior and medial sides of the ribs. Polyester resin can also be used to reinforce the medial side of the rib. The weight of the ribs should not be borne by the diapophyses because they may break. The ribs should be wired and/or epoxied to their own supports, and the armature bolted or spot welded to the vertebral armature. Only if lightweight cast ribs are used can they

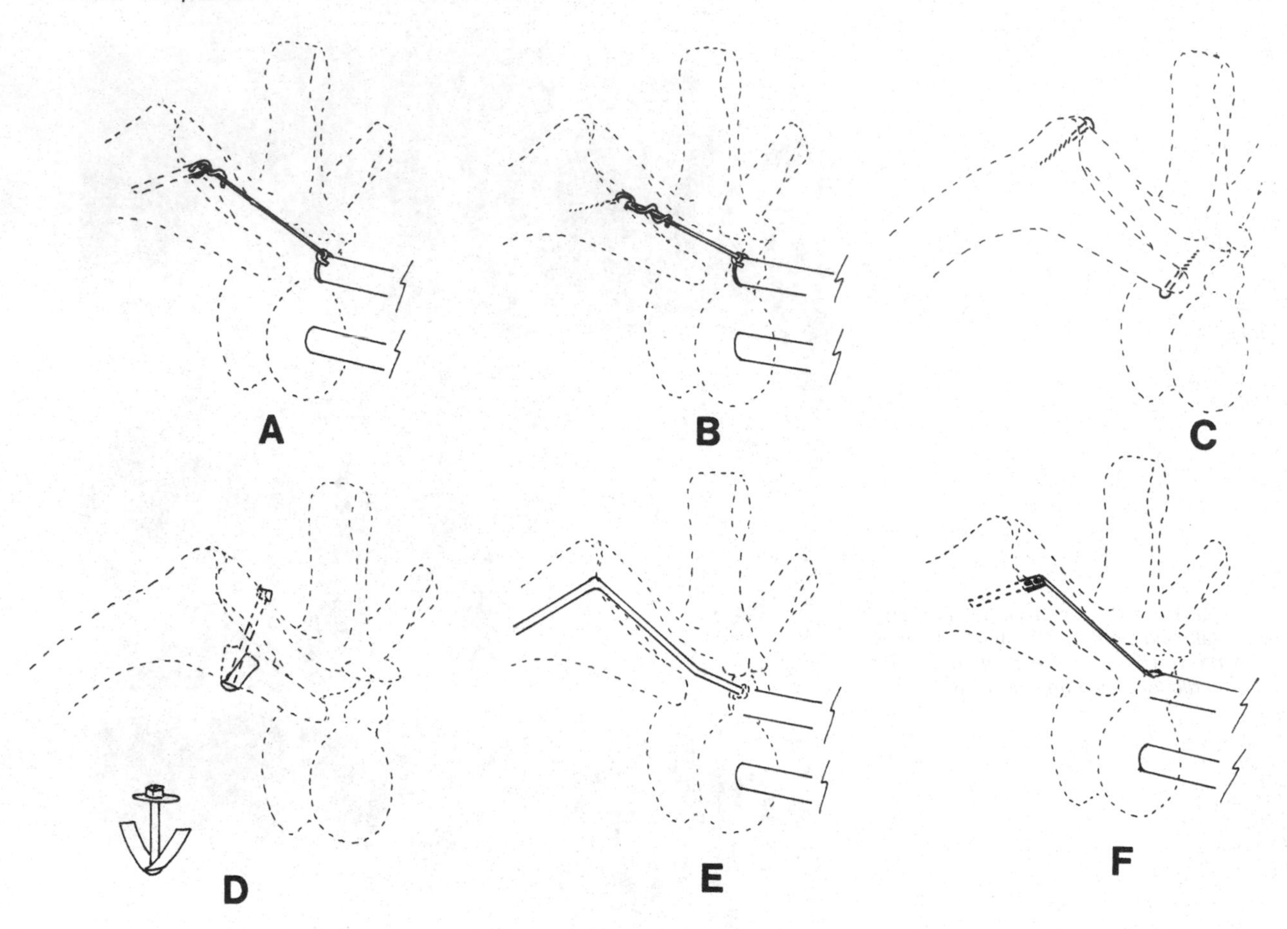

Figure 11.32. Various methods of attaching a rib to a vertebra. (A) Wire loop to a strap embedded in the rib (see Figure 11.23D for how the wire is attached to the strap). (B) Wire looped through the eye of a hook. (C) Screws used to attach plastic cast ribs. (D) A bolt and cradle, in which the bolt is inserted through a hole drilled through the diapophysis. (E) Oblique rear view showing a steel rod that extends down the posterior surface of the rib. The rib is epoxied to the rod. (F) Short rod or strap attached to a rib strap (see Figure 11.23D for this method of attachment).

be attached directly to the diapophyses. Various options for attaching the ribs are shown in Figure 11.32.

The scapula is supported medially by steel bars (see Figure 11.17C and 11.20C). In dinosaurs with a horizontal, elongate scapula, the armature is attached by two points to the vertebral armature: anteriorly between the last cervical and first dorsal and posteriorly between the third and fourth dorsals. The supports follow the posterior surface of the ribs to the vertebral armature. In mammals the scapula is nearly vertical, and the armature can be attached by a single pin. If the bone is strong enough (e.g., Pleistocene bones from La Brea tar pits), then no armature is necessary and the bone may be pinned to the humerus (see Figure 11.33).

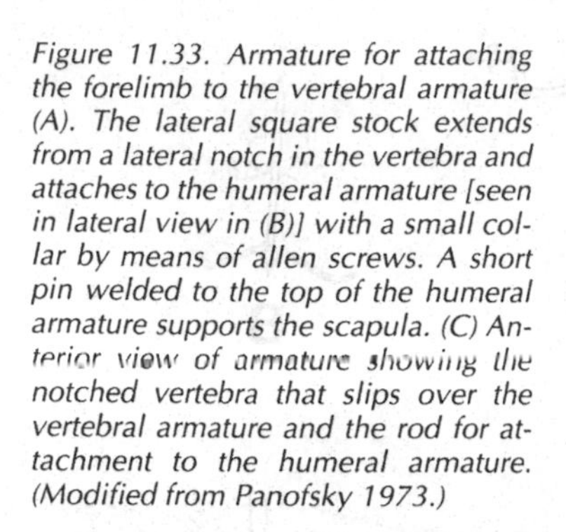

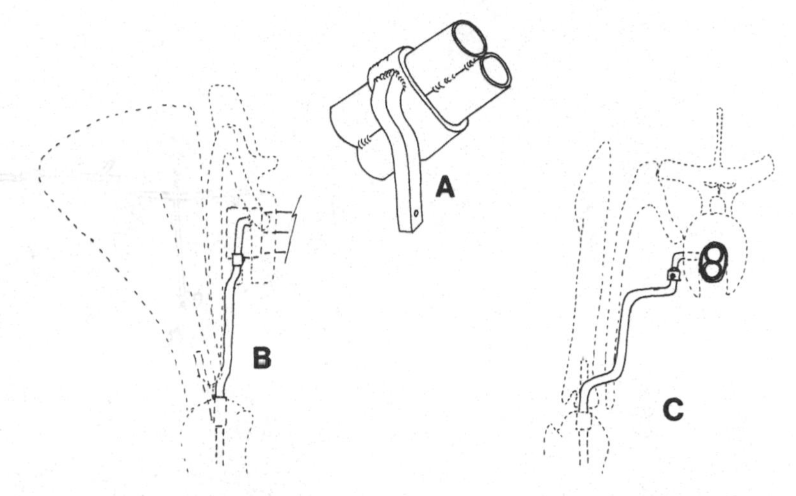

Figure 11.33. Armature for attaching the forelimb to the vertebral armature (A). The lateral square stock extends from a lateral notch in the vertebra and attaches to the humeral armature [seen in lateral view in (B)] with a small collar by means of allen screws. A short pin welded to the top of the humeral armature supports the scapula. (C) Anterior view of armature showing the notched vertebra that slips over the vertebral armature and the rod for attachment to the humeral armature. (Modified from Panofsky 1973.)

The scapulae of all animals are attached to the armature by round-headed (stove) bolts or clamps that grasp the edge of the scapula, especially the ventral edge. With a heavy scapula, a short peg from the armature may be necessary at the glenoid on which the scapula rests. For dinosaurs, the armature supporting the coracoid can be attached to the scapular armature near the glenoid. Sternal plates, if preserved, can then be attached to the coracoid armature. Articulated dinosaur skeletons show that coracoids and sternal plates are not in contact, but are separated by a gap (see Figure 11.9).

The forearm armature can be attached to the scapular armature, or to the upright, if the skeleton is heavy (see Figure 11.17B). The humeral armature should extend down the posterior side of the humerus to carry the weight of the bone. Near the olecranon groove, the upper arm armature joins the forearm armature.

If the forearm bones are separate, the ulna should be mounted first. The ulnar armature should wrap around the medial side of the olecranon where it joins the humeral armature. It should then extend down the posterior or posteromedial side, whichever is least conspicuous and least difficult to mount. In quadrupeds, the ulnar armature must be long enough to accommodate the hand before being welded to a steel base plate or attached to the base with nuts and washers. The length of steel needed for the ulnar armature is greater if the elbow is flexed because the humerus is farther from the olecranon. The radius may be pinned to the ulna proximally and distally or held by steel bands to the ulnar armature. If the ulna and radius are heavy, they may be supported by short pegs or brackets welded to the armature near the distal end of the bones.

As with the hind foot, the phalangeal and carpal elements of the manus may be threaded and epoxied onto wire and epoxied into the distal ends of the radius and ulna. Another method is to make a small armature for the underside of the phalanges.

The skull can be mounted on a cradle (see Figures 11.30 and 11.34), or if well

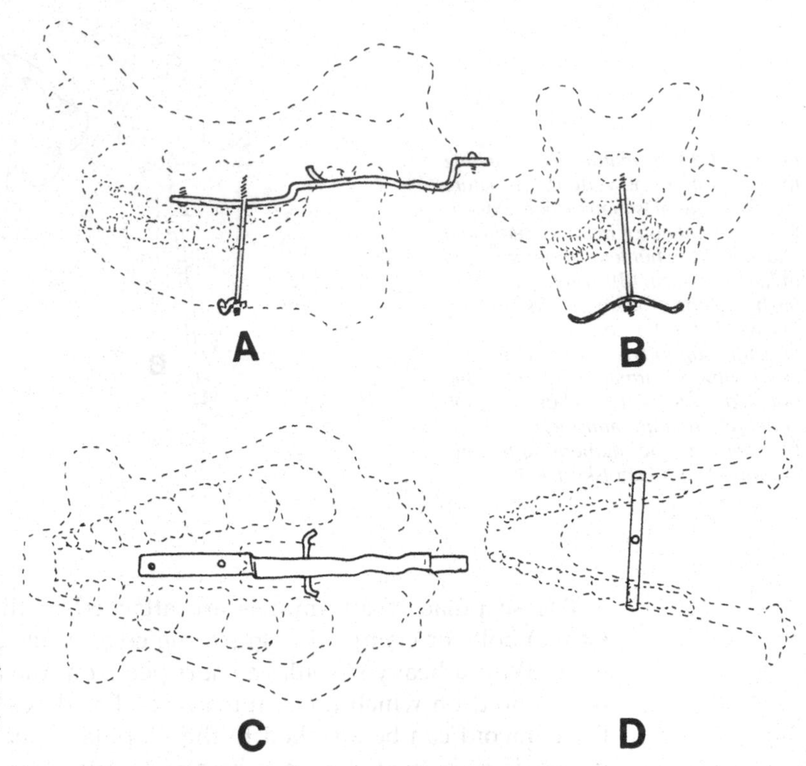

Figure 11.34. Example of a skull cradle that attaches to the vertebral armature in lateral view (A), anterior view (B), and ventral view (C). (D) is dorsal view of armature for supporting the lower jaws.

preserved and lightweight (e.g., La Brea tar pit skull), it can be slipped onto the vertebral armature through the foramen magnum (opening into the braincase). If hard matrix fills the brain cavity, a hole can be drilled through the foramen magnum and a bar or rod epoxied in place. This bar will be used to attach the skull to the neck. With exceptionally heavy skulls, or moderately heavy skulls on a long neck, a separate upright post may be needed (see Figure 11.20A, B).

Mounting casts

Many of the steps used for mounting original bone skeletons can also be used for casts – with a few exceptions. Before mounting a cast skeleton, prepare it by removing all flash, the thin pieces of casting material formed along the seams of the molds. This may be done with saws, diagonal wire cutters, a small knife (X-acto) or utility knife, or a hand drill using a burr or router bit (the Foredom drill is preferred over a Dremmel drill because the Dremmel is limited to light use).

Touch-up and most restoration of casts can be done with automotive body filler (e.g., Bondo), available in most automotive departments of large stores. Body fillers have a polyester resin base and are therefore compatible with polyester and WEP casts. Final touch-ups after restoration of large blemishes with body filler can be made with premixed spackling compound (or Synkaloid) or paper-mâché. Holes are filled and the area smoothed or textured with a finger dipped in water. If done carefully, little or no sanding of the repair is necessary.

Casts can be drilled with spade or steel twist bits. However, if the cast resin

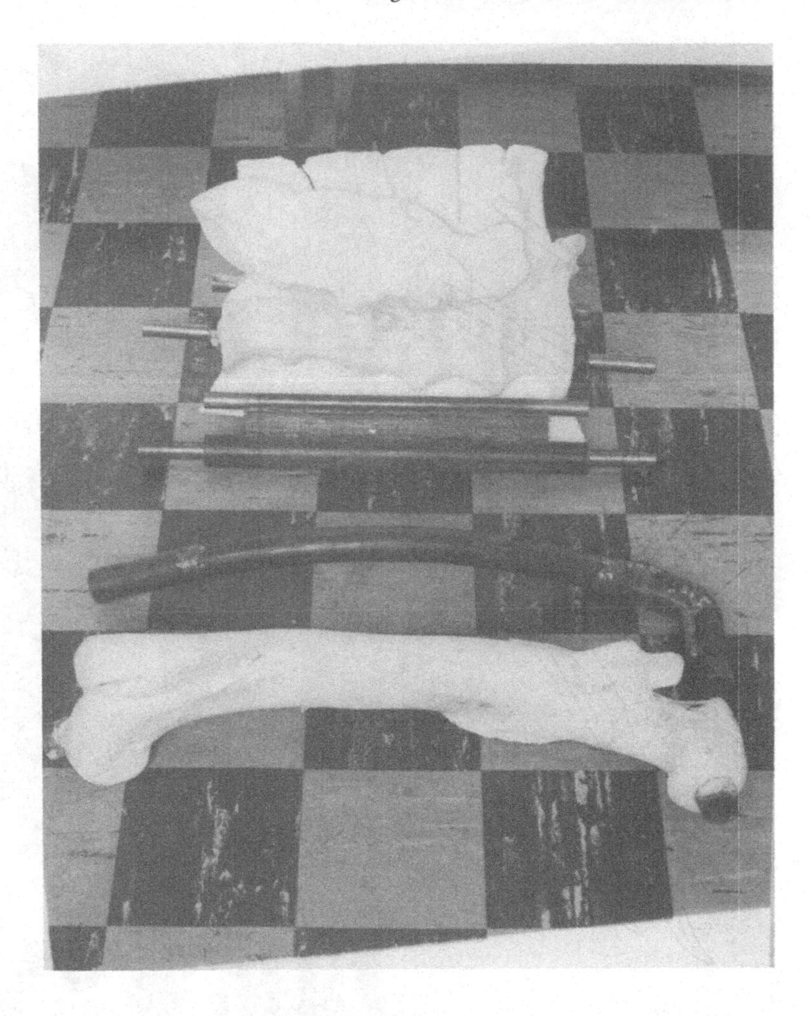

Figure 11.35. Examples of armature cast within bones and the armature used: sacrum (top) and femur (bottom).

is filled with marble dust (actually calcium carbonate powder that is used to reduce shrinkage in WEP casts), a carbide masonry bit is necessary. A hole saw can be used on hollow polyester casts and thin WEP casts. Small, tapered, tungsten carbide die-grinder burrs may also be used on polyester and WEP casts. An electroplated burr made by American Vermont is very useful for marble dust-filled casts, because conventional and high carbon steel burrs quickly dull.

Mounting should proceed in the same order as for original bone. Armature pieces can often be cast in place (see Figure 11.35), but this requires a considerable amount of planning in constructing the mold. The mold must accommodate the armature, yet be snug enough to prevent leakage of the resin or plaster around the armature. If the armature is not cast internally, the casts can be cut open or notched to accommodate the armature (see Figure 11.36). With large skeletons placing the armature internally makes the elements heavier. It may be necessary, therefore, to make short segments of vertebrae or parts of the legs, which are then welded, pinned, or bolted together.

Pigments may be added to the cast resin, but the difficulties in getting a good fossil bone color and getting all the casts to have the same shade sometimes

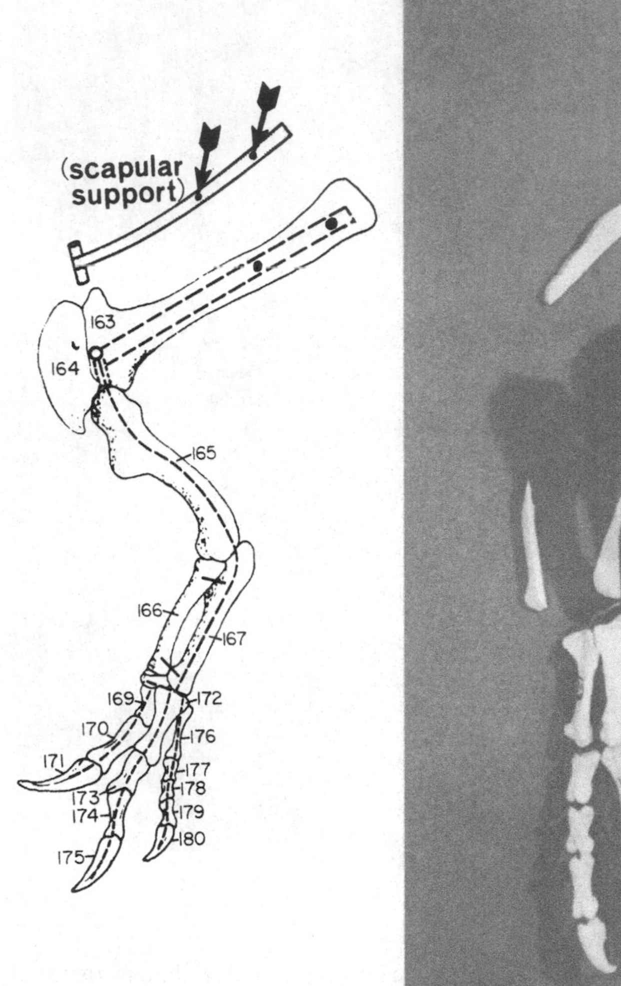

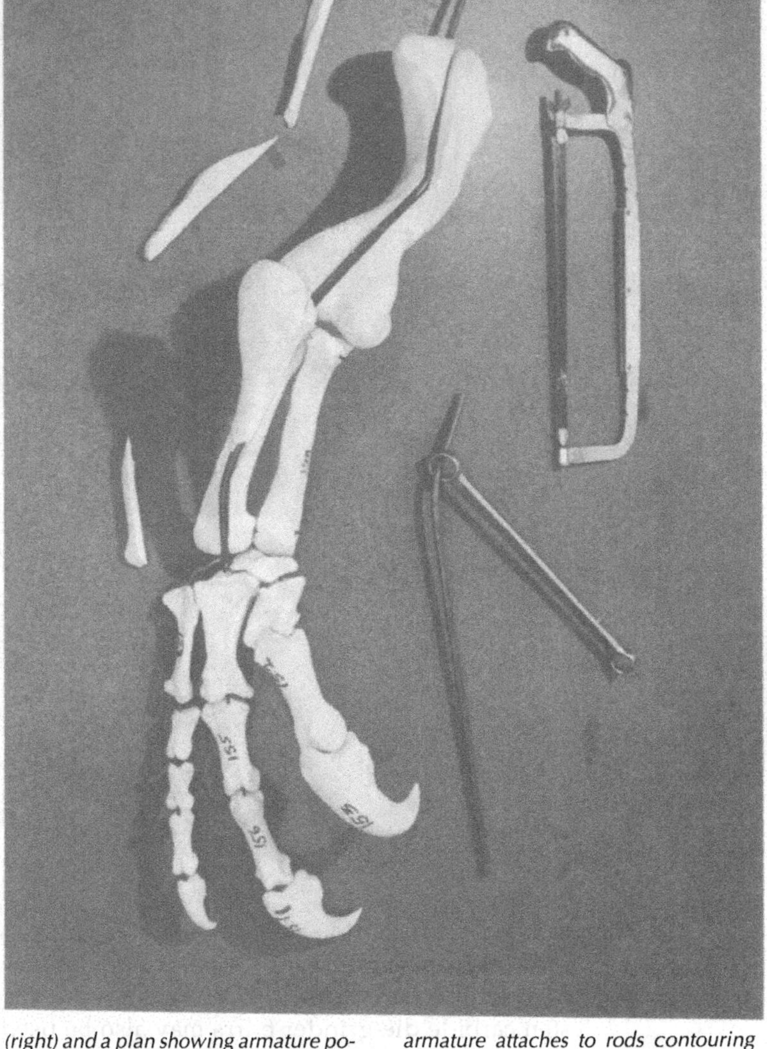

Figure 11.36. Example of cutting a cast to mount the armature internally (right) and a plan showing armature position. Arrows show where the scapular armature attaches to rods contouring the ribs from the vertebral column.

make this more trouble than it is worth. Coloring a cast should be done after the skeleton is mounted because then nicks and blemishes can be easily hidden. Painting is often an art unto itself, and the best results can be obtained by dabbing paint with cotton balls. Acrylic paints are preferable to oil paints because of their relatively quick drying time, which allows for the successive buildup of washes, and because of the ease of cleanup.

Before painting, all restored parts must be dry and dust-free. Acrylic aerosol paints are preferred for the base coat because of their ease in application. An antique effect is made by the application of a thin coat of dilute acrylic or latex paint of a contrasting color (often pale gray or brown) that is quickly wiped off with a damp cloth or water-wet paintbrush. The result is to leave the contrasting color in the pits, grooves, cracks, and other blemishes of the original fossil. Wood stains can also be used to color casts or to antique paint applied as a base coat. For a good imitation of North American Late Cretaceous bone color, Jacobean stain is recommended.

Special problems

Ossified tendons are often removed during the preparation of ornithischian dinosaurs. These long, thin bones actually served an important role in the living animal because they kept the back and tail stiff and horizontal. This kept the body counterbalanced over the hind limbs in a bipedal pose. Such tendons can be reconstructed on a skeleton using plexiglass or acrylic rods of the appropriate diameter. The rods may be bent to conform to any bend in the vertebral column with a hot air gun or hair drier. They may be pinned or epoxied in place. The arrangement of these tendons varies among the different dinosaurs, so original scientific descriptions or photographs of the skeletons may be required (e.g., crosshatched bundles in hadrosaurs, parallel diagonal bundles in ceratopsians).

Exceptionally small skeletons (e.g., *Hypertragulus* or rodents) can be glued together. Clear acrylic rods can be used to support the weight of the skeleton. These rods can be formed to contour with the limbs or serve as upright supports (see Figure 11.37). Heavy wire or a braid of twisted wire can also be epoxied on the inner side of the limbs to function as an armature.

A unique method of mounting skeletons has been used at the Senckenberg Natural History Museum in Frankfurt, Germany. The skeleton is mounted within a life-size silhouette of the living animal cut out of a coarse wire mesh or large sheet of smoked plexiglass. Another variation of this technique involves mounting the bones on one side of a silhouette of the animal as a panel mount (see Figure 11.38). Then missing bones may be painted on the silhouette. The advantage of this method is that it conveys to the museum visitor how incomplete fossil skeletons may be.

Fossil fish skulls can be reconstructed as three-dimensional exhibits using thin wire and Carbowax. Large fish skulls may require heavier wire or steel rods and epoxy putty. If a fish skeleton is to be mounted three dimensionally, the vertebrae can be drilled and epoxied onto a steel rod. Multiple ribs can be cast from a single mold of a rib. If cast in plastic, their curvature can be altered with a heat gun and their length shortened by cutting them distally. Distal ends of the altered ribs can also be used to reconstruct missing neural and hemal spines.

Conclusions

We have presented some guidelines and details for the mounting of fossil skeletons, keeping in mind that other methods and techniques may be as good or better. We certainly do not have the final word on the subject. The three of us have improved our mounting skills over the years by an open exchange of ideas with colleagues and appreciate that improvement comes only with experience. And to get that experience, we have mixed mistakes with our successes along the way.

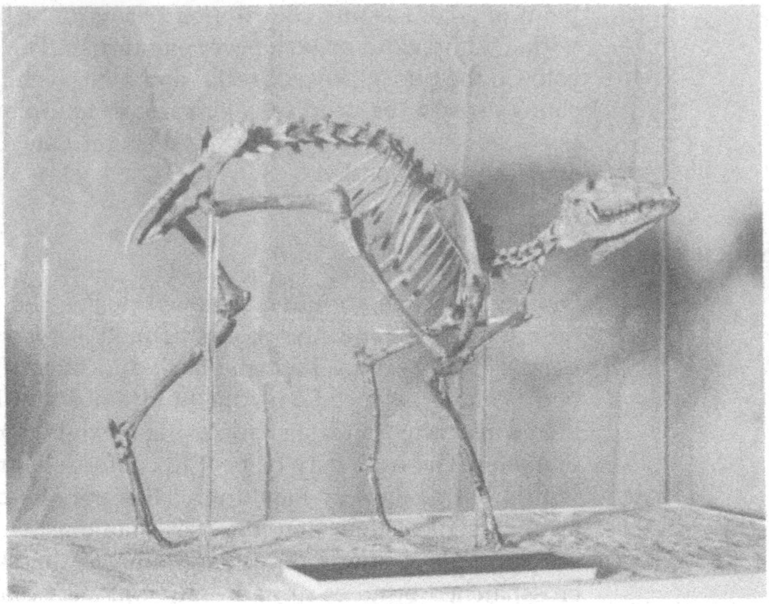

Figure 11.37. A skeleton of the small deer Hypertragulus *mounted as if scratching itself behind the ear. Bones were glued together with epoxy and clear plexiglass rods were used for vertical supports. (University of Colorado Museum.)*

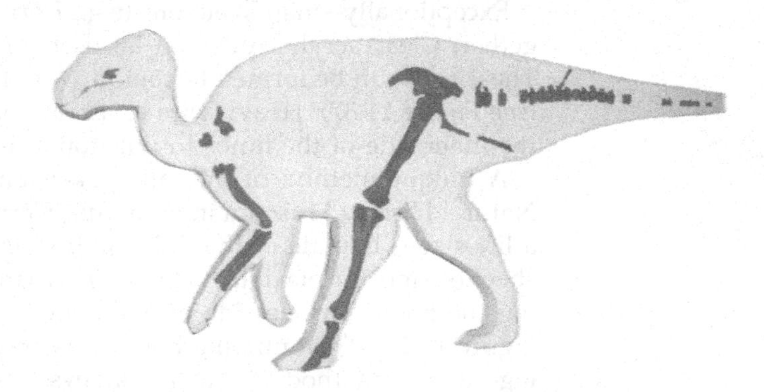

Figure 11.38. Life-size cutout of Hadrosaurus foulkii *with casts of bones mounted in place. Silhouettes of the missing bones were later painted. Contrast this with the first attempted mount of this skeleton in Figure 11.1. (Academy of Natural Sciences, Philadelphia.)*

More than a hundred years have passed since the first fossil skeleton was mounted for display. During that interval, there has been dramatic refinement of mounting techniques. Even the mounts themselves have become more sophisticated with the elimination of distracting external armature. In recent years the development of strong, lightweight, plastic casts has kindled a debate about whether original fossil bone or cast replicas should be mounted. At least one of us (Madsen) believes that only casts should be mounted because the accessibility and scientific integrity of the original specimen may be jeopardized. Whether museums accept this notion and mount more replica skeletons remains to be seen.

Acknowledgments

This chapter is dedicated to all the preparators who have ever struggled with the problems of mounting a fossil vertebrate skeleton. The caiman used in the locomotion study is courtesy of Peter Johnson, Monel Chemical Senses Institute, Philadelphia. Comments by Dan Chaney (Smithsonian Institution) and Bryan Small (Denver Museum of Natural History) improved the original manuscript.

References

Alexander, E. 1983. *Museum masters: their museums and their influence.* Nashville: American Association for State and Local History.

Bathel, K. 1966. Mounting a skeleton of *Smilodon californicus* Bovard. *Curator 9*:119–124.

Bessom, L. 1963. A technique for mounting skeletons with fiberglass. *Curator 6*:231–239.

Burke, A., M. Anderson, A. Weld, and E. Gaffney. 1983. The reconstruction and casting of a large extinct turtle, *Meiolania. Curator 26*:5–25.

Carpenter, K. 1984. Skeletal reconstruction and life restoration of *Sauropelta* (Ankylosauria: Nodosauridae) from the Cretaceous of North America. *Canadian Journal of Earth Sciences 21*:1491–1498.

———. 1989. Common mistakes in the mounting of fossil skeletons. *Journal of Vertebrate Paleontology 9*, Supplement to No. 3: p. 15A.

Chamberlain, F. 1943. Atlas of avian anatomy. *Michigan Agricultural Experiment Station Memoir Bulletin 5.*

Colbert, E. 1968. *Men and dinosaurs: The search in field and laboratory.* New York: Dutton.

Converse, H. 1984. *Handbook of paleo-preparation techniques.* Gainesville: Florida Paleontological Society.

Dollo, M. 1883. Note sur la presence les oiseaux du "Troisieme trochanter" des dinosauriens et sur la fonction de celui-ci. *Bulletin du Musee Royal d' Histoire Naturelle de Belgique* 2:13–20.

Dong, Z. 1988. *Dinosaurs from China.* London: British Museum of Natural History.

Drevermann, F. 1930. Wie man Skelette fossiler Wirbeltiere montiert. *Natur und Museum 60*:469–473.

Ellenberger, W. 1956. *An atlas of animal anatomy for artists.* New York: Dover.

Flower, W., and R. Lydekker. 1891. *Mammals living and extinct.* London: Adam and Charles Black.

Gilbert, S. 1968. *Pictorial anatomy of the cat.* Seattle: University of Washington Press.

Gilmore, C. 1925. A nearly complete articulated skeleton of *Camarasaurus*, a saurischian dinosaur from the Dinosaur National Monument. *Memoirs of the Carnegie Museum 10*: 347–384.

Gilpin, O. 1959. A free-standing mount of Gorgosaurus. *Curator* 2:162–168.

Gray, H., and C. Goss. 1973. *Anatomy of the human body.* Philadelphia: Lea and Febiger.

Hermann, A. 1909. Modern laboratory methods in vertebrate paleontology. *American Museum of Natural History Bulletin 26*:283–331.

Hildebrand, M. 1974. *Analysis of vertebrate structure.* New York: Wiley.

———. 1985. Walking and running. In M. Hildebrand et al. (Eds.), *Functional vertebrate morphology.* Cambridge: MA: Harvard University Press.

———, D. Bramble, K. Liem, and D. Wake. 1985. *Functional vertebrate morphology.* Cambridge, MA: Harvard University Press.

Keller, T. 1973, Diplodocus – Fossil und Leiche. *Natur und Museum 103*:320–333.

Kellogg, R. 1936. A review of the Archaeoceti. *Carnegie Institute of Washington Publication 482*:1–366.

Kunkel, R. 1981. *Elephants.* New York: Abrams.

Leidy, J. 1865. Cretaceous reptiles of the United States. *Smithsonian Contribution to Knowledge 14.*

Lockley, M., and D. Gillette. 1989. Dinosaur tracks and traces: An overview (pp. 3–10). In M. Lockley and D. Gillette (Eds.), *Dinosaur tracks and traces.* New York: Cambridge University Press.

Lockley, M., K. Houk, and N. Prince. 1986. North America's largest dinosaur trackway site: Implications for the Morrison Formation paleoecology. *Geological Society of America Bulletin 97*:1163–1176.

Lull, R. 1933. Revision of the Ceratopsia or horned dinosaurs. *Memoirs of the Peabody Museum of Natural History 3*, pt. 3.
Madsen, J. 1973. On skinning a dinosaur. *Curator 16*:224–266.
Majumdar, P. 1974. A free-standing mount of an Indian rhynchosaur. *Curator 17*:50–55.
Muybridge, E. 1887. *Animal locomotion*. Philadelphia.
———. 1957. *Animals in motion*. New York: Dover.
Norman, D. 1980. On the ornithischian dinosaur *Iguanodon bernissartensis* from the Lower Cretaceous of Bernissart (Belgium). *Institut Royal des Sciences Naturelles de Belgiue Memoire 178*:1–103.
———. 1985. *Illustrated encyclopedia of dinosaurs*. New York: Crescent Books.
Osborn, H. 1912. Integument of the iguanodont dinosaur *Trachodon*. *American Museum of Natural History Memoirs* 1:33–50.
Ostrom, J. 1978. The osteology of *Compsognathus longpipes* Wagner. *Zitteliana* 4:73–118.
Panofsky, A. 1973. Current techniques for fossil skeleton mounting. *Stanford Linear Accelerator Center Technical Note 73–7*.
Rixon, A. 1976. *Fossil animal remains: Their preparation and conservation*. London: Athlone.
Romer, A., and T. Parsons. 1977. Philadelphia: Saunders.
Sarjeant, W. 1975. Fossil tracks and impressions of vertebrates. In R. Frey (Ed.), *The study of trace fossils*. Amsterdam: Springer-Verlag.
Schultz, C., and H. Reider. 1943. Modern methods in the preparation of fossil skeletons. *Compass 23*:268–291.
Simpson, G. 1942. The beginnings of vertebrate paleontology in North America. *Proceedings of the American Philosophical Society 86*:130–188.
Sisson, S., and J. Grossman. 1938. *The anatomy of domestic animals*. Philadelphia: Saunders.
Sternberg, C. 1951. Complete skeleton of *Leptoceratops gracilis* Brown from the Upper Edmonton Member on Red Deer River, Alberta. Annual Report of the National Museum of Canada. *Bulletin 123*:225–255.

12

Methods and use of CT scan and X-ray

CT scan of fossils

Sandy Clark and Ian Morrison

Computerized axial tomography, commonly known as CAT scanning, is a radiological technique defined by Hounsfield (1973) and later developed by Ledley et al. (1974). It was not until 1984 that computer tomography (CT) was used on a complete fossil skull of a Miocene ungulate *Stenopsochoerus* (Conroy and Vannier 1984). With this technique structures can be seen that are not normally seen using more conventional preparation methods, without possible damage to the specimen. The risk of damaging a specimen during preparation of a fragile or difficult to reach area has always been a major concern; in some cases the risk is so great that these specimens remain unprepared on the shelf. Thus CT has provided researchers with a unique avenue for the study of vertebrate fossil material. One of its first uses in the paleontology field was in paleoanthropology (Conroy and Vannier 1985). It was only a matter of time before vertebrate paleontologists started using CT scanning in their research – for example, McGowan (1989a, 1989b, 1990, 1991) and Haubitz et al. (1988), to name a few. In this chapter we hope to provide readers with a basic understanding of computer tomography and point out some of its uses in the vertebrate paleontology field.

CT scanning process

X-rays travel through fossil specimens with varying degrees of absorption, depending on mineralized bone density, composition of the rock matrix, thickness of the slice taken, X-ray energy value, and interference. The X-rays are picked up by xenon gas detectors enclosed in a gantry, an apparatus shaped like a square donut that surrounds the examining table. From here the information is interpreted by the data acquisition system (DAS) and sent as digital information to a computer, which processes it into an image that can be displayed on a viewing screen and stored on magnetic tape or optical disk for later use. The

CT scanner we use is a General Electric 9800 Quick. Initial photography of the CT image is done using a multiformat camera system, on a 14" × 17" single emulsion film that is fed into an automatic processor for developing. The film format used depends on the number of slices taken per fossil and the ultimate display of information. One to 12 images per film can be displayed, depending on one's requirements

There are certain restrictions on the size and weight of fossil specimens to be scanned, determined by the make and model of the CT scanner used. The General Electric 9800 Quick scan table can normally hold up to 350 pounds and guarantees accurate table incrementation to 1 mm. The gantry aperture is 70 cm in diameter, and specimens should not measure more than 150 cm in circumference, to ensure their entire visualization. Once these criteria are met and a prescan plan has been discussed with the technician, the specimen is positioned in such a way as to include all the areas to be scanned in as central a location on the table bed as possible. Doing this will eliminate the need for large compensatory *X* and *Y* scan coordinates while ensuring equal distribution and absorption of applied X rays. For safety, the specimen is secured using tape and/or sponges. Securing the specimen also ensures it will not move, thus eliminating any unwanted interference called "motion artifact." Artifact can also be reduced by placing the specimen in water while scanning. The "water emulates the soft tissue coverings of the skull in life for which CT scanners are optimized" (Conroy and Vannier 1984). However, check that the fossil can be safely submerged in a water bath. The stability of the matrix may be compromised if it is soluble in water, thus undermining the integrity of the fossil.

At the control panel, after identification of the specimen – name, catalog number, institution, and so on – a scout or initial planning view is obtained. The scout is similar to an X ray and encompasses all the areas that are to be scanned. This initial view may be up to a maximum of 480 cm long. Consequently, larger specimens may need more than one scout view. The scout view is used to determine a flexible start and end point. A mid-image scan location is chosen as a preliminary test slice to determine scan parameters. The parameters include the *X* and *Y* coordinates needed to centralize the image and an appropriate scan field of view, which is used to size the entire fossil or portion of which is to be viewed on screen. The scan values needed for optimum visualization include kilovolts (kV), milliamperes (mA), and time (s) an algorithm, which is used as a filter to enhance different structures; and a choice of matrix size, which provides for degrees of volume visualization.

CT scan was developed for use in a medical setting as a diagnostic tool for the human head (Hounsfield 1973). Differences in density of various soft tissues and bone in humans are greater than the differences in density between mineralized fossil bones and the enclosing rock matrix. A bone algorithm is a function that allows enhancement of the image on the screen to show as much bony detail as possible. It acts much like a filter on a camera in that it can increase or decrease the quality of the image. Because bone is usually infilled with the rock matrix, the algorithm becomes an important function for utilizing CT scanners in paleontological research.

Matrix size is a variable value that controls the resolution on the screen. Three are available on the 9800 Quick: 256, 320, and 512. The matrix size corresponds to the number of lines available on the viewing screen to reproduce the scanned image. (It works just like a television screen.) Therefore a low matrix size corresponds to a lower number of lines, which in turn decreases the resolution. A high matrix size corresponds to a higher number of lines, which in turn increases the resolution.

The most commonly used scan values for fossilized specimens on the 9800 Quick are 120 kV, 140–200 mA, and 3 or 4 s. This set of parameters is only a starting point, and they relate specifically to this machine. The radiology technologist should be able to derive a comparable set of parameters for your machine using these figures. These values will vary according to the size of the specimen, its condition, and the composition of the rock matrix. Also of consequence is the mount used for its display and or preparation. This could affect the quality of the image produced and ultimately decide if scanning is possible. Often what can or cannot be scanned is not obvious until a scan is attempted. The composition of the rock matrix will usually be the determining factor in the quality of the scanned images. If heavy metallic minerals are present in the sediment (e.g., pyrite), the possibility of producing good quality images may diminish (Zangerl and Schultze 1989). As far as we know, there has been no comprehensive study on the variable response of CT scans to the composition of the rock matrix.

Throughout the process, adjustment of the scan values may be necessary due to the specimen's varying dimensions. The optimum results are often obtained by trial and error, always remembering that if the settings are too low the image will be grainy with a low contrast; too-high settings produce artifact (unwanted interference). In general a high resolution scan is desirable – a thin slice (1.5–3mm in thickness) in conjunction with a bone algorithm and 512 matrix size. This type of scanning allows for the greatest amount of bony detail to be imaged.

A postprocessing function (called "reformat") can be used to display the axial image in different planes. The selections most commonly used are the sagittal and coronal reformats. Coronal, sagittal, and axial planes can be transposed into *X*, *Y*, and *Z* planes of a graph. The *X* axis represents the coronal plane, the Y axis represents the sagittal plane, and the *Z* axis represents the axial plane. The orientation of these planes, for example, can be placed in relation to a person lying on his or her back, moving head first through the gantry aperture. The axial or transverse plane is the "slice" or section taken through the body. The coronal plane moves anterior to posterior and the sagittal plane moves medial to lateral. These two last planes use the information gathered from contiguous axial slices to produce their images. Both the coronal and sagittal reformats are a postprocessing function. When considering a reformat of the information, keep in mind that the thickness of the slice plays an important role. The thinner the slice, the more information is available and thus a clearer image. The thicker the slice, the less information is available, and thus a "fuzzier" image. The sagittal and coronal selections provide for one reformat per manipulation. A coronal orientation allows for anterior to posterior reformatting and a sagittal orientation allows medial to lateral reformatting. A "batch" selection can also be done; this allows multiple consecutive sagittal or coronal reformats.

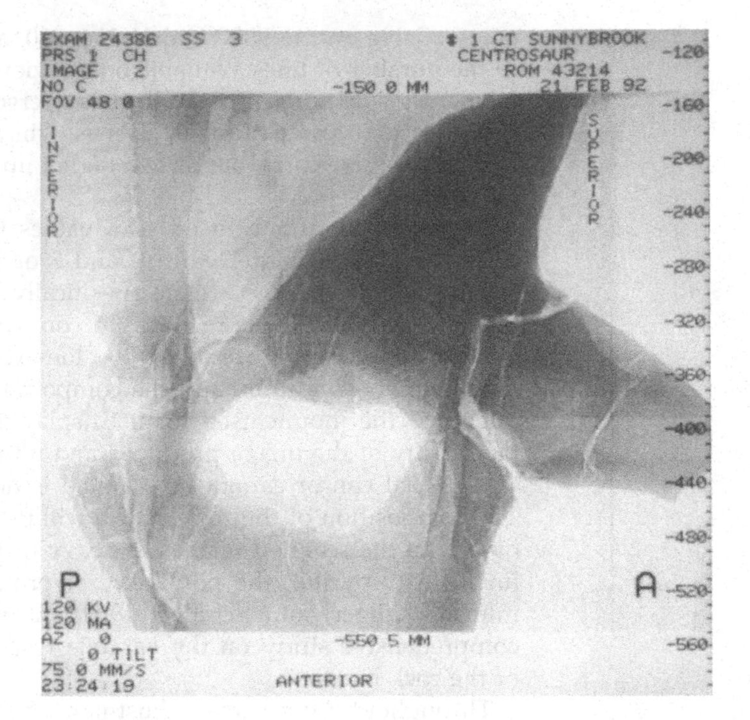

Figure 12.1. A scout image of Centrosaurus *skull ROM 43214.*

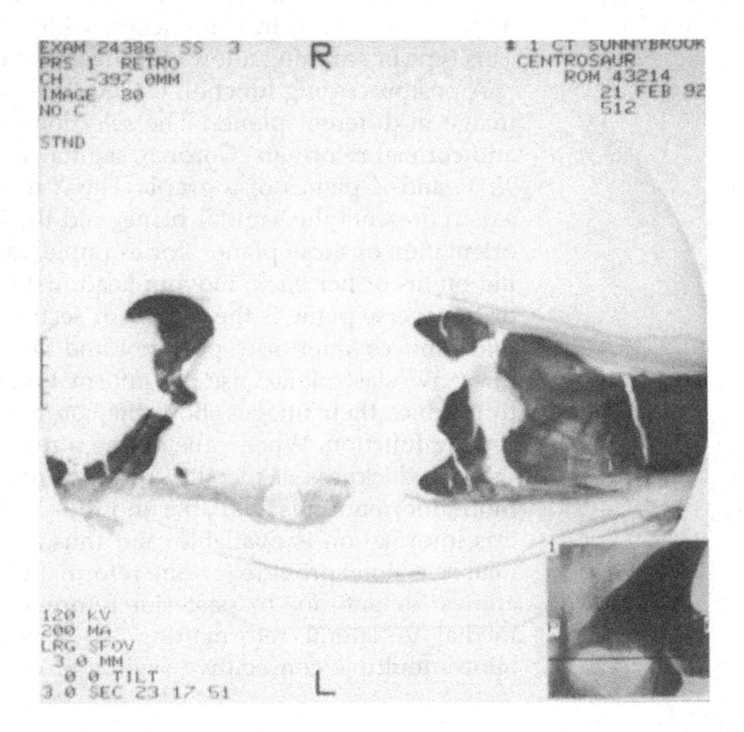

Figure 12.2. Axial image of ROM 43214. The image on the lower right shows the location of the axial scan on the scout image.

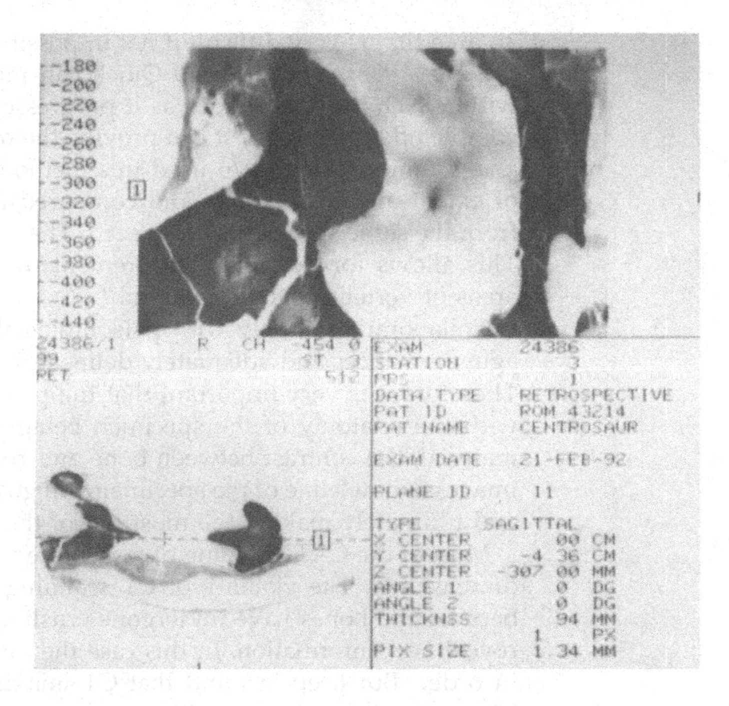

Figure 12.3. A sagittal reformat image of ROM 43214. The image on the lower left shows axial image 99 with the location of the reformat 1.

As an example we scanned a ceratopsian skull from the Royal Ontario Museum (ROM 43214). We scanned a total of 150, 3 mm slices at 120 kV, 200 mA and 3 s, in conjunction with a 512 × 512 matrix. Figures 12.1–12.3 are examples of the images we collected. The photographic reproductions are contact prints made from the original CT images. Therefore the bone in the photographs is black against a white background.

CT scan uses

For the best three-dimensional results, contiguous slices are needed. The smaller the scan increment the smoother and more accurate the reconstruction will be. The three-dimensional imaging system we use is the ISG Allegro 3D imaging workstation. The original scan data must be reconstructed using the same *X* and *Y* coordinates and the same 512 matrix, and requires changing the algorithm to standard. The data are stored on a magnetic tape, which is then transferred to the ISG unit for both two- and three-dimensional imaging. The tape can be stored and information accessed any time.

Due to the precise technique of CT scanning, accurate measurements can be taken of linear and volumetric elements. Computer tomography images provide a centimeter scale with every scanned image. Thus measurements can be taken off an image and measured against this scale for an accurate reading. Another method is to measure from an enlarged photographic print of a scanned image (McGowan 1989a, 1989b). This also allows for a test of accuracy for measurements taken from the fossil (McGowan 1989b). The distance between two points in a CT image can also be measured, providing an accurate measurement from an area that is very difficult if not impossible to measure accurately (Conroy and Vannier 1986). The GE 9800 Quick can measure the volume of a specific area within each individual scan as it progresses through the specimen. When it has completed the imaging, it can provide the measurement of the total area. Conroy and Vannier (1985, 1986) used this method to measure the endocranial volume of extant and fossil skulls and compared the results with measurements taken from the same specimens using seed. The conclusion was a 1%–3% accuracy rate. This allows for possible measurements of previously inaccessible matrix-filled areas of vertebrate fossil material.

Collaborating closely with paleontologists is essential to ensure that the imaging enhances and adequately delineates the unique anatomy of the specimen. Therefore it is very important that the paleontologist and technician be familiar with the anatomy of the specimen being scanned. Due to the nature of fossil material, the contrast between bone and rock matrix will be low in the scanned images. Knowledge of the specimen will go a long way in interpreting the images and ultimately making the most out of the information.

Most of the CAT scanning of vertebrate fossil material has been relatively three dimensional. The viability of CT scanning two-dimensional material is limited because the bones have undergone crushing and the images produced may not reveal new information. In this case the more traditional type of X rays may be in order. But keep in mind that CT imaging has more depth perception in its images, and this may provide better resolution of the fossil. Also remember the reformat function can "fillet" the specimen in 1 mm slices, thus giving access to otherwise inaccessible areas.

Once the decision has been made to scan the fossil material, there are two factors to consider: the availability of a CT scanner in your area and the costs involved in using one. In our situation a local hospital has allowed us to use its facility after hours at no expense, but we do pay the radiographer for running the machine. We also pay for the film, the cost of developing, and for the magnetic tape on which the information is stored. Availability of CT scanners and user cost will vary from place to place, but if you are interested in CAT scanning fossil material, a good place to inquire is at your local hospital.

Computerized axial tomography is a relatively new informational tool in the vertebrate paleontology field. It provides new avenues of study of previously inaccessible fossil material while preserving the integrity of the specimen. Information from CT scans of fossil preservation is limited, but with increased use of CT scanning in paleontological research, a more comprehensive study of the viability of CAT scanning fossil material will emerge.

Acknowledgments

We would like to thank Sunnybrook Health Science Centre, Toronto, for the use of its scanner. We would also like to thank Dr. Chris McGowan of the Royal Ontario Museum for his permission to CT scan the *Centrosaurus* skull ROM 43214. We thank Kevin Seymour for reviewing the manuscript through all its phases and for his support and encouragement.

References

Conroy, G. C. and M. W. Vannier. 1984. Noninvasive three-dimensional computer imaging of matrix-filled fossil skulls by high resolution computer tomography. *Science 226*: 456–458.

———. 1985. Endocranial volume determination of matrix-filled fossil skulls using high resolution computer tomography (pp. 419–426). In P. V. Tobias (Ed.), *Hominid evolution: Past, present and future*. New York: Alan R. Liss.

———. 1986. Three dimensional computer imaging: Some anthropological applications. In Else and Lee (Eds.), *Primate evolution*. Proceedings of the 10th. Congress of the International Primatological Society, Vol. 1. Cambridge: Cambridge University Press.

———. 1987. Dental development of the Taung skull from computerized tomography. *Nature 329*:625–627.

Haubitz, B., M. Prokop, W. Dohring, J. H. Ostrom, and P. Wellenhofer. 1988. Computer tomography of Archaeopteryx. *Paleobiology 14*:206–213.

Hounsfield, G. N. 1973. Computerized transverse axial scanning tomography. *British Journal of Radiology 46*:1016.

Ledley, R. S., G. DiChiro, J. J. Luessenhop, and H. L. Twigg. 1974. Computerized transaxial x-ray tomography of the human body. *Science 186*:207–212.

McGowen, C. 1989a. Computer tomography reveals further details of *Excalibosaurus*, A putative ancestor for the swordfish-like ichthyosaur *Eurhinosaurus*. *Journal of Vertebrate Paleontology 9*:269–281.

———. 1989b. The ichthyosaurian tailbend: A verification problem facilitated by computer tomography. *Paleobiology 15*:429–436.

———. 1990. Computer tomography confirms that Eurhinosaurus (Reptilia: Ichthyosauria) does have a tailbend. *Canadian Journal of Earth Science 27*(11):1541–1545.

———. 1991. An ichthyosaur forefin from the Triassic of British Columbia exemplifying Jurassic features. *Canadian Journal of Earth Science 28*(10):1553–1560.

Zangerl, R., and Hans-Peter Schultze. 1989. X-radiographic techniques and applications (pp. 165–177). In R. M. Feldmann, R. E. Chapman, and J. T. Hannibal (Eds.), *Paleotechniques*. Paleontological Society Special Publication No. 4.

Radiography of fossils

Jorg Harbersetzer

Radiography of fossils is an invaluable aid in the preparation of specimens enclosed within rock, mud stone, slate, and the like. Also, X rays are very important for scientific evaluation of fossils, because they can show internal details without destruction. Further, radiographs provide comparative morphological data.

Just a few months after the 1895 discovery of X rays by Wilhelm Conrad Roentgen, radiographs of fossils were prepared in Paris and Berlin. However, the use of X rays in paleontology was more or less occasional. One of the large radiographic collections of Hansruck slate fossils came from Lehmann in the 1980s. The largest collection of 22,000 radiographs was established by the late Professor

Sturmer, which even today is a source of research on Devonian fossils. This collection and another 4000 radiographs of Messel fossils and extant specimens represent a 30-year X-ray experience of the Senckenberg Museum and Research Institute. The spectrum of actually applied methods ranges from standard radiographs for screening purposes to the sophisticated newly developed digital microradiography for three-dimensional imaging.

The objective of this chapter is to explain briefly the most important X-ray physics (detailed information is indicated with the references) and to show with some examples how to obtain excellent prints from standard radiographs. Special methods are described separately to keep the text easy to read for beginners in the field. The chapter notes whenever our experiences with paleontological specimens differ from technical advances in textbooks for medical or industrial radiography.

Radiography on films

Conventional X-ray films

X-ray films for medical applications are normally double-coated and used with a film cassette with intensifying screens on the front and back sides. They are processed in a darkroom with special chemicals to give high contrast (as compared to normal photographic chemicals). Film processing can be done by hand in a tank (special tank developer) or automatically in a machine (the latter is the normal procedure in a radiology facility).

The front plate of the cassette and the physical properties of the intensifying screens may considerably reduce the contrast and definition in the radiographs of small and thin fossils, in comparison to a radiograph of the same specimen when it is X-rayed on single emulsion film in a black paper envelope. When using this modified technique, one has to consider that the radiation dose (mA × s = mAs) must be enlarged by a factor of 12–20(!) according to the type of fossil.

For very small objects or residues of bones, the mammography technique may be used because of the better contrast by soft radiation. For mammography single-coated films (with only one intensifying screen) are also used. Again the resolution can be increased using the mammography film without screen but the mAs factor will be lower than 15.

With both modified techniques, the facilities of a radiology facility can be fully utilized. For fossil plates, photo timing techniques can be used by taking multiple shots on the same film (e.g., 15 shots for the factor of 15). The results of this method are good enough to avoid the false diagnosis "not suited for radiographical studies."

NDT films

Nondestructive testing (NDT) films are special industrial films with very thick emulsions and high silver content. As a consequence their dynamic range (number of possible gray steps) is extraordinarily high (6–7 log units). There are double coated high speed films, which can be enlarged approximately 6× and single-coated low-speed (1/8th) films that can be enlarged up to 20×. For paleontological objects these films should not be used with salt or ledscreens. The disadvantage is the long development time compared to other films. This requires a special machine or hand processing.

If the fossil plate varies in thickness, if there are different large steps in contrast by thick overlapping skeletal elements, and if there are additional pure traces of bone to be shown (e.g., end of a phalanx or ribs) or even residues of soft tissue, the results on NDT films are the best compared to any other films. This is because of the high definition, the high contrast, and the large dynamic range. For less critical applications, a large range for over- and underexposure is advantageous whenever very different specimens have to be studied. Underexposed NDT films are similar in density to normal photographic films. However, when the full range has to be utilized, a special processing is required for prints (see the discussion on processing radiographs for prints). NDT films have been successfully combined with all of the various X-ray apparatus.

Other films

Excellent radiographs can be achieved on conventional photographic film. We have tested a large variety of such film and have gotten optimum results when orthochromatic or panchromatic film was developed in highly concentrated, fine grain developers. From this procedure radiographs of high contrast, medium density, and high definition result. Prints with an enlargement of more than 30× can be taken from the negatives. For extreme enlargements (approximately 400×) high-resolution plates or special emulsions normally used in autoradiography can be used. Negatives must be preenlarged by a microscope. These are extremely low-speed emulsions and require very long exposure times.

Radiographs on Polaroid film are of low dynamic range, medium contrast, and medium definition. They are recommended for a quick screening of small series when there are no facilities for film processing.

X-Ray physics and techniques

X-ray spectrum

For X-ray studies of paleontological specimens a range of 10–130 kV is sufficient for the vast majority of applications. For very small and delicate structures, radiation with only a few kilovolts is preferable. When the target (anode) of the X-ray tube is hit only by slowly accelerated electrons, a soft X-ray spectrum (long wavelength) with low penetration and thus high contrast for tiny structures will result. Mammography X-ray tubes have a target and an X-ray window of molybdenum, an element characterized by a high peak of soft radiation. Conventional medical tubes have a tungsten target and a harder X-ray spectrum. If conventional tubes have to be used for soft-ray radiographs, the spectrum can be considerably elongated to soft portions when the collimator windows and all aluminum filters, which are necessary in medicine to reduce skin doses, are disengaged. The soft spectrum of NDT X ray is between these extremes. A beryllium X-ray window or a very thin aluminum filter is often combined with a tungsten target.

For the radiography of very thick specimens high tube voltages should be used, because the wavelength of radiation becomes smaller with increasing kilovolts and thus the penetration increases. However, a higher portion of X-ray photons now hits the film and this reduces contrast. Therefore, in practice, use one always looks for a compromise of better penetration (high kV) and higher contrast (low kV). Despite this correlation of voltage and spectrum additional filtering may be necessary to minimize the soft radiation – for example, to reduce disturbing scattering radiation (fogging on the radiograph) – when very thick fossil plates are X-rayed. Aluminum as well as copper filters are suited to harden the X-ray spectrum.

Geometrical unsharpness

As long as the specimen is very close to the film, the size of the focal spot of the X-ray tube does not play an important role. However, fossils within a plate cannot be attached close to the film. Because of the parallax of X-ray beams originating from different points of the target, the projected image on the film will become distorted, and as a result of magnification, specimen proximity is higher from the film.

For carefully prepared radiography this unsharpness can be easily calculated. The magnification is the ratio of the focus to film distance (e.g., 1.2 m) and the focus to object distance (e.g., 0.8 m). This magnification (1.5×) is used as a factor for the focal spot size to obtain geometrical unsharpness on the radiograph (e.g., 0.5 mm spot size × 1.5 = 0.75 mm unsharpness). A good practice is either to change the geometry (e.g., move the tube or the film) or choose an appropriate film speed (that means a certain grain size with a related so-called "inner unsharpness" of the film) according to the geometrical unsharpness.

X-ray apparatus

Some properties of conventional X-ray apparatus in medicine, mammography and NDT tubes have already been mentioned. The spot sizes vary from 0.1 to 0.5 mm up to 1.2 mm. However there are also special microfocus tubes with spot sizes of 50 to even 2 μm. The 10–60 w power of the apparatus is very low compared to the power of conventional X-ray apparatus, which provides from a few hundred watts up to several thousand (for very short exposure times). This lack of low power can be compensated for by moving the tube very close to the specimen. By the same emitted radiation, the dose is four fold for half the former distance and eightfold for a quarter of the distance. The increase in geometrical unsharpness caused by this procedure is often unimportant because of the microfocus. However, it must be pointed out that the statements in some textbooks and manuals that the depth of the measuring field is infinite and the magnification with microfocus devices is unrestricted are definitely wrong. Of course, geometrical unsharpness is calculated by the same formula and must also be considered with microfocus tubes when high-definition films are used. In general, a microfocus is of advantage when radiographs should or must be pre-enlarged.

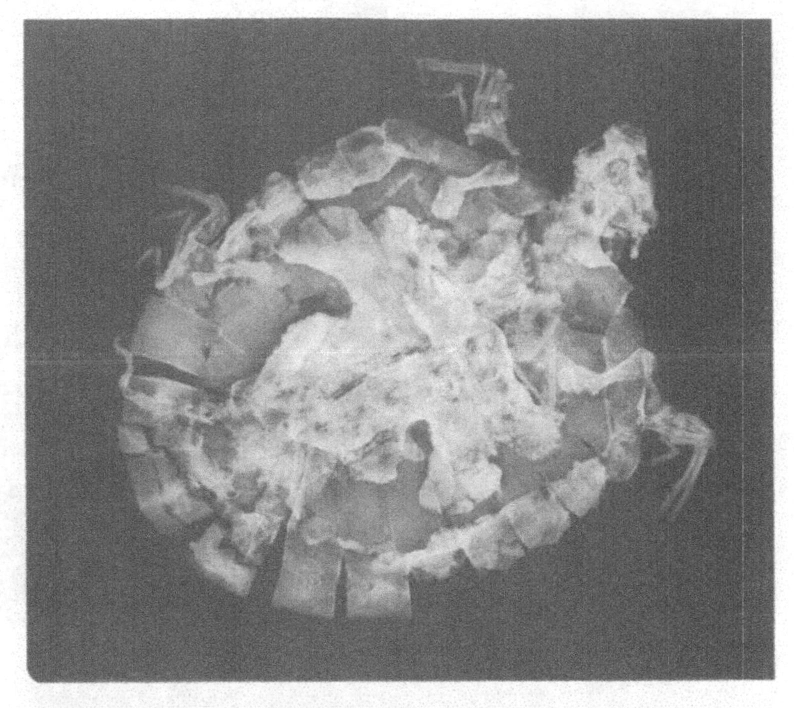

*Figure 12.4. Tortoise (*Allaeochelys crassessculpta*). Du Pont NDT 45, 90 kV, 400 mAs, FFA110 cm × 1.27. (Senckenberg Museum Frankfurt am Main; paleontological find from Messel Mine.)*

Processing radiographs for prints

General problems

The purpose of special processing is to get really striking prints from radiographs. In industrial and medical radiography the radiologist is looking for indications on the original (negative) films that most likely will not reproduce in prints. In addition, radiographs of fossils should also be suitable for publications and exhibitions. Development of the X-ray film is not difficult and can be done with machines or

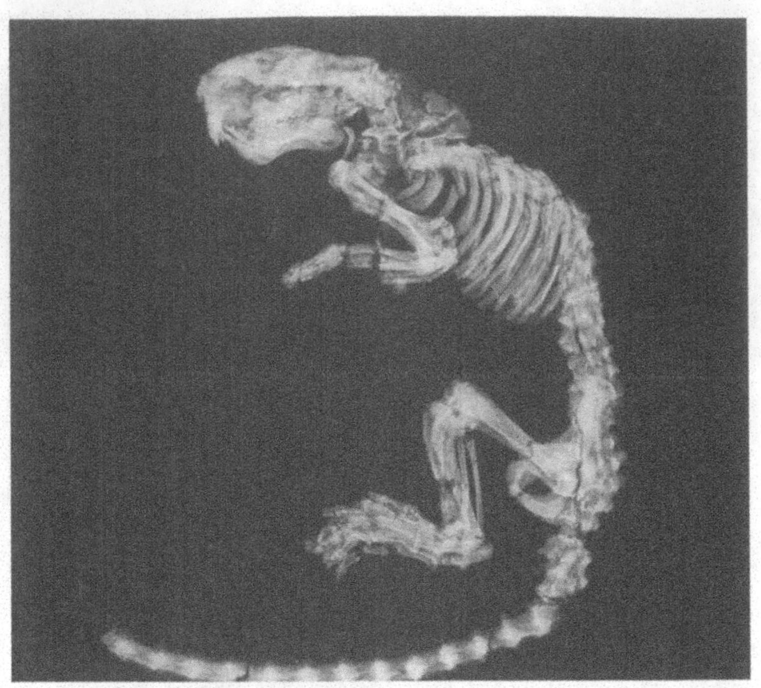

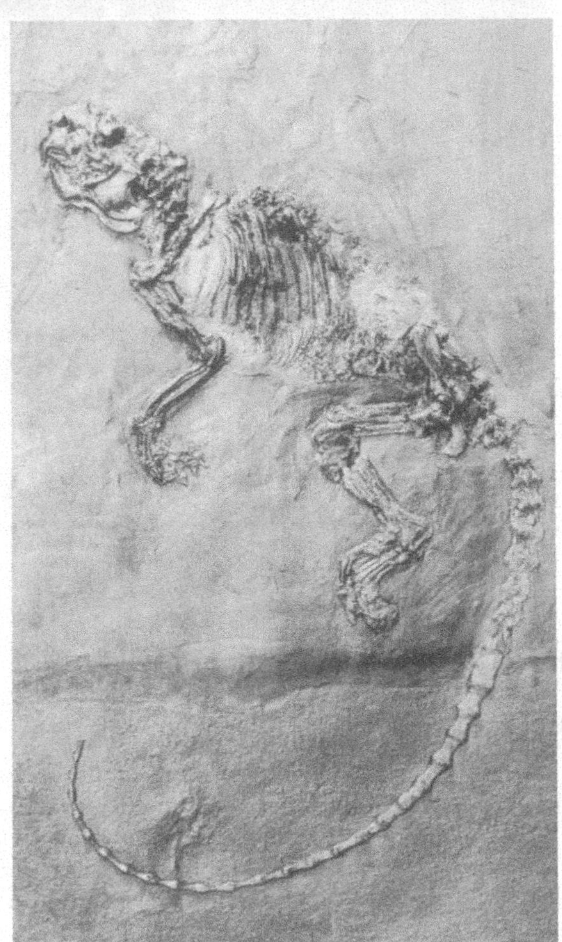

*Figure 12.5. Rodent (*Ailuravus macrurus*). Du Pont NDT 45, 90 kV, 400 mAs, FFA110 cm × 1.13. (Senckenberg Museum Frankfurt am Main; paleontological find from Messel Mine.)*

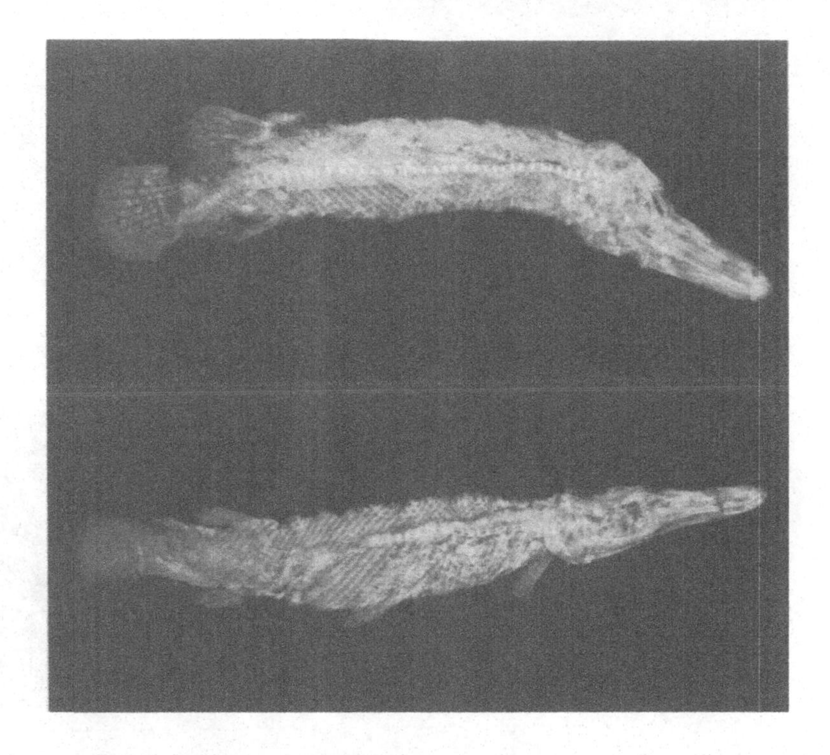

*Figure 12.6. Bony pike (*Atractosteus strausi*). A Du Pont NDT 45, 50 kV, 500 mAs, FFA10 cm × 1.30;* B *Du Pont NDT 35, 35 kV 800 mAs, FFA80 cm × 9.13. (Senckenberg Museum Frankfurt am Main; paleontological find from Messel Mine.)*

by hand. However, the production of good prints from film radiographs is a complex, skilled procedure. We discuss here examples of prints obtained from the radiography of fossils using the different techniques already described.

Sources and preparation of specimens

The high water content (40%) of oil shale poses special problems: Exposed to the air, the fossil dries and begins to decay. Fossils can be preserved in water for a

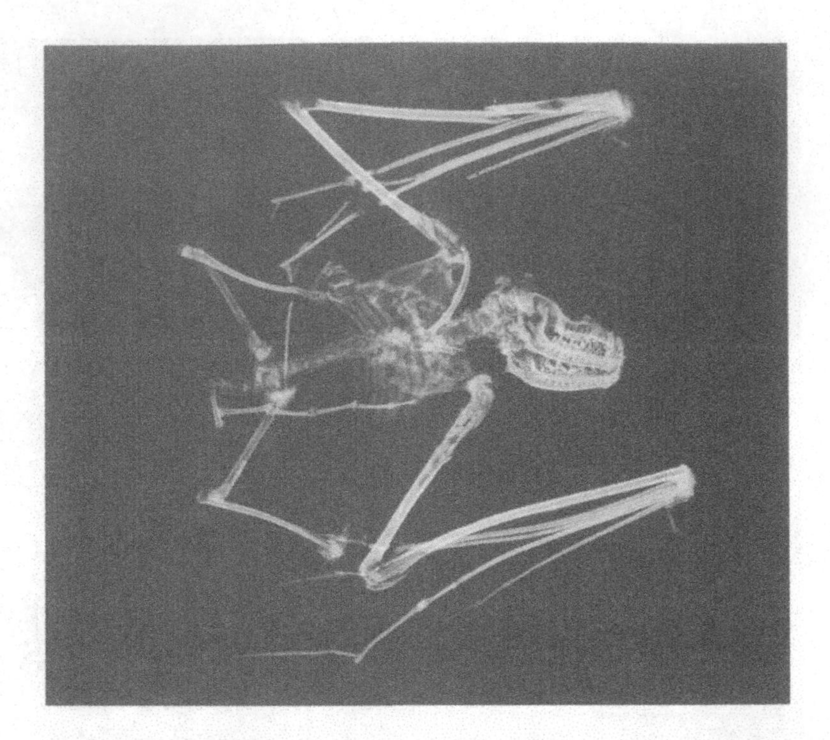

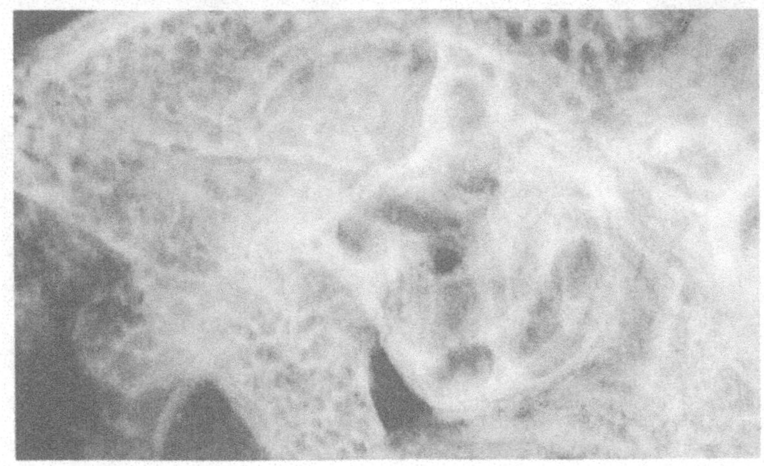

*Figure 12.7. Bat (*Palaeochiropteryx tupaiodon*). A Du Pont NDT 35, 35 kV, 600 mAs, FFA90 cm × 4.39.* B *Du Pont NDT 35, 45 kV, 55 mAs, FFA 38 cm × 28.0. (Senckenberg Museum Frankfurt am Main; paleontological find from Messel Mine.)*

certain period. Previously fossil plates were immersed in glycerin or mounted on support plates of wax or chalk. Today they are mounted on synthetic resin plates after all traces of oil shale are removed under a microscope, with the result that the skeleton and remnants of soft parts and hair rest freely on the plate.

Although skeletal structures may appear to be complete, details are often destroyed by pressure of the sediment in which they rested; or the ferric salts that have become embedded in the fossils prevent X-ray, thereby concealing additional minute structure. Often, samples must be X-rayed in order to reveal all the inner details of the fossil. For example, in order to produce radiologically 5 intact inner ears, a total of 170 bats had to be X-rayed.

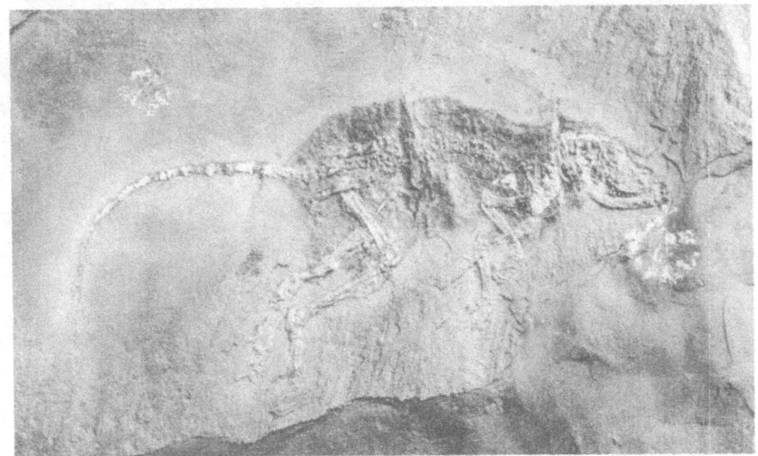

Figure 12.8. Insectivore (Macrocranion tupaiodon). *Du Pont NDT 35, 60 kV, 170 mAs, FFA90 cm × 2.21. (Senckenberg Museum Frankfurt am Main; paleontological find from Messel Mine.)*

Combination of X-ray and film techniques

Even though the soft X-ray specimen from a tube with a molybdenum anode (mammography tubes) or tungsten anode and 0.02 mm aluminum filteration is used, special techniques are necessary to show the thinnest bones and retain a high contrast. Film emulsions permitting great enlargement are necessary. At the Senckenberg Research Institute we use a combination of high-resolution non-screen film (double-coated Du Pont NDT-45 and single-layer NDT-35) with a microfocus X-ray tube having a focal spot diameter of 10 μm. A projective enlargement of 1.5×–3× onto the film permits a final enlargement of 40×–60×;

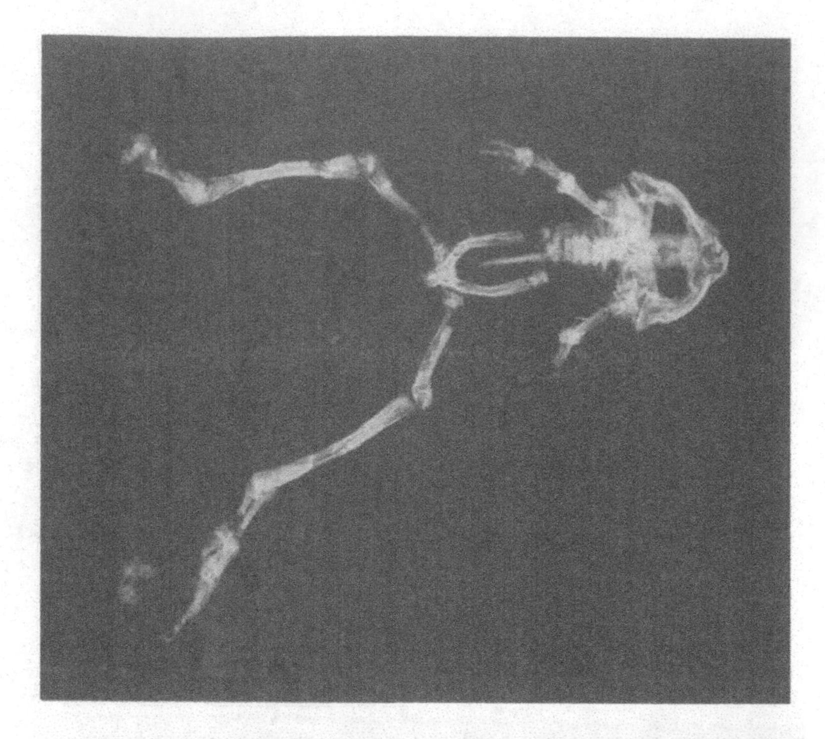

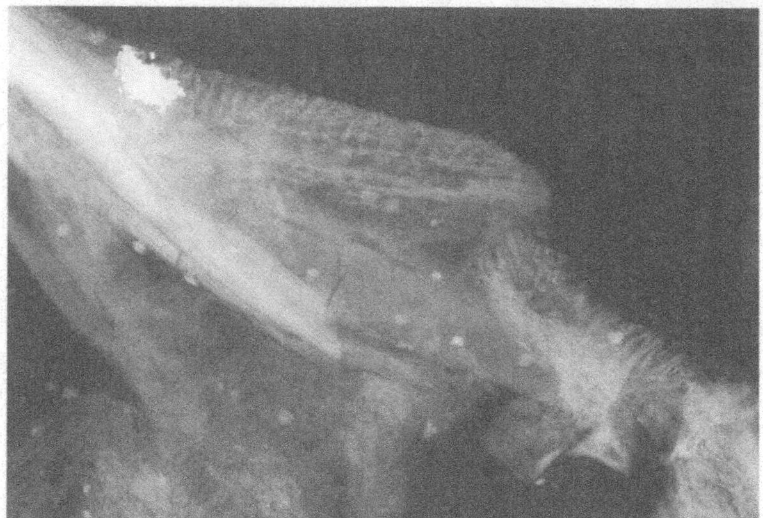

*Figure 12.9. Frog (*Pelobatidae*). A Du Pont NDT 35, 55 kV, 250 mAs, FFA80 cm × 2.31. B Du Pont NDT 35, 45 kV, 30 mAs, FFA38 cm × 13.8. (Senckenberg Museum Frankfurt am Main; paleontological find from Messel Mine.)*

and even the tiny teeth of frogs and the details of the inner ear of bats can be seen on the radiographs. Larger fossils are radiographed by more conventional techniques without projective enlargement but with aluminum filters removed.

To achieve a very high image contrast, in addition to reducing the X-ray beam filtration to a minimum, the dynamic spectrum of the NDT films was fully utilized. The maximum density was between 4.8 and 5.8 log units; the minimum was just over 1.0. The classic masking technique of combining an unsharp intermediate film mask with the original radiograph is difficult to do with radiographs having very high densities. An alternative procedure, designed to maintain high-contrast detail with a shortened film density range that could be accommodated on a paper print, has been developed.

The nonlinear base region (densities less than 1.0), which does not contribute to the image, is etched away with potassium ferricyanide. Then the regions with minimal object thickness (i.e., high optical density on the film) are partly etched so that the density changes in the various regions of the fossil are reduced, but not removed. Lastly, regions with very little film density are covered with a conventional glaze to produce a higher optical density. Thus the density range of the radiograph is reduced while maintaining maximum detail contrast.

In order to perform this procedure it was necessary to take additional radiographs using the hard-ray technique. However, these lower density radiographs were used only for reference purposes, and the final paper prints were made from single, high-density radiographs with a modified density range. By this method of density compression, artifacts such as pseudosolarization effects were completely avoided.

The modified radiographs, having a density range of about 2.5 density units, were additionally processed during printing with a conventional laser scanner. Examples of the resulting radiographic prints are shown in Figures 12.4–12.9.

Index